Stance and Motion

Facts and Concepts

Stance and Motion
Facts and Concepts

Edited by

V. S. Gurfinkel

Institute for Problems of Information Transmission
USSR Academy of Sciences
Moscow, USSR

M. E. Ioffe

Institute of Higher Nervous Activity and Neurophysiology
USSR Academy of Sciences
Moscow, USSR

J. Massion

Laboratory of Functional Neurosciences
CNRS
Marseille, France

and

J. P. Roll

Laboratory of Human Neurobiology
University of Provence
Marseille, France

Plenum Press • New York and London

Library of Congress Cataloging in Publication Data

Soviet–French Roundtable Meeting on Neurobiology (3rd: 1986: Moscow, R.S.F.S.R. and Leningrad, R.S.F.S.R.)
Stance and motion: facts and concepts / edited by V. S. Gurfinkel . . . [et al.].
p. cm.
"Proceedings of the Third Soviet–French Roundtable Meeting on Neurobiology, held May 19–24, 1986, in Moscow, Leningrad, USSR"—T.p. verso.
Bibliography: p.
Includes index.
ISBN 0-306-43108-4
1. Locomotion—Regulation—Congresses. 2. Posture—Congresses. 3. Sensory-motor integration—Congresses. I. Gurfinkel, V. S. II. Title.
QP301.S755 1986 88-37435
612′.76—dc19 CIP

Proceedings of the Third Soviet–French Roundtable Meeting on Neurobiology, held May 19–24, 1986, in Moscow and Leningrad, USSR

A Division of Plenum Publishing Corporation
233 Spring Street, New York, N.Y. 10013

Printed in the United States of America

PREFACE

This book dealing with stance and motion was planned in June 1986 at a meeting held in Moscow and Leningrad between a group of Soviet and French scientists interested in motor control. This meeting took place in the framework of an exchange program between the USSR Academy of Sciences and the French Centre National de la Recherche Scientifique. It was very successful event and was greatly appreciated by all those who attended it. Several participants put forward the proposal that the possibility of publishing a book was worth exploring.

What were the reasons for publishing a book on stance and motion ? The interest aroused in the participants by each others contributions was not a sufficiently decisive argument. It was felt, however, that a large proportion of the orginal material presented at the meeting, especially in the field of posture and locomotion but also on other aspects covered by the book could be presented in a summarized form which should appeal to a larger audience because the facts and hypotheses they contained especially those from the Soviet participants, were not very familiar among international circles, and that many scientists would appreciate having a single volume containing a survey of the current state of research in this field. This was also the opinion of Plenum Press, who agreed to publish the book. Each participant at the meeting submitted a paper which was examined by two referees before being accepted. The papers were arranged in five chapters dealing with peripheral and central aspects of motor control, locomotion, posture and modelling.

The preparatory work was carried out in Marseilles with the help of serveral co-workers who have done an excellent job. We would like to express our thanks to Y. Coiton, K. Mattei, C. Quoniam and G. Riviére for their help in editing the manuscripts. Jessica Blanc was responsible for revising the English, and the editors express here their thanks for her very crucial contribution to the book.

We hope that the readers will share our enthusiasm for the present volume and will agree that this venture was worth attempting.

V.S. G., M.E. I., J. M., J.P. R.

CONTENTS

PERIPHERAL MECHANISMS OF MOTOR CONTROL

Encoding of Muscle Contractile Tension by Golgi Tendon Organs 1
G. Horcholle-Bossavit, L. Jami, J. Petit, R. Vejsada, and D. Zytnicki (Paris)

Modulating Processes of Single Fusimotor Fibre Discharge in Man 11
J.P. Vedel and J.P. Roll (Marseille)

Extraocular Proprioception and Body Postural References 23
J.P. Roll and R. Roll (Marseille)

Gravitational Mechanisms in the Motor System.
Studies in Real and Simulated Weightlessness 37
I. Kozlovskaya, I. Dmitrieva, L. Grigorieva, A. Kirenskaya, and Yu. Kreidich (Moscow)

CENTRAL MECHANISMS OF MOTOR CONTROL

Synaptic Influences on Motoneurones of the Spinal Cord in Vertebrates 49
A.I. Shapovalov and N.P. Vesselkin (Leningrad)

Neuronal Organization of the Corticorubral System 61
V.V. Farnardjian and V.L. Gorodnov (Yerevan)

Role of the Monkey Substantia Nigra Pars Reticulata in Sensorimotor Control of Eye-Head-Arm Coordination 71
F. Lestienne, P. Cailler, and P. Liverneaux (Paris)

The Participation of the Cerebral Cortex and Basal Ganglia in Motor Act Programming 83
A.S. Batuev, A.A. Orlov, and V.I. Shaefer (Leningrad)

CONTROL OF LOCOMOTION AND RHYTHMIC ACTIVITY

The Stance Phase and its Control by Sensory Information During Walking in Crustacea 93
F. Clarac and A. Chrachri (Arcachon)

Neuronal Mechanisms Controlling Rhythmic Movements in Gastropod Molluscs 107
Yu. I. Arshavsky, T.G. Deliagina, I.M. Gelfand, G.N. Orlovsky, Yu. V. Panchin, and G.A. Pavlova (Moscow)

On the Bulbospinal Locomotor Column in the Cat 123
O.V. Kazennikov, V.A. Selionov, and M.L. Shik (Moscow)

Analysis of the Pattern of Activity in "Knee Flexor" Motoneurons During Locomotion in the Cat 133
C. Perret, J.M. Cabelguen, and D. Orsal (Paris)

Modulation of Locomotor Activity Induced
by Natural Stimulation of the Vestibular System 143
V.V. Marlinsky, D.A. Vasilenko, and Th.I. Tsyntsabadze (Kiev)

Presynaptic Control of Afferent Input During Real Locomotion in Rats 153
I.A. Yakhnitsa, A.I. Pilyavsky, and N.V. Bulgokova (Kiev)

Role of Motor Cortex in Control of Locomotion 163
I.N. Beloozerova and M.G. Sirota (Moscow)

CONTROL OF POSTURE AND MOVEMENT

Anticipatory Postural Adjustments and Dynamic Asymmetry
of Voluntary Movement 177
S. Bouisset and M. Zattara (Orsay)

Body Scheme in the Control of Postural Activity 185
V.S. Gurfinkel, Yu.S. Levik, K.E. Popov, B.N. Smetanin,
and V.Yu. Shlikov (Moscow)

Forward and Backward Axial Movements: Two Modes of Central Control 195
P. Crenna, C. Frigo, J. Massion, A. Pedotti, and A. Deat (Marseille and Milan)

The Role of Nigrostriatal and Mesolimbic Dopaminergic Brain Systems
in the Control of Voluntary Movements and Postural Adjustement in Dogs 203
K.B. Shapovalova and A.F. Yakimovsky (Leningrad)

On the Role of Motor Cortex in the Learned Rearrangement
of Postural Coordinations 213
M.E. Ioffe, N.G. Ivanova, A.A. Frolov, E.V. Birjukova,
and N.V. Kiseljova (Moscow)

On the Influence of Movement Kinematics on the Support Pressure Pattern
During Postural Adjustment of Quadrupeds 227
A.A. Frolov, E.V. Birjukova, and M.E. Ioffe (Moscow)

MODELLING OF MOTOR ACTIVITY CONTROL

The Effect of Repetitive Stimulation on Firing Motoneurons.
Computer Simulation Based on a Motor Unit Study 239
M. Piotrkiewicz, R. Person, and L. Kudina (Warsaw and Moscow)

Topological Aspects of Sensori-Motor Control 251
J. Droulez (Paris)

Rapid One-Joint Movements:
a Qualitative Model and its Experimental Verification 261
R.M. Abdusamatov, S.V. Adamovich, M.B. Berkinblit,
A.V. Chernavsky, and A.G. Feldman (Moscow)

Contributors 271

Index 275

ENCODING OF MUSCLE CONTRACTILE TENSION BY GOLGI TENDON ORGANS

G. Horcholle-Bossavit *, L. Jami *, J. Petit *,
R.Vejsada **, and D. Zytnicki **

* Laboratoire de Neurophysiologie,
Collège de France, 75231 Paris, Cedex 05, France

**Institute of Physiology,
Czechoslovak Academy of Sciences,
Videnska 1083, 142 20 Prague 4, Czechoslovakia

Summary

The discharges of tendon organs of peroneal muscles were recorded in deeply anaesthetized cats during unfused contractions of single motor units. The tension developed during unfused contractions displays oscillations, in phase with the stimulation, superimposed on a steady level of static tension.

In responses to such tension profiles, the discharge frequency of individual tendon organs provided information about the frequency of tension oscillations but none about their amplitude and the level of static tension. However, observations made on several tendon organs activated by the same motor unit showed that such information is present in the ensemble discharge of these receptors. The average instantaneous frequency was calculated from the pooled discharges of all the activated tendon organs. It showed variations which followed the variations of contractile tension.

Upon combined stimulation of two motor units, there was approximately linear summation of tensions. The ensemble discharges of tendon organs did not display linear summation although they still followed the variations of the tension profile. The dynamic components of contractile force, rather than its static level, prevail in the information provided to the Central Nervous System by the total Ib afferent discharge.

Introduction

In addition to spindles that are length sensors, mammalian skeletal muscle contains contraction sensitive mechanoreceptors, the Golgi tendon organs, innervated by fast-conducting Ib afferent fibres. Tendon organs are mostly located at myo-tendinous or myo-aponeurotic junctions rather than in tendons proper (Barker 1974).

A tendon organ is made of a collagen bundle surrounded by a capsule and connected at one end with the tendons of a small fascicle of muscle fibres, while the other end is in continuity with the whole muscle tendon or aponeurosis. The Ib fibre enters the capsule and divides in terminal branches interwoven among the collagen strands. Contraction of the muscle fibres attached to the tendon organ exerts strain on the collagenous bundle, which causes deformation of sensory terminals and conductance changes in their membrane. The resulting receptor potential propagates to an impulse initiating site and from that point, the discharge travels towards the Central Nervous System. In addition to the muscle fibres attached in-series with the tendon organ, a large number of fibres run in-parallel with the receptor, inserting around its tendinous or aponeurotic end. The contraction of these fibres tends to reduce the strain on the collagenous bundle and on the sensory terminals, i.e., to unload the tendon organ.

Table 1. Numbers of specific mechanoreceptors in cat peroneal muscles

	Numbers of Motor Units[1]	Spindle	Tendon Organs[2]	Ratios of motor units to Spindles	Tendon organs
Muscles:					
Tertius	34	14	10	2.4	3.4
Brevis	78	40	36	1.9	2.1
Longus	104	17	16	6.1	6.5

1. Means calculted from the data of Horcholle-Bossavit et al. 1985
2. From Scott and Young 1987

It is generally held that muscles contain more spindles than tendon organs, although the exact numbers of tendon organs present are known for a few muscles only (Barker 1962; Scott 1987; Swett and Eldred 1960; Wohlfart and Eriksson 1960). Rather than the number of receptors, the ratio of motor units to tendon organs might be a functionally significant figure, as earlier proposed by Cooper (1966) for spindles. Table 1 shows that in a given muscle of the cat peroneal group, the ratios for spindles and for tendon organs are similar. Differences between muscles suggest that motor unit activity could be more closely monitored by muscle receptors in peroneus brevis than in peroneus longus, since the former has less motor units per receptor.

The adequate stimulus for a tendon organ is muscle contraction and, more precisely, the contraction of those motor units contributing fibres to the fascicle attached in-series with the receptor. For this stimulus, the tendon organ has a very low threshold, and can respond to a single motor unit (Houk and Henneman 1967). In cat leg muscles, the set of motor units activating a given tendon organ usually comprises 10-15 units belonging to different physiological types (Reinking et al. 1975). Since, in addition, a given motor unit usually activates several tendon organs, it is very likely that the contraction of every motor unit in a muscle is monitored by at least one tendon organ (Jami and Petit 1976a).

The set of motor units activating an individual tendon organ displays the same range of mechanical properties as the total muscle population. For an example, Fig 1 shows the distribution of tetanic tensions in a group of 15 motor units that were found to activate a single peroneus longus tendon organ. The weakest of these units developed only 1.2g while the strongest developed 75g. If contractile force were the parameter encoded by individual tendon organs, one would expect their discharge frequency to display some clear relation with the strength of the contracting motor unit. Experience never confirmed this assumption in any of the muscles examined so far (Houk and Henneman 1967; Reinking et al. 1975; Stauffer and Stephens 1975; Jami and Petit 1976b; Gregory and Proske 1979). Slow-contracting motor units developing small forces were often found to elicit more discharges from a given tendon organ than large fast-contracting units, not only in responses to maximal tetanic contractions but also in responses to single twitches (see Fig 1 in Gregory and Proske 1979). To overcome this discrepancy, the hypothesis was proposed that the ensemble responses from all the tendon organs in a muscle could provide the Central Nervous System with a good measure of the total forces being produced by contraction (Houk and Henneman 1967; Reinking et al. 1975).

In experiments on cat peroneal muscles, which contain relatively limited populations of tendon organs (Table 1), we were able to examine this hypothesis because about 10 single Ib afferent fibres could be prepared for recording in dorsal root filaments, so that activation of several tendon organs by the same motor unit could be monitored. Stimulation patterns eliciting unfused tetanic contractions were used because such contractions are likely to result from the usual firing rates of alpha-motoneurones under physiological conditions (see the review by Henneman and Mendell 1981).

Methods

The experiments were carried out on adult cats (2.5-3.5kg) anaesthetized with an initial intraperitoneal dose of 45mg/kg pentobarbital sodium. The level of anaesthesia was

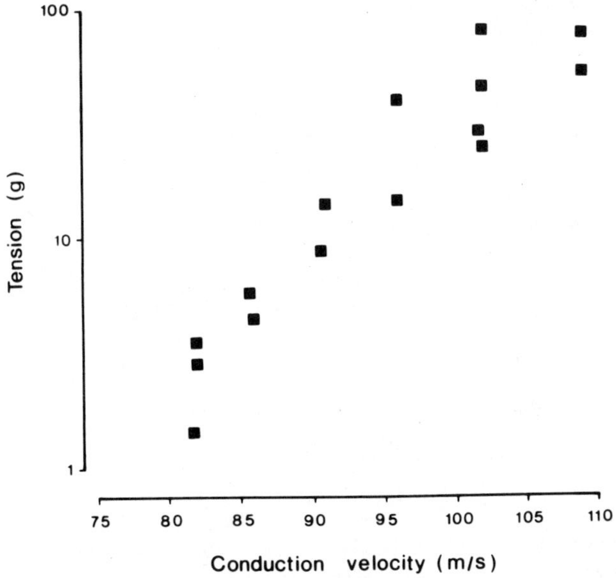

Fig. 1. Distribution of the tetanic tensions of 15 motor units activating the same tendon organ. Peroneus longus muscle. The motor unit tensions are plotted on a logarithmic scale against their axonal conduction velocities. Maximal tetanic tensions were measured upon stimulation at 150/s. (Modified from Jami and Petit 1976a).

subsequently maintained by additional intravenous injections of the drug (4mg/kg) whenever necessary. Blood pressure was monitored, and body temperature was maintained at 38°C.

The methods for isolating and identifying single Ib-afferent fibres and single alpha-axons have already been described in details, as well as the arrangement for recording afferent discharges and contractile tension (Jami and Petit 1976a). In each experiment 8-10 Ib-afferent fibres were functionally isolated in dorsal root filaments. The filaments were placed on the branches of a multi-wired electrode, each of which could be connected to an amplifier. The effect of stimulating single motor units was tested on each afferent and each of the studied tendon organs was found to be activated by several (up to 10) motor units of different physiological types. The muscle tendon was attached to a load cell with low compliance and the muscle length was set systematically within 1mm of the physiological maximum, i.e. the length for which twitch tension was maximal.

Data were collected on an IBM-PC XT microcomputer fed with the output of the amplifiers used to record Ib-afferent discharges and contractile tension. The computer measured the times of arrival of action potentials in dorsal roots and stored them on a floppy disk together with records of contractile tension digitized at a rate of either 500 or 1000 points/s. Records were also kept of the stimulation patterns. The instantaneous frequencies of discharges and stimulation were calculated and displayed, with the corresponding tension records, either on the screen of an oscilloscope or on a X-Y plotter.

Results

The unfused contractions of a motor unit develop a characteristic tension profile displaying oscillations superimposed on a steady level of static tension. In the example illustrated by Fig 2, a fast-contracting motor unit was stimulated with a "step" pattern in which stimulation started at 10/s for 1s, was then abruptly increased to 20/s and after another 0.5s, raised to 40/s by a further step; stimulation frequency was maintained at this level for 1s before returning, by successive steps, to 20 and 10/s. In the motor unit response, tension oscillations appeared in phase with stimulation and, upon augmentation of stimulation frequency, their amplitude decreased while the steady level of static tension

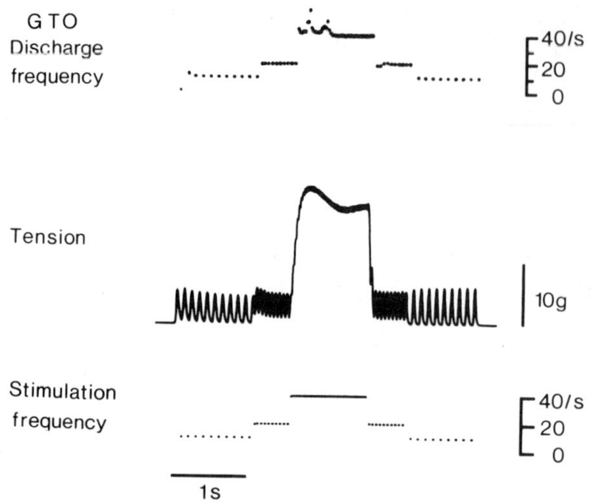

Fig. 2. Response of a peroneus tertius tendon organ to unfused contractions of a fast-contracting motor unit. The discharge of the tendon organ (GTO) and the stimulation are represented in instantaneous frequency. Stimulation pattern described in the text. Conduction velocities of the Ib afferent fibre, 88m/s and of the motor axon, 92m/s.

increased. The tendon organ activated by this motor unit fired one impulse in response to each oscillation of the tension, whatever the amplitude of these oscillations and regardless of the steady level upon which they where superimposed. Comparison of upper and lower records in Fig 2 shows that, except for a brief period at the onset of the 40/s step, the frequency of discharge in the tendon organ response reproduced almost exactly the stimulation pattern. This type of discharge, locked on the stimulation frequency, is termed "1:1 driving" by analogy with the responses of some spindles to fusimotor stimulation, as described by Kuffler et al. (1951). Driving is very common in the responses of tendon organs to unfused contractions of motor units of the fast-contracting types. It was already described in details elsewhere and shown to depend on the dynamic sensitivity of tendon organs (Jami et al. 1985). At any rate, it is clear from Fig 2 that in 1:1 driven responses, the information encoded in the frequency of the tendon organ discharge concerns only the frequency of tension oscillations and not their amplitude or the mean level of tension.

Fig 3 illustrates observations made during an experiment in which the actions of single motor units on several tendon organs could be recorded. Here a fast-contracting motor unit activated 4 tendon organs and their responses to the "step" pattern of stimulation are shown. The first tendon organ (GTO1) was strongly activated and responded with relatively high frequency discharges; note, however, that the maximal frequency of response did not appear when the tension was maximal, i.e. during the 40/s step. The responses of the second and third tendon organs (GTO2 and GTO3) were very similar to that shown in Fig 2. Finally, the fourth tendon organ (GTO4) gave an intermediate response: it was more activated than GTO2 and GTO3 during the first two steps at 10 and 20/s, but less activated during the 40/s step. The differences observed between the individual responses of the four tendon organs suggest that the contraction of this motor unit did not provide an equally efficient stimulus for each receptor. The strongest activation was that of GTO1, but in the response of this tendon organ, no correlation appeared between the variations of discharge frequency and of contractile force. In fact, none of these tendon organ responses, by itself, gave an appropriate account of the tension profile.

However, the pooled discharges of the four tendon organs suggested that this information might be found in the ensemble responses of the receptors. The "average instantaneous frequency" of the four discharges was established as follows: first, for each individual tendon organ, the period during which the interspike intervals were

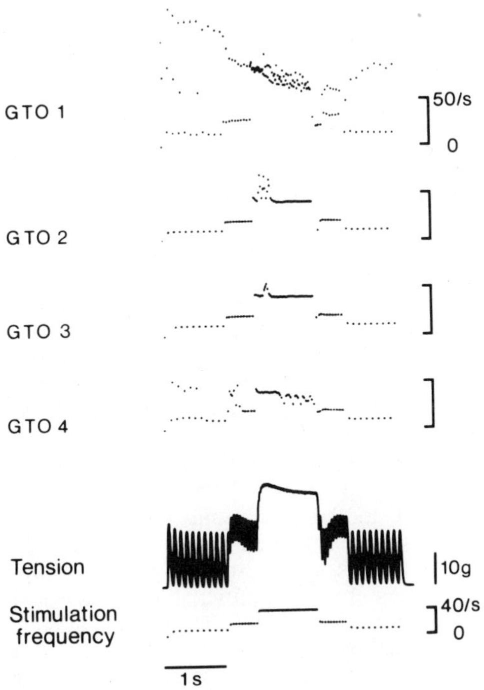

Fig. 3. Responses of four peroneus tertius tendon organs activated by the same motor unit.Same arrangement as in Fig 2. Conduction velocities of the Ib afferent fibres ranged between 88 and 96m/s. Conduction velocity of the motor axon 92m/s.

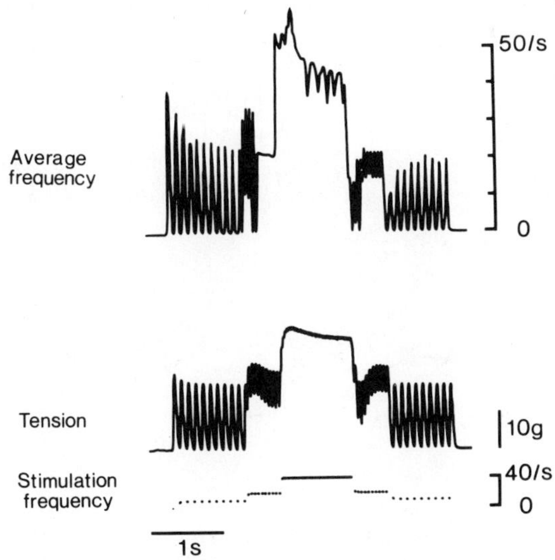

Fig. 4. Instantaneous average frequency of the discharges of the four tendon organs shown in Fig 3. The method used to calculate the average frequency is explained in the text.

collected was divided in bins of equal durations. Different bin widths were tested, ranging from 10 to 50 ms. Then, within each bin, the "average instantaneous frequency" (i.e. the average of reciprocal intervals) was calculated for each tendon organ. Four figures were thus obtained in each bin, i.e., one average frequency per tendon organ. Second, the average of these four figures was calculated within each bin, and this gave the average instantaneous frequency of the pooled responses of the four tendon organs. The result obtained with a bin width of 25 ms is shown in Fig 4. The average frequency of the four tendon organ discharges displayed a profile comparable to the tension profile, with variations reflecting the variations of tension. Modulations of average frequency occurred in correspondence with tension oscillations (see the steps at 10 and 20/s in Fig 4) and a steady level appeared also in correspondence with that of the tension (see the steps at 20 and 40/s in fig 4). In the 40/s step, the average frequency slowly decreased from about 55/s to 40/s in parallel with a slow decrease of tension from about 33g to 30g. It is not certain whether, in the Central Nervous System, the processing of impulses from muscle receptors includes operations comparable to averaging, as proposed by Matthews and Stein (1969). Whatever may be the case, the average frequency of the collective discharge of the tendon organs activated by a single motor unit was found here to provide informations about that unit tension which were not detectable in the instantaneous discharge frequency of individual tendon organs.

In the same experiment, another motor unit was found to act on six tendon organs, of which four were the same as those activated by the unit illustrated in Figs 3 and 4. It was therefore possible to compare the ensemble responses of tendon organs to both motor units and to observe the effects of their combined stimulation. This is shown in Fig 5, where the two units are termed *a* and *b* respectively. Upon stimulation at 10 and 20/s, the amplitudes of tension oscillations were smaller for unit *a* than for unit *b*, and this difference was evident also in the average frequencies of tendon organs. During the step at 40/s, although the two units developed similar tensions, the average frequency of the four tendon organs activated by unit *b* was higher (from 55/s to 40/s) than that of the six receptors activated by unit *a* (from 35/s to 25/s). This example shows that two motor units developing comparable tensions, as recorded at the muscle tendon, can elicit quite different degrees of tendon organ activation, at least in terms of avarage discharge frequency. If the total Ib afferent traffic was considered, in terms of absolute number of impulses, less difference would appear between the effects of unit *a* and unit *b* because six tendon organs firing at about 30/s will produce in total as many impulses as four tendon organs firing at about

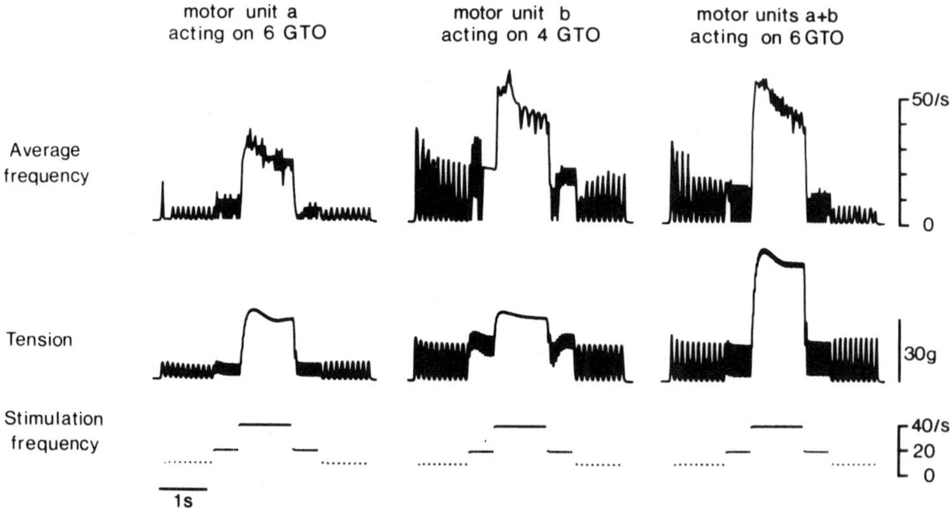

Fig. 5. Instantaneous average frequencies of the tendon organs activated by two motor units stimulated either separately (records on the left and in center) or together (records on the right). The six tendon organs activated by motor unit a included the four tendon organs activated by motor unit b, which were the same as those shown in Figs 3 and 4.

45/s. It remains that the total amount of tendon organ activity produced by a given motor unit does not seem to depend on the number of receptors activated but rather on the efficiency of coupling between the motor unit and each receptor.

When motor units *a* and *b* were stimulated simultaneously (right hand records in Fig 5), summation of tensions occurred, resulting in a slight increase of the amplitude of tension oscillations during the 10 and 20/s steps and a much larger augmentation of the static tension level during the 40/s step. At the end of this step, the mean tension was about 30g for each unit when stimulated separately, and 57g when both units were stimulated together, which indicates an almost linear summation of tensions at the muscle tendon. In contrast, summation of tendon organ discharges was hardly visible. The discrepancy between the increase in tension and the virtual absence of change in tendon organ responses is very clear during the 40/s step, where average frequency decreased slowly from about 55/s to 40/s, i.e., almost the same frequency as for stimulation of unit *b* alone. In this case the ensemble responses of tendon organs still displayed variations reflecting the variations of tension but did not account for the increase in mean level of tension. In other terms, the information carried by the total Ib afferent discharges appeared more appropriate for signaling the dynamic component of contractile force than for measuring its static component.

Discussion

Golgi tendon organs are thought to monitor muscle contraction but, when considering the responses of individual receptors activated by one or several motor units, it is difficult to find which parameter of contraction is encoded in their discharge (see the review by Proske 1981). The force developed in unfused contractions is a composite stimulus for tendon organs, containing a static component (the mean tension) and a dynamic component (the superimposed oscillations). Individual tendon organs respond essentially to the dynamic component but without giving much information about the amplitude of tension variations (Figs 2 and 3; see Jami et al. 1985). The pooled discharges of several tendon organs activated simultaneously were shown to provide additional information on this point and also on the static component of the stimulus.

This was possible because activation of different tendon organs by a single motor unit is not uniform. The same motor unit can be more efficient with some receptors than with others (compare the response of GTO1 with that of GTO3 in Fig 3). Morphological factors may account for these differences. Each of the motor units activating a tendon organ usually contributes a single muscle fibre to the fascicle attached in series with the receptor, but in some instances two fibres may be supplied from the same unit and a stronger activation of the tendon organ can be expected in such instances (Table 1 in Spielmann and Stauffer 1986). Assuming that all the fibres in a motor unit develop the same tension, it seems obvious that if a motor unit contributes more fibres to the fascicle inserted on one tendon organ and less fibres to the fascicles of other receptors, this unit will elicit higher discharge frequencies from the receptor "receiving" more fibres than from any of the other receptors. However, this apparently simple idea implies a further assumption, namely, that the mechanical coupling between muscle fibres and sensory terminals is similar in all tendon organs, which is not always the case. In rat tendon organs, Zelena and Soukup (1983) have shown that some collagenous bundles, within the capsule, are supplied with numerous sensory terminals while others are poorly innervated. As suggested by these authors, a motor unit connected to the tendon organ through a richly innervated bundle is likely to evoke strong responses whereas a unit connected through a bundle receiving only occasional sensory terminals should be less efficient.

The present observations verified only partly the hypothesis that the collective discharge of the tendon organ population in a muscle would provide information on the total amount of contractile force produced in the muscle (Houk and Henneman 1967; Reinking et al. 1975). An acceptable account of the force developed by a single motor unit could be found in the pooled discharges of several tendon organs activated by this unit (Fig 4 and unit a in Fig 5) but when two motor units were stimulated simultaneously, the ensemble discharge of tendon organs did not increase linearly with contractile tension (Fig 5). This was not an artefactual consequence of the particular type of processing chosen to represent tendon organ activity. Similar non-linearities (with all the tested bin widths) were observed when calculating the instantaneous frequency of the pooled discharges without averaging and also when simply counting the numbers of impulses in the pooled responses.

Several causes of non-linear summation may be found in the properties of individual tendon organs. Non-linearity is a characteristic of the receptor potential, as shown by studies of isolated tendon organs (Fukami and Wilkinson 1977; Wilkinson and Fukami 1983) and this is a first reason why linear discharge frequencies cannot be expected. Second, non-linear summation is likely to appear in tendon organs because of overlap between the actions of two different stimulus. It is known that there are slightly more tendon organs in a muscle than Ib fibres in this muscle nerve (Barker 1974) and it may happen that two motor units act separately on two tendon organs innervated by the same afferent fibre. In such instances, interaction between the discharges from the two receptors rarely produces a linear summation (Fukami 1980). Similar interactions may occur within a single tendon organ between the multiple myelinated branches of the Ib fibre (Gregory et al. 1985) and there is also a possibility of mechanical interactions between collagen strands within the capsule (Gregory and Proske 1979; Zelena and Soukup 1983).

Finally, non-linear summation of the effects of two motor units on a group of tendon organs may result from unloading influences exerted by one of the motor units upon some of the tendon organs activated by the other unit (Stuart et al. 1972). A relatively simple example of how this can occur is given in Fig 6 with a tendon organ which discharged spontaneously, in the absence of any contraction, when the muscle length was kept close to its physiological maximum. Stimulation of motor unit *a* interrupted the tendon organ discharge, a clear case of unloading due to the fact that some fibres of this unit ran in-parallel with the receptor (Houk and Henneman 1967). Motor unit *b* developed slightly less tension than unit *a* and strongly activated the tendon organ. When stimulation of unit *a* was superimposed on that of unit *b*, the discharge frequency of the tendon organ was reduced while contractile tension increased. It cannot be inferred from this observation whether the ensemble discharge of all the tendon organs activated by these two units was actually reduced when unit *a* was superimposed on unit *b*, since, besides its unloading effect on this tendon organ, unit *a* certainly activated several other receptors. At any rate, this example suggest that unloading effects can limit the increase in tendon organ discharges when contractile force augments upon recruitment of motor units.

In conclusion, the causes of non-linear relations between the tension developed during muscle contraction and the discharges of tendon organs appear inherent to the structure and to the location of the receptor. It is therefore not surprising to find that there is

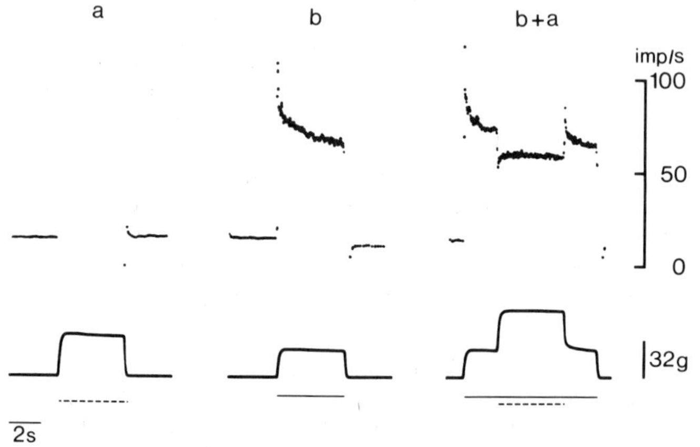

Fig. 6. Unloading of a peroneus longus tendon organ by the contraction of a motor unit. Upper trace, instantaneous frequency of the tendon organ discharge. Lower trace, isometric tension recorded at the muscle tendon. The horizontal lines indicate the duration of the stimulation of motor unit a (dashed line) and b (full line). Both units were stimulated at 75/s. Note that before stimulation of motor units, this particular tendon organ fired continuously. (Jami, Petit, Proske and Zytnicki, unpublished observation).

no linear summation of the ensemble discharges of tendon organs when several motor units are contracting together. It remains that the Ib afferent traffic from a given muscle provides reliable information about the variation of tension, which is likely to be the most significant information on tension used by the Central Nervous System for motor control.

Acknowledgements

The authors wish to thank Pr Y. Laporte for very helpful comments on the manuscript. Thanks are also due to Mrs D. Couton for excellent technical assistance.

This work was supported by C.N.R.S. (ATP n° 960140 and a fellowship to R. Vejsada).

References

Barker, D., 1962, The structure and distribution of muscle receptors, in: Barker, D., ed., Symposium on muscle receptors, Hong-Kong University Press, Hong-Kong, 227-240.

Barker, D., 1974, The morphology of muscle receptors, in: Hunt, C.C., ed., "Muscle Receptors", Springer, Berlin Heidelberg New York, Handbook of Sensory Physiology, Vol III/2: 1-190.

Cooper, S.,1966, Muscle spindles and motor units, in: Andrews, B.L., ed., "Control and Innervation of Skeletal Muscle", Livingstone, Edinburgh, London, 9-16.

Fukami, Y., 1980, Interaction of spike activities originating from individual Golgi tendon organs innervated by branches of a single axon., J. Physiol., 298: 483-499.

Fukami, Y.,1981, Responses of isolated Golgi tendon organs of the cat to muscle contraction and electrical stimulation, J. Physiol., 318: 429-443.

Fukami, Y., and Wilkinson, R. S., 1977, Responses of isolated Golgi tendon organs of the cat, J. Physiol., 265: 673-689.

Gregory, J. E., and Proske, U., 1979, The responses of Golgi tendon organs to stimulation of different combinations of motor units, J. Physiol., 295: 251-262.

Gregory, J. E., and Proske, U., 1981, Motor unit contractions initiating impulses in a tendon organ in the cat, J. Physiol., 313: 251-262.

Gregory, J. E., Morgan, D. L., and Proske, U., 1985, Site of impulse initiation in tendon organs of cat soleus muscle, J. Neurophysiol., 54: 1383-1395.

Gregory, J. E., Morgan, D. L., and Proske, U., 1986, The discharge of cat tendon organs during unloading contractions, Exp. Brain Res., 61: 222-226.

Henneman, E., and Mendell, L., 1981, Functional organization of motoneuron pool and its inputs, in: "The Nervous System", Motor Control. Am. Physiol. Soc, Bethesda Md., Handbook of Physiology, Section 1, Vol II, Part 1, 423-507.

Houk, J. C., and Henneman, E., 1967, Responses of Golgi tendon organs to active contraction of the soleus muscle of the cat, J. Neurophysiol., 30: 466-481.

Horcholle-Bossavit, G., Jami, L., Thiesson, D., and Zytnicki, D., 1985, Anatomy of motoneuronal pools of peroneal muscles in cat spinal cord, Neurosci. Lett., Suppl. 22, S266.

Jami, L., and Petit J., 1976a, Heterogeneity of motor units activating single Golgi tendon organs in cat leg muscles, Exp. Brain Res., 24: 485-493.

Jami, L., and Petit, J., 1976b, Frequency of tendon organ discharges elicited by the contraction of motor units in cat leg muscles, J. Physiol., 261: 633-645.

Jami, L., Petit, J., Proske, U., and Zytnicki, D., 1985, Responses of tendon organs to unfused contractions of single motor units, J. Neurophysiol., 53: 32-42.

Kuffler, S. W., Hunt, C. C., and Quilliam, J. P., 1951, Function of medullated small nerve fibers in mammalian ventral roots: efferent muscle spindle innervation, J. Neurophysiol., 14: 29-54.

Matthews, P. B. C., and Stein R. B., 1969, The sensitivity of muscle spindle afferents to sinusoidal stretching, J. Physiol., 200: 723-748.

Proske, U., 1981, The Golgi tendon organ. Properties of the receptor and reflex actions of impulses arising from tendon organ, Int. Rev. Physiol. 25: 127-171.

Reinking, R. M., Stephens, J. A., and Stuart, D. G., 1975, The tendon organs of cat medial gastrocnemius: significance of motor unit type and size for the activation of Ib afferents, J. Physiol., 250: 491-512.

Scott, J. J. A., and Young, H., 1987, The number and distribution of muscle spindles and tendon organs in the peroneal muscles of the cat, J. Anat., 151: 143-155.

Spielmann, J. M., and Stauffer, E. K., 1986, Morphological observations of motor units connected in-series to Golgi tendon organs, J. Neurophysiol., 55: 147-162.

Stauffer, E. K., and Stephens, J. A., 1975, The tendon organs of cat soleus: static sensitivity to active force, Exp. Brain Res., 23: 279-291.

Stuart, D. G., Mosher, C. C., Gerlach, R. L., and Reinking, R. M., 1972, Mechanical arrangement and transducing properties of Golgi tendon organs, Exp. Brain Res., 14: 274-292.

Swett, J. E., and Eldred, E., 1960, Distribution and numbers of stretch receptors in medial gastrocnemius and soleus muscles of the cat, Anat. Rec., 137: 453-460.

Wilkinson, R. S., and Fukami, Y., 1983, Responses of isolated Golgi tendon organs of cat to sinusoidal stretch, J. Neurophysiol., 49: 976-988.

Wohlfart, G., and Henriksson, K. G., 1960, Observations on the distribution, number and innervation of Golgi muscular tendinous organs, Acta Anat. 41: 192-204.

Zelena, J., and Soukup, T., 1983, The in-series and in-parallel components in rat hindlimb tendon organs, Neuroscience, 9: 899-910.

MODULATING PROCESSES OF SINGLE FUSIMOTOR FIBRE DISCHARGE IN MAN

J.P. Vedel* and J. P. Roll**

* Laboratoire de Neurosciences Fonctionnelles,
CNRS, U2, 31, Chemin Joseph Aiguier
13402 Marseille Cedex 9, France

** Département de Psychophysiologie, Laboratoire de Neurobiologie
Humaine, UA CNRS 372, Université d'Aix-Marseille
13397 Marseille Cedex 13, France

Summary

Experiments were performed on awake human subjects in which single nerve fibre activity was recorded in the lateral peroneal nerve using tungsten microelectrodes as described by Hagbarth & Vallbo (1967).

This discharge of thirteen single efferent fibres innervating the tibialis anterior muscle (TA) or the extensor digitorum longus muscle (EDL) was recorded. On the basis of their functional activity, seven fibres were identified as fusimotor fibres.

Their efferent nature was demonstrated by the fact that the various tests used to identify afferent fibres elicited no response of these fibres. These efferent fibres were considered as fusimotor because their discharges were uncorrelated with any activation of extrafusal muscle fibres. Several means were used to detect activation of extrafusal fibres: surface EMG electrodes, tungsten electrodes deeply implanted in the muscle and especially the use of a high-sensitivity tension transducer (0.1 mN) placed on muscle tendons. Fusimotor fibres were generally spontaneously active with some fluctuation in the discharge frequency.

The activity in fusimotor fibres could be either elicited or modulated under the following conditions: clenching of the fists, pinna twisting, voluntary isometric contraction, passive phasic stretch of the muscle, mental computation, environmental disturbances, subject laughing, the sound of hand clapping, and subject listening to manoeuvre instructions. Moreover, during spontaneous fusimotor fiber activity two subjects were able to voluntarily stop the unit discharge.

The results are compared to those obtained in animal studies and discussed with reference to the notion of alpha-gamma linkage, static and dynamic gamma motoneuron activities, and to another available data concerning the effects of various stimulations on muscle spindle afferent activities in man.

Introduction

The structure and function of mammalian muscle spindles have been extensively studied, including the activity of fusimotor neurons in anaesthetized and awake animals (see Matthews, 1972; 1981a,b; Murthy, 1978; Laporte, 1979; Boyd, 1981; Hulliger, 1984; Prochazka & Hulliger, 1983).

In humans, using the technique of recording single fibre activity with tungsten microelectrodes introduced by Hagbarth and Vallbo (1967), the sensory discharge from muscle spindles has been investigated during voluntary movement (see Vallbo et al., 1979; Burke, 1981; Vedel & Roll, 1983, Hulliger, 1984; Prochaska & Hulliger, 1983). From changes in the afferent spindle discharge as well as muscle tension and electromyography activity, the activity in fusimotor neurons has been indirectly assessed. Surprisingly, no information has been published on activity in gamma fusimotor neurons although the microneurographic technique has been used to study activity in sympathetic axons (Burke et al., 1977; Vallbo et al., 1979) which are considerably smaller.

Since spindle afferent discharge depends not only on muscle length changes and concurrent fusimotor and skeletomotor activity, but also on after-effects of prior fusimotor discharge and stretch (Emonet-Dénand et al., 1985b) a more direct measure of fusimotor activity would be highly desirable. In the present study, conducted on human subjects, efferent activity has been recorded from axons which appeared to be clearly fusimotor. The results have been extensively published in the Journal of Physiology (Ribot, Roll and Vedel, 1986).

Methods

Experiments were performed on healthy human subjects aged 19-30, all of whom had given informed consent to the experimental procedures as required by the Helsinki declaration (1964). The subjects were seated in a comfortable armchair. Their legs were placed in cushioned grooves permitting the maintenance of a stereotyped relaxed position without any muscular activity. The right foot was placed on a stationary plate and the left foot fixed to an electronically driven rotating pedal (Roll & Vedel, 1982). Another mechanical device made it possible to move the toes. Ankle and toe passive and active movements were recorded using linear potentiometers.

Activity from single nerve fibres was recorded from the lateral peroneal nerve (LPn) of the leg at the level of the popliteal fossae. Single unit recording was performed using insulated tungsten microelectrodes (impedance 300-1000 Kohm at 1000 Hz) sterilized with formaldehyde. The microelectrode was inserted transcutaneously until the detection of physiological noise induced by various types of sensory stimulations on the pretibial and dorsal side of the foot which is innervated by the lateral peroneal nerve. The microelectrode was then guided to an appropriate site in the nerve where single unit activity was sought by moving the microelectrode in minute steps. The electrode was always manipulated manually and left "free-floating" with no rigid external fixation. Microelectrodes recording was continuously monitored on an oscilloscope and a loudspeaker.

The activities of the tibialis anterior (TA), extensor digitorum longus (EDL) and soleus-gastrocnemius (SG) muscles were recorded by EMG surface electrodes and sometimes using insulated tungsten elctrodes deeply implanted transcutaneously in the muscle. Moreover, a highly sensitive transducer was placed on the TA and toe extensor muscle distal tendons at the level of the dorsal side of the ankle.

Electrophysiological recordings, transducer recordings and signals indicating the onset and offset of the reinforcement manoeuvres were stored on tape during experiments and then filmed on a multibeam oscilloscope. Spike frequency discharge was analysed using an instantaneous frequency meter.

Results

Thirteen unitary fibres were recorded in the lateral peroneal nerve, all of which were characterized by impulses with a satisfactory signal-to-noise ratio. The usual forms of natural stimulation used to identify afferent fibres, i.e., pressure applied to the skin, tendon, joint and muscle, muscle stretch at low velocity, passive joint movements, response to mechanical vibrations applied to the skin and to the the distal tendon of the muscle involved, produced no activation or modulation of the discharge of these fibres. The fibres tested were thus identified as efferent units and on the basis of their response to the various reinforcement manoeuvres and their discharge characteristics, seven of them were considered as fusimotor fibres.

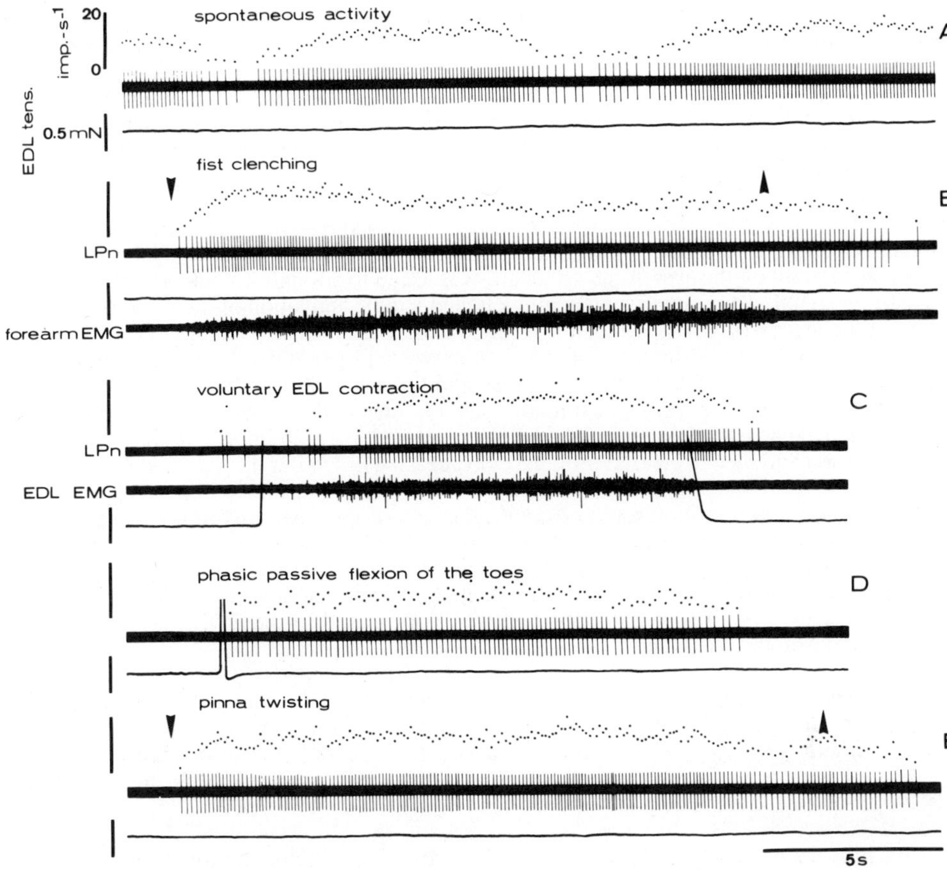

Fig. 1. Modulation of the discharge of a fusimotor fibre innervating the EDL muscle, caused by various reinforcement manoeuvres. Instantaneous frequency curve (upper recording), unitary discharge (LPn, medial recording), muscle tension transducer (lower recording) and occasionally EMG activity were simultaneously recorded. Modulation of the unit discharge frequency was successively observed during spontaneous discharge (A), clenching of the fists (B), during voluntary EDL muscle contraction (C), passive plantar flexion of the toes (D) and pinna twisting (E). The maximal discharge frequency observed during spontaneous activity and the various manoeuvres fluctuated from 10 to 18 imp.s^{-1}. (Spikes were retouched.)

These seven fusimotor fibres recorded all showed a conspicuous spontaneous discharge (5-20 imp.s^{-1}) in completely relaxed subjects. Here the spontaneous discharge fluctuated from time to time, sometimes stopping momentarily and then firing again for a long period although the subject was still relaxed and the environment relatively quiet and steady in terms of mental stimulation (Fig. 1A). This spontaneous discharges was never related to the EDL or TA muscle activity monitored by EMG surface electrodes and particularly by the high-sensitivity transducer. Such an efferent activity suggested that these unitary discharges were related to fusimotor fibres.

Various manoeuvres were then made in order to test this assumption. Of the seven fibres tested, five yielded only partial information about their modalities of activation because their recording time was short (10-20 min). The two others were recorded during more than one hour, so that it was possible to test several times all the various reinforcement manoeuvres likely to activate the fusimotor system.

When the subject was asked to perform a particular task although he was required to rest and remain relaxed, the verbal description of the task usually caused a variable enhancement of the discharge frequency of the unit recorded which was probably related to change of the subject's attention (Fig. 2B). The starting signal was given only when the unit activity was stable (or nil). No difference in the fibre discharge modulation was observed whether or not the subject received feedback information about the unitary activity provided by a loudspeaker and an oscilloscope.

1) Excluding the Jendrassik's manoeuvre, which frequently triggered muscle activity, a motor reinforcement manoeuvre was performed in which the subject, whose forearms were lying comfortably on the arm of the armchair, was asked to strongly clench both fists. When the starting signal was given, fist clenching elicited a tonic activation of the fusimotor fibres after a short latency (Fig. 1B) without any TA or EDL muscle tension change. This activation persisted throughout the duration of the manoeuvre and generally continued for several seconds after the instruction to stop was given and forearm EMG activity disappeared. This kind of manoeuvre had very conspicuous effects on all the units tested. Under these conditions, the discharge frequency of the units increased to 15-25 imp.s^{-1} at the begining of the stimulation and then decreased progressively, ceasing only after the signal to stop.

2) With regard to motor and reflex effects, six of the seven fusimotor units studied were activated during slight isometric muscle contraction. It was thus possible to identify the location of the muscle spindle supplied by each fibre: four were found to supply the EDL muscle responding to its selective isometric contraction, and the other two were attributed to the TA muscle. The seventh unit recorded was also related to the EDL muscle but only responded to isotonic contraction.

The activity of only one fusimotor fibre began regularly to increase before isometric muscle contraction, one fired poorly under the same condition (Fig. 1C), and the discharge of four other fibres (3 EDL, 1 TA) increased after the skeletomotor fibre activation, sometimes with a latency of several seconds. At the end of the contraction, fusimotor discharge stopped immediately or persisted for only a short time.

Since the existence of an autogenic excitation of the gamma motoneurons by the Ia sensory fibres of its own muscle has been confirmed by various authors (Ellaway & Trott, 1976, 1978; Ellaway et al., 1976; Appelberg et al., 1983), reflex activation of the recorded human fusimotor fibres was tested using muscle tendon vibration. No activation was obtained under these conditions, only in one unit did activity (TA) increase slightly when vibration induced a tonic vibration reflex. Nevertheless, noticeable selective activation of five fusimotor fibres was observed (4 identified as EDL units, and 1 responding only to the EDL isotonic contraction) when a brief EDL muscle stretch was induced by phasic plantar flexion of the toes (Fig. 1D). In all five cases the increase in the discharge frequency persisted for a long time after the end of phasic stimulation (5-20 s). The two units innervating the TA muscle could not be submitted to the phasic reflex test. This failure resulted from the displacement of the electrode during the manoeuvre causing the loss of the fibre. Trials involving TA tendon percussion did not produce any activation of these two units.

3) As previously demonstrated by Granit et al. (1952) in animals, pinna twisting was also able vigorously to excite the fusimotor neurons (Fig. 1E). This effect, called "pinna reflex" by Granit et al. (1952), was obtained without any muscle activation. The increase in fusimotor discharge frequency began either immediately or after a short latency, continued as long as stimulation was maintained and generally persisted several seconds after pinna twisting stopped.

4) In addition to these three reinforcement manoeuvres, fusimotor fibre activation was also observed in various situations such as hand clapping behind the subject's head (Fig. 2A), subject laughing (Fig. 2D), subject speaking and changes in environmental conditions, such as a person entering the experimental room. Discharge frequency of the fusimotor fibres in these situations was very variable but never exceeded the values obtained in the previous reinforcement manoeuvres. In fact, none of the activating processes ever elicited a firing rate higher than the maximum spontaneous discharge of these fusimotor fibres.

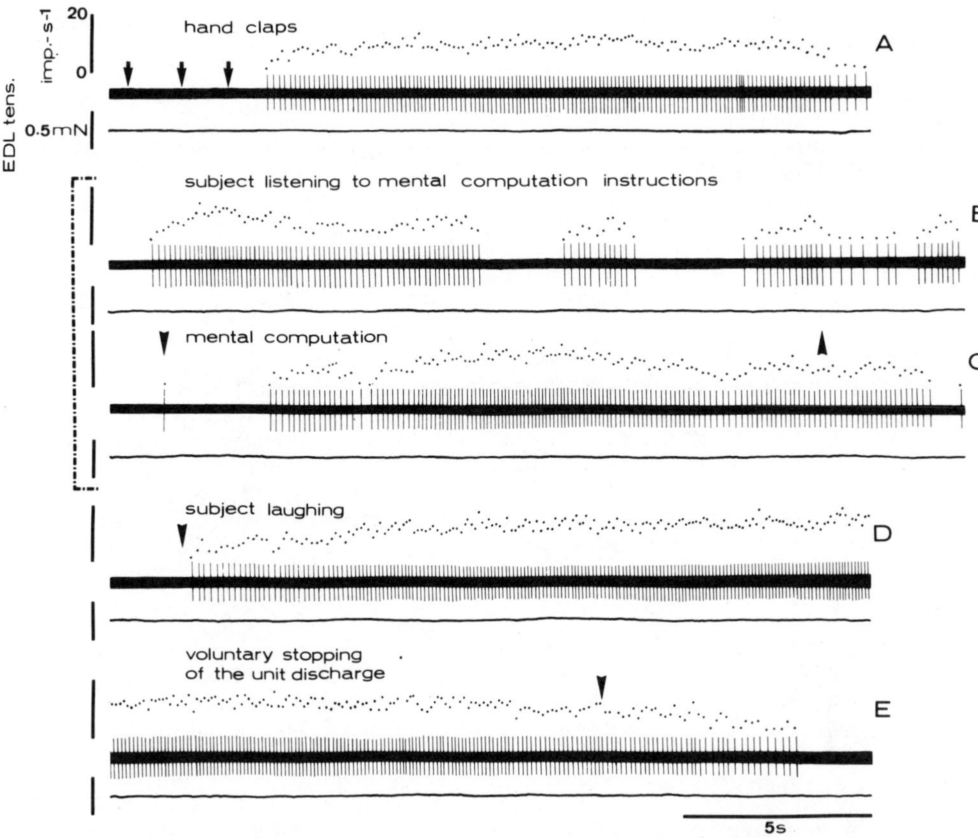

Fig. 2. Modulation of the same fusimotor fibre activity as in Fig. 1 but under different experimental conditions. Modulation of unit discharge frequency (maximal discharge frequency: 8-18 imp.s^{-1}) by hand clapping (A), subject listening to the mental computation instructions (B), mental computation (C), subject laughing (D) and during voluntary control by the subject of the fibre discharge causing its activity to stop (E). (Spikes were retouched.)

5) Mental computation (for example: addition of all the odd numbers from 34 to 72) was also a very efficient means to modulate fusimotor activity. This manoeuvre was tested on four of the seven units recorded. In this case, providing verbally the instructions to the subject produced always an increase of the unit discharge. After the starting signal, the latency of the unit discharge facilitation was very variable (sometimes for several seconds). Both intertrial and intersubject variability were observed. Discharge frequency under these conditions often reached a lower value than during other manoeuvres and was found to be somewhat variable but never stopped during the mental computation. At the stopping signal, the activity progressively decreased, but increased again when the experimenter asked the subject to give the result. Figures 2B and 2C illustrate on one EDL unit activity the effect of mental computation.

6) In two experiments, subjects were asked during spontaneous firing of EDL fusimotor fibres to try to stop the unit discharge by entering a state of deep mental and somatic relaxation. During this test, subjects received feedback information concerning their own fibre activity, which was provided by an oscilloscope and a loudspeaker. After several trials, one subject consistently succeeded in achieving this state, completely silencing

fusimotor activity with a relatively short latency (Fig. 2E). The other subject suceeded only in decreasing the firing rate.

7) Finally, various other reinforcement manoeuvres known to activate gamma motoneurons in animals (see, Hagbarth, 1952; Granit et al., 1952; Hulliger, 1984) never modified the human fusimotor fibre activity (pressure on the eye bulb, pinching or scratching of the skin covering the EDL and TA muscles, moving the head and slightly contracting the contralateral leg).

Discussion

A) Methodological considerations

The present study deals with the recorded activity from seven fusimotor units. This may appear to be a limited pool of neurons although five years has passed since we first recorded efferent activity. The reason why the activity of these fibres was so difficult to record in man with the electroneurography method could be either that the muscular efferent fibre bundles are difficult to reach with a microelectrode (e.g. they may be located on the posterior side of the nerve), or that the efferent fibres are not arranged in bundles as are the sensory units and so they mingle with the afferent fibres. It should be noted that the recordings of efferent activities reported here have mainly been made by chance and we are still unable to describe any systematic method for locating such fibres.

The activity recorded from the seven single axons are considered to be efferent discharges because of the total lack of response to natural sensory stimulation. In particular the possibility that some of them, mainly activated by muscle contraction, could be Golgi tendon organ afferents was excluded because of the lack of response to stimuli which normally excite these receptors such as large amplitude muscle stretch and strong pressure applied on muscle and tendon (Burke et al.,1976a,b; Vallbo et al., 1979; Vedel & Roll, 1983).

The main purpose of this study was to demonstrate that the seven unitary activities recorded could be ascribed to fusimotor fibres. Their distinction from the skeletomotor fibres was mainly based on the existence of spontaneous activity and on the increase in discharge produced by various reinforcement manoeuvres, both observed without any concomitant muscle contraction. Another important point was to distinguish the fusimotor activity from that of sympathetic fibres which were the only efferents previously recorded and which were often activated by similar reinforcement manoeuvres (Delius et al., 1972a; Normell & Wallin, 1974). The first argument in favour of identifying these units as fusimotor fibres is the relatively large size of the unitary nerve impulses recorded. Indeed, Vallbo et al. (1979) have reported that in most of their experiments the C fibre firing was characterized by mass activity and only occasionally by discharge of single small-amplitude units (Hallin & Torejork, 1974). Moreover, the authors specify that "in muscle nerve fascicles the resting activity of the sympathetic fibres is characterized in man by fairly regular pulse-synchronous bursts of impulses, sometimes occurring in short sequences separated by periods of neural silence, sometimes in more staccato patterns" (Hagbarth & Vallbo, 1968; Delius et al., 1972b; Sundlof & Wallin, 1977). The spontaneous efferent activities recorded in our experiments never manifested such patterns of discharge. Confusion with beta fibres might be possible during isometric contraction, but the fact that the same unitary discharge was elicited by other manoeuvres, even when there was no concomitant muscular activity, rules out this possibility.

B) Functional considerations

In the light of the present results, it is interesting to consider certain data and hypotheses relating to fusimotor fibre properties and functions in man.

With regard to the gamma motoneuron activity, its occurrence and the modulation of its discharge were only inferred indirectly in previous studies on man where variations in muscle spindle sensory discharge were recorded. Recording of the activity of fusimotor fibres gives more direct answers to several questions.

a) Are skeletomotor fibres and fusimotor fibres coactivated or independent?

It has recently been demonstrated (Taylor & Appenteg, 1981; Prochaska & Wand, 1981; Appenteg et al., 1983) that in anaesthetized as well as in freely moving cats, strict alpha-gamma linkage does not necessarily occur during all motor activities. In man, the possibility of alpha-gamma independence has also been suggested (Vallbo & Hulliger, 1981; Burke et al., 1980). Our observations do not provide any evidence on the possibility that such alpha-gamma independence occurs during movement, but show that in the relaxed and awake subject, selective activation of fusimotor fibres occurs spontaneously or can be induced. This observation supports the view that alpha-gamma independence can occur under some behavioural conditions.

b) Skeletomotor fibre-fusimotor fibre coactivation: difference in onset time.

With regard to the difference in onset time between skeletomotor and fusimotor activation, Vallbo (1971) observed at the onset of a voluntary isometric contraction, "that, at best, the fusimotor neurons were activated at the same time, but not before alpha motoneurons were". This suggestion, extrapolated from observations of spindle afferent activity, has also been put forward by other authors (Hagbarth et al., 1975; Burke & Eklund, 1977; Burke et al., 1978a,b), and is confirmed by our more direct results, since under similar experimental conditions, only one of six fusimotor fibres fired before the onset of the EMG burst. In this case our observations on the temporal relations between alpha and gamma discharge do not seem to comply with the motor fibre recruitment law (the size principle) established by Henneman and his collaborators (Henneman et al., 1965a,b; Henneman & Medel, 1981).

c) Are static or dynamic fusimotor axons activated in relaxed human subjects?

The similarity of fusimotor firing patterns observed during various manoeuvres in the present study suggests that the recorded activity resulted from the same kind of unit, probably dynamic gamma motoneurons. In support of this argument is the fact that when the same tests were applied to twenty fibres identified as Ia afferent fibres, none of these exhibited an increase of firing rate (Vedel & Roll, unpublished data). Since Ia fibre tonic discharge is increased to a greater extent by gamma static neurons than by the gamma dynamic neurons (Crowe & Matthews, 1964; Brown & Matthews, 1966) one would have expected tonic activation of spindle afferents if the reinforcement manoeuvres activated static fusimotor fibres. The lack of spindle afferent static activation in our experiments might be attributable to the fact that dynamic gamma motoneurons selectively innervate the bag1 intrafusal muscle fibre (Boyd et al., 1977; Laporte, 1978), which contracts almost without shortening, unlike the bag2 and chain intrafusal muscle fibres, which shorten considerably during static gamma stimulation (Boyd, 1981; Boyd et al., 1981). Hence, it may be suggested that this quasi-isometric intrafusal contraction acting only on the bag1 spiral ending of the Ia fibre may not suffice, under our experimental conditions, to increase the primary ending static discharge since the latter seems to be mainly attributable to the Ia spiral endings located on the bag2 and chain fibres.

Although the above considerations provide support for our hypothesis concerning the selective activity of dynamic gamma motoneurons in relaxed human subjects, they do not provide final confirmation. A strong argument would be provided if a significant increase could be observed in primary ending dynamic sensitivity using the same reinforcement manoeuvres as for the fusimotor fibres in our experiments. Unfortunately, only negative results on dynamic sensitivity modulation in relaxed human subjects have been reported. For example, complete block of the peroneal nerve produced by anaesthesia (Burke et al., 1976b) or pressure (Burke et al., 1979; Burke, 1981) has no effect on the response of the primary endings to muscle stretch, tendon percussion and vibration (Vallbo, 1974; Burke et al., 1979).

Indirect confirmation of our findings and deductions may be found in previous studies on the human monosynaptic spinal reflex (Paillard, 1955, 1959) comparing its amplitude variation when it is elicited by electrical nerve stimulation (Hoffman's "H" reflex) or by tendon percussion (tendon jerk reflex, "T" reflex). With this method it has been clearly demonstrated that T reflex amplitude can be selectively increased, without H reflex change, by reinforcement manoeuvres similar to those used in our microneurography experiments: e.g. loud sounds, fist clenching and mental computation. Since T reflex amplitude is determined by the level of the dynamic sensitivity of spindle primary endings, it can now be suggested that its increase might result from specific activation of dynamic gamma motoneurons. From these observations, Paillard (1955; 1959) had already deduced that in relaxed human subjects, modification of the spindle dynamic reactivity could result from changes in the "intrafusal myofibril stiffness" without increase in the receptor spontaneous discharge.

d) Fusimotor fibre reflex activation

Phasic plantar flexion of the toes was the only peripheral manoeuvre investigated which caused activation of the EDL fusimotor fibres. This effect might be due to the autogenetic reflex of gamma motoneurons which has now been firmly established (Fromm & Noth, 1976; Fromm et al., 1974; Trott, 1976; Ellaway et al., 1976). Phasic flexion of the toes indeed elicits a brief EDL muscle stretch, mainly activating the Ia fibres in a phasic burst, which is a prerequisite for gamma motoneuron autogenetic reflex activation (Lundberg & Winsburg, 1960; Stuart et al., 1970). Furthermore, the long duration of the increase in the fusimotor discharge characterizing this reflex effect in our experiments could suggest a reflex action occurring through polysynaptic pathways as demonstrated by animal experiments (Murthy, 1978).

In conclusion, although we successfully recorded from seven fusimotor fibres in man, our investigation has yielded mainly qualitative data. Presumably, greater experimental efficiency would allow a more quantitative study of the functional properties of the fusimotor fibres in man. It would also allow investigation of their role in a wider variety of voluntary motor tasks in order to disclose the involvement of these fibres in the preparation for movement and its regulation.

Nevertheless, these preliminary data demonstrate the great variety of situations able to modulate the fusimotor discharge, these situations being motor as well as mental, behavioural as well as environmental.

References

Appelberg, B., Hulliger, M., Johansson, H., and Sojka, P. C., 1983, Actions on gamma-motoneurons elicited by electrical stimulation of muscle group I afferent fibres in the hindlimb of the cat, J. Physiol., 335: 237-253.

Appenteg, K., Hulliger, M., Prochazka, A., and Zangger, P., 1983, Fusimotor action during movement, deduced by matching the pattern of afferent discharge in freely moving and anaesthetized cats, J. Physiol., 334: 29-30.

Boyd, I. A., 1981, The action of the three types of intrafusal fibres in isolated cat muscle spindle on the dynamic and length sensitivities of primary and secondary sensory endings, in: Taylor, A., Prochazka, A, Eds., "Muscle receptors and movement", MacMillan, London, pp. 17-32.

Boyd, I. A., Gladden, M. H., McWilliam, P. N., and Ward, J., 1977, Control of dynamic and static nuclear bag fibres and nuclear chain fibres by gamma and beta axons in isolated cat muscle spindles, J. Physiol., 265: 133-162.

Brown, M. C., and Matthews, P. B. C., 1966, On the subdivision of efferent fibres to muscle spindle into static and dynamic fusimotor fibres, in: Andrew, B.L., Ed., "Control and innervation of skeletal muscle", Truex, Oxford, pp. 18-31.

Burke, D., 1981, The activity of human muscle spindle endings in normal behaviour, Int. Rev. Physiol., 20: 91-136.

Burke, D., Hagbarth, K. E., and Lofstedt, L., 1978a, Muscle spindle responses in man to changes in load during accurate position maintenance, J. Physiol., 276: 159-164.

Burke, D., Hagbarth, K. E., and Lofstedt, L., 1978b, Muscle spindle activity in man during shortening and lengthening contractions, J. Physiol., 277: 131-142.

Burke, D., Hagbarth, K. E., and Skuse, N., 1979, Voluntary activation of spindle endings in human temporarily paralysed by nerve pressure, J. Physiol., 287: 329-336.

Burke, D., MacKeon, B., and Westerman, R. A., 1980, Induced changes in the threshold for activity in preparation for a voluntary contraction, J. Physiol., 302: 171-181.

Burke, D., Sundlof, G., and Wallin, G., 1977; Postural effects on muscle nerve sympathetic activity in man, J. Physiol., 272: 399-414.

Burke, D., Hagbarth, K. E., Lofstedt, L., and Wallin, B. G., 1976a, The response of human muscle spindle endings to vibration of non-contracting muscles, J. Physiol., 261: 673-693.

Burke, D., Hagbarth, K. E., Lofstedt, L., and Wallin, B. G., 1976b, The response of human spindle endings to vibration during isometric contraction, J. Physiol., 277: 131-142.

Burke, D., McKeon, B., Skuse, N.F., and Westerman, R. A., 1980, Anticipation and fusimotor activity in preparation for a voluntary movement, J. Physiol., 306: 337-348.

Crowe, A., and Matthews, P. B. C., 1964, The effects of stimulation of static and dynamic fusimotor fibres on the response to stretching of the primary endings of muscle spindles, J. Physiol., 174: 109-131.

Delius, W., Hagbarth, K. E., Hongell, A., and Wallin, B. G., 1972a, General characteristics of sympathetic activity in human nerves, Acta Physiol. Scand., 84: 65-81.

Delius, W., Hagbarth, K. E., Hongell, A. and Wallin, B. G., 1972b, Manoeuvres affecting sympathetic outflow in human skin nerves, Acta Physiol. Scand., 84: 177-186.

Ellaway, P. H., and Trott, J. R., 1976, Reflex connections from muscle stretch receptors to their own fusimotor neurons, Prog. Brain Res., 44: 113-122.

Ellaway, P. H., and Trott, J. R., 1978, Autogenic reflex action onto gamma motoneurons by stretch of triceps surae in the decerebrate cat, J. Physiol., 276: 49-66.

Ellaway, P. H., Pascoe, J. E., and Trott, J. R., 1976, The effects upon fusimotor neurons on small, brief stretches of their muscles, J. Physiol., 258: 48-49.

Emonet-Dénand, F., Hunt, C. C., and Laporte, Y., 1985a, Fusimotor after-effects on responses of primary endings to test dynamic stimuli in cat muscle spindles, J. Physiol., 360: 187-200.

Emonet-Dénand, F., Hunt, C. C., and Laporte, Y., 1985b, Effects of stretch on dynamic fusimotor after-effects in cat muscle spindles, J. Physiol., 360: 201-223.

Fromm, C., and Noth, J., 1974, Autogenetic inhibition of gamma motoneurons in the spinal cat uncovered by DOPA injection, Pflügers Archiv für die ges. Physiol., 349: 247-256.

Fromm, C., Haase, J., and Noth, J., 1974, Length-dependent autogenetic inhibition of extensor gamma motoneurons in the decerebrate cat, Pflügers Archiv für die ges. Physiol., 363: 81-86.

Granit, R., Job, C., and Kaada, B. R., 1952, Activation of muscle spindle in pinna reflex, Acta Physiol. Scand., 27: 161-168.

Hagbarth, K. E., 1952, Excitatory and inhibitory skin area for flexor and extensor motoneurons, Acta Physiol. Scand. (suppl. 94), 26: 1-58.

Hagbarth, K. E., 1979, Exteroceptive, proprioceptive and sympathetic activity recorded with microelectrodes from human peripheral nerves, Mayo Clinic Proc., 54: 353-365.

Hagbarth, K. E., and Vallbo, A. B., 1967, Mechanoreceptor activity recorded percutaneously with semimicroelectrodes in human peripheral nerves, Acta Physiol. Scand., 69: 121-122.

Hagbarth, K. E., and Vallbo, A. B., 1968, Pulse and respiratory grouping of sympathetic impulses in human muscle nerves, Acta Physiol. Scand., 74: 96-108.

Hagbarth, K. E., Wallin, G., and Lofstedt, L., 1975, Muscle spindle activity in man during voluntary fast alternating movements, J. Neurol. Neurosurg. and Psychiat., 38: 1143-1153.

Hallin, R. G., and Torebjork, H. E., 1974, Single unit sympathetic activity in human skin nerves during rest various manoeuvres, Acta Physiol. Scand., 92: 303-317.

Henneman, E., Somjen, G., and Carpenter, P. O., 1965a, Functional significance of cell size in spinal motoneurons, J. Neurophys., 28: 560-580.

Henneman, E., Somjen, G., and Carpenter, P. O., 1965b, Excitability and inhibitibility of motoneurons of different sizes, J. Neurophys., 28: 599-620.

Henneman, E., and Mendell, L. M., 1981, Functional organization of motoneuron pool and its inputs, in: Brookhart, J.M., Mountcastle, V.B., and Brooks, V.B., Eds., "Motor control (part 1), Handbook of Physiology", sect. 1, vol. 2, American Physiological Society, Bethesda, pp 423-507.

Hulliger, M., 1984, The mammalian muscle spindle and its central control, Rev. Physiol. Bioch. Pharmacol., 101: 110 pp.

Jendrassik, E., 1883, Beiträge zur Lehre von den Sehnenreflexen, Statist Archiv für Klin. Med., 33: 177-199.

Laporte, Y., 1978, The motor innervation of the mammalian muscle spindle, in: Porter R., Ed., Studies in Neurophysiology presented to McIntyre A.K., Cambridge University Press, pp 45-59.

Laporte, Y., 1979, Innervation of cat muscle spindles by fast-conducting skeletomotor fibres, in: Asanuma H., Wilson V.J., Eds., "Integration in the nervous system", Igaku-Shoin, Tokyo, pp 3-12.

Lundberg, A., Winsbury, G., 1960, Selective adequate activation of large afferents of muscle spindles and golgi tendon organs, Acta Physiol. Scand., 49: 155-164.

McCloskey, D. I., 1978, Kinesthetic sensibility, Physiol. Rev., 58: 768-820.

Matthews, P. B. C. 1972, Mammalian muscle receptors and their central action, Monographs of the Physiological Society, Edward Arnold Ldt., London, 630 pp.

Matthews, P. B. C., 1977, Muscle afferents and kinaesthesia, British Med. Bull, 33: 137-142.

Matthews, P. B. C., 1981a, Evolving views on the internal operation and functional role of the muscle spindle, J. Physiol., 320: 1-30.

Matthews, P. B. C., 1981b, Muscles spindles: their messages and their motor supply, in: Brookhart, J.M., Mountcastle, V.B., and Brooks, V.B., Eds., "Motor control, part 1, Handbook of Physiology", sect 1, vol. 2, American Physiological Society, Bethesda, pp 189-228.

Murthy, K. S. K., 1978, Vertebrate fusimotor neurons and their influences on motor behavior, Prog. in Neurobiol., 11: 249-307.

Normell, L. A., and Wallin, B. G., 1974, Sympathetic skin nerve activity and skin temperature changes in man, Acta Physiol. Scand., 91: 417-426.

Paillard, J., 1955, Réflexes et régulations d'origine proprioceptive chez l'Homme, Etude neurophysiologique et psychophysiologique, Thèse de Doctorat d'Etat, Arnette, Paris, 293 pp.

Paillard, J., 1959, Functional organization of afferent innervation studies in man by monosynaptic testing, Am. J. Phys. Med., 38: 239-247.

Prochazka, A., and Wand, P., 1981, Independence of fusimotor and skeletomotor systems during voluntary movement, in: Taylor, A., and Prochazka, A., Eds., "Muscle receptors and movement", McMillan Ldt., London, pp 229-243.

Prochazka, A., and Hulliger, M., 1983, Muscle afferent function and its significance for motor control mechanisms during voluntary movement in cat, monkey and man, in: Desmedt, J.E., Ed., "Motor control mechanisms in health and disease", Raven Press, New York, pp 93-132.

Ribot, E., Roll, J. P., and Vedel, J. P., 1986, Efferent discharges recorded from single skeletomotor and fusimotor in man, J. Physiol., 375: 251-268.

Roll, J. P., and Vedel, J. P., 1982, Kinaesthetic role of muscle afferents in man, studied by tendon vibration and microneurography, Exp. Brain Res., 47: 177-190.

Ruffini, A., 1898, On the minute anatomy of the neuromuscular spindles of the cat, and on their physiological significance, J. Physiol., 23: 190-208.

Stuart, D. G., Mosher, C. G., Gerlach, R. L., and Reinking, R. M., 1970, Selective activation of Ia afferents by transient muscle stretch, Exp. Brain Res., 10: 177-187.

Sundlof, G., and Wallin, B. G., 1977, The variability of muscle nerve sympathetic activity in resting recumbent man, J. Physiol., 272: 383-397.

Taylor, A., and Appenteg, A., 1981, Distinctive modes of static and dynamic fusimotor drive in jaw muscles, in: Taylor, A., Prochazka, A., Eds., "Muscle receptors and movement", McMillan, London, pp 179-192.

Trott, J. R., 1976, The effect of low amplitude vibration on the discharge of fusimotor neurons in the decerebrate cat, J. Physiol., 255: 635-650.

Vallbo, A. B., 1971, Muscle spindle response at the onset of voluntary isometric contractions in man, Time difference between fusimotor and skeletomotor effects, J. Physiol., 218: 405-431.
Vallbo, A. B., 1974, Human muscle spindle discharge during isometric voluntary contractions, Amplitude relations between spindle frequency and torque, Acta Physiol. Scand., 90: 303-318.
Vallbo, A. B., and Hulliger, M., 1981, Independence of skeletomotor and fusimotor activity in man, Brain Res., 223: 176-180.
Vallbo, A. B., Hagbarth, K. E., Torebjork, H. E., and Wallin, B. G., 1979, Somatosensory, proprioceptive and sympathetic activity in human peripheral nerves, Physiol. Rev., 59: 919-957.
Vedel, J. P., and Roll, J. P., 1983, Muscle spindle contribution to the coding of motor activities in man, in: Massion, J., Paillard, J., Schultz, W., and Wiesendanger, M., Eds., "Neural coding of motor performance", Springer Verlag, Berlin, Heidelberg, New York, Exp. Brain Res., Suppl. 7, pp 253-265.

EXTRAOCULAR PROPRIOCEPTION AND BODY POSTURAL REFERENCES

J.P. Roll and R. Roll

Laboratoire de Neurobiologie Humaine
U.A. C.N.R.S. 372 - Université de Provence
Avenue Escadrille Normandie Niemen
13397 Marseille, France

Summary

Painless, well patterned vibrations, applied with an increasing frequency from 10 to 80 Hz, to either the medial or lateral rectus of a subject's eye(s) (right, left or both), were found to induce directional perceptual and motor effects which were closely related to the postural context. The same was true with the superior and inferior recti. Illusory and/or actual movements of head, trunk or whole body were produced depending on the postural constraints (head free or fixed, standing or seated subject). Likewise, illusory directional shift of a visually fixed target in darkness during extraocular muscle vibration was reported by subjects. The contribution of extraocular proprioception to the coding of eye, head and body position in relation to postural and environmental conditions is discussed.

Introduction

The existence of reciprocal relationships between the visual and postural system has now been clearly established (De Cyon, 1911; Baron, 1955; Nashner, 1970; Berthoz, 1978; Lacour, 1981). Both vision and plantar support, closely linked with vestibular information, are responsible for maintaining postural equilibrium. Thanks to the eyeball's great mobility, the eye is constantly exploring the environment and informing the organism, so that it can act or react by adopting a postural configuration which is appropriate to the ongoing or forthcoming event. On the other hand, a stabilized head seems to be a prerequisite for efficient visual information collection.

All these processes serve to connect two classically dissociated spaces: body and extracorporeal space. The former is constantly being organized into a series of static and dynamic postural configurations corresponding to the whole range of body activities. This organization involves the use of always available basic body references, which result from the subject's knowledge of the respective positions of the mobile linked body segments, namely the eyes, head, trunk and lower limbs.

Extracorporeal space is the physical space surrounding objects, within which their geometrical relationships are defined and our actions take place. The successful accomplishment of these actions mainly depends on their specific coding by the CNS, which interrelates the two types of space during spatially oriented behavior. Since the gaze plays a leading part in body and limb orientation (Paillard and Beaubaton, 1978), we assumed that extraocular proprioception might make a non-negligible contribution to this type of interaction between body and environment.

The visual localization of an object in extracorporeal space is known to involve both retinal and extra-retinal position cues: the former provide information about the degree of object eccentricity, and the latter to the position of the eye itself in space. Most of the previous studies on the interactions between vision and posture have focused, however, on retinal factor analysis, particularly on the decisive role of the peripheral retina in the dynamic control of posture (Berthoz, 1974; Lee and Lishman, 1975; Dichgans and Brandt, 1978). Few have dealt with the relationships between postural regulation and oculomotricity as such, or more specifically, with the contribution of extraocular proprioception.

Extraocular proprioception, although often rejected by those favouring the "efferent copy" hypothesis (Brindley and Merton, 1960), has been shown to exist in vertebrates in studies identifying its specific sensory receptors (Cooper and Daniel, 1949; Cooper et al., 1955; Bach y Rita and Ito, 1966; Barker, 1973; Mukuno, 1986), nervous pathways (Batini and Buisseret, 1974), and central projections (Milleret, 1985; Ashton et al., 1984; Lacour et al., 1987). In addition, as Buisseret and Gary-Bobo (1979) have shown, extraocular proprioceptive afferents play a decisive role in the development of directional selectivity in the visual neurons. At the behavioural level, early suppression of these afferents in kittens seriously affects the development of visually guided activity (Hein and Diamond, 1982). In man, kinaesthetic sensations and directional postural displacements can be elicited by vibrating muscles in the eyeball region (Roll and Roll, 1987), which suggests that a close relationship may exist between extraocular proprioception and postural regulation and coding.

At skeletal level, the fact that a vibratory stimulus applied to the muscle tendon of blindfolded subjects elicits illusory sensations of movement without any real joint movement taking place, strongly supports the idea that these signals are of muscle proprioceptive origin.

At our laboratory, using both neurophysiological and psychophysiological approaches, we have shown directly, by means of microneurographic recordings on human sensory nerves, that the nervous messages elicited by tendon vibration leading to illusory sensation of movement were almost exclusively of muscle spindle origin (Roll and Vedel, 1982). The specificity of the vibratory stimulus used to activate the primary sensory endings of muscle spindles, and the fact that the latter respond one to one to the mechanical stimulus between 1 and 100 cycles per second, make this method particularly suitable for eliciting artificial proprioceptive patterns which are very similar to those produced by performance of an actual movement. Any modulation of the vibration frequency consistantly entails a corresponding change in the firing rate of most of the muscle spindles in the vibrated muscle. Vibration therefore constitutes an efficient means of obtaining true copies of actual nervous messages, and thus of eliciting "at will" illusory movements the direction, speed and duration of which can be preselected. By peripherally specifying the spatial distribution (agonist and/or antagonist muscles), the time course (duration, simultaneity, alternation) and the frequency (constant, increasing or decreasing) of the vibratory stimulation, it is possible to elicit complex proprioceptive messages which are identified by the cortical decoding structures as if they resulted from real movements. These messages can thus be called "neuromimetic messages".

This kind of illusory movement can extend to the whole body when vibration is applied to muscles involved in postural stance. For instance, vibration of the two Achilles tendons or the Tibialis anterior muscles tendons gives rise to a persistent sensation of forward or backward body movement, respectively. As in the case of a single limb, the perceived speed of whole body illusory movements depends on the vibration frequency (Fig. 1 E). This parallels the finding that recordings of real movements performed at increasing speed show an increase in the discharge frequency of the muscle spindles (Roll and Vedel, 1982).

The data as a whole suggest that there may exist, at cortical level, a kind of repertory of "kinaesthetic shapes", the various components of which can be activated by specific peripheral messages which are usually associated with the performance of motor activities. In addition, several experimental findings indicate that these cortical maps might be built up during motor learning, on the basis of muscle and associated visual and tactile cues. Whole body or single limb illusory movements can be obtained from muscle proprioceptive (Roll, 1981) as well as visual (Lee and Aronson, 1974; Tardy-Gervet et. al., 1984) or tactile (Berthoz, 1978) patterns.

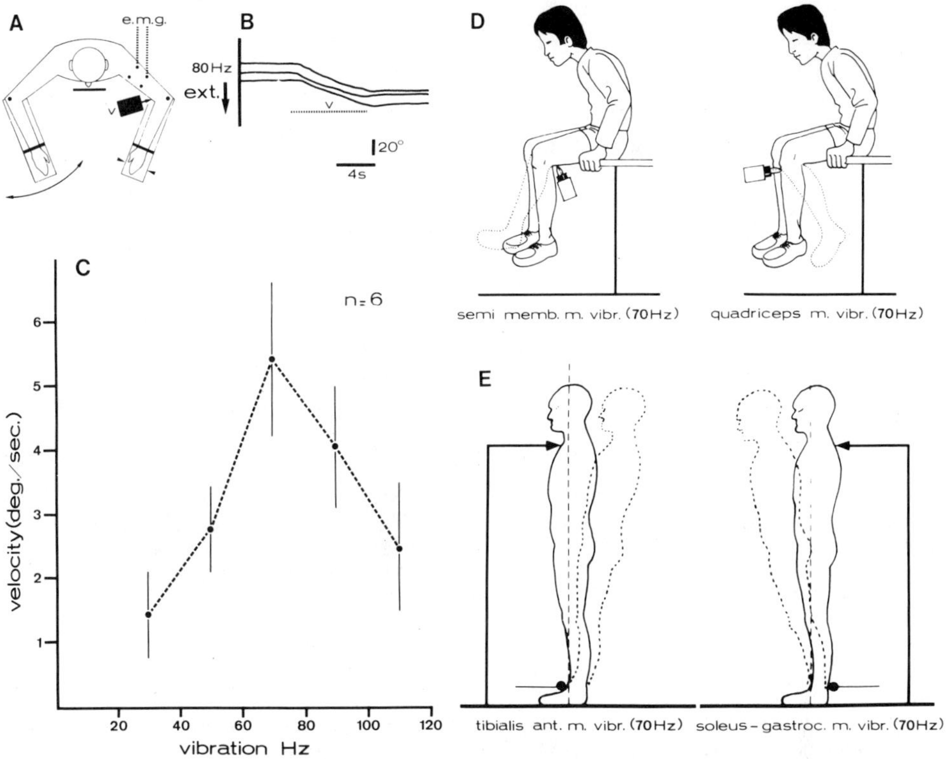

Fig. 1. From A to C: Under isometric conditions an illusory sensation of forearm extension was reported by subject when mechanical vibrations were applied to the distal part of his Biceps Brachii tendon. The perceived movement was actively reproduced by subject with his contralateral arm so that it could be assessed by the experimenter. B: Recordings of contralateral movements (3 trials, 80 Hz). C: Vibration frequency versus speed of illusory movement (each point is the mean of 18 trials). D and E: Diagram of vibration-induced illusory movements in leg (D) and foot (E).

Some recent findings it is now proposed to outline seem to indicate that muscle information plays a leading role in the kinaesthetic coding of the head and eye position separately, as well as in relation to the whole body. Besides the visual and vestibular contribution, extraocular and neck proprioception might play a decisive role in postural control and body orientation. These two functions are based on egocentric references without which it would be impossible to successfully carry out activities within, or directed towards, extrapersonal space. Since the spatial detection of a target-objet, as well as whole body or limb orientation towards it, usually result from "eye grasping", the proprioceptive information associated with the gaze direction presumably plays a leading role in the interrelating of static and dynamic body space with extracorporeal space. This hypothesis is supported at neurophysiological level by two facts: that muscle spindles have been found to exist in the eye muscles (Barker, 1973; Buisseret, 1978; Mukuno, 1986); and that receptors are sensitive to low amplitude mechanical vibration, as demonstrated in the case of the skeletal muscles (Burke et al., 1976; Roll and Vedel, 1982).

Method

Mechanical vibrations with an amplitude of 0.1 to 0.2 mm peak to peak (rectangular pulses: 3ms) were applied to various points on the periphery of the eyeball, by means of a mini electromagnetic vibrator (L.D.S. type 101) on which small probes could be adapted to the subject's eye morphology. Their length and shape varied according to whether medial or

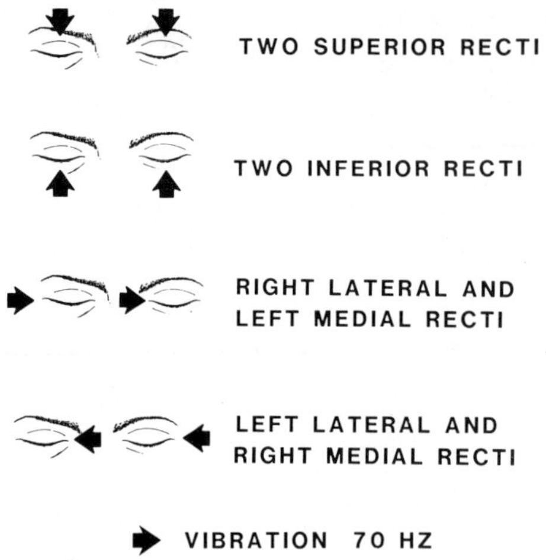

Fig. 2. Conditions of stimulation applied to both eyes. The arrows correspond to the simultaneous application of vibrations to both eyes, leading probably to the selective activation of proprioceptors of different extraocular muscle pairs. Single eye vibration is not represented here.

external muscles were to be vibrated. The contacting surface was polished and had a concave shape which could be easily adapted to the eyeball.

The vibration train duration was 5 seconds; the frequency could vary from 1 to 100 Hz but was usual by 70 Hz or an increasing frequency from 10 to 70 Hz.

In the standing situation, postural sways were recorded from four strain gages, which were enclosed in the stabilometer, set under each subject's foot.

In the seated situation, the subject's head was free, and his trunk was fastened to a fixed seat back. Horizontal head movements were recorded with a high precision potentiometer mounted on a head helmet.

In the opposite situation -trunk free and head restrained- the subject was seated in a rotating chair on the axis of which was also fixed a potentiometer for recording the horizontal body rotations. A head and chin-rest device was added to prevent any head movements.

Lastly, in the target viewing situation, the subject was seated in the dark, monocularly facing a small electroluminescent diode located 57 cm ahead of him, at eye level.

The subject's eyes were closed in all situations except for the last. Either one eye or both eyes alternately were vibrated. When both eyes were stimulated, vibrations were applied either to the same muscle of each eye (two inferior and two superior recti), or to the synergist muscles, i.e. the lateral rectus of one eye together with the medial rectus of the other eye (Fig. 2).

Before testing, the apparent lengthening of a visual target in an expected direction perceived by the subject in response to application of low frequency vibration to a given muscle, was used to check that the vibrator was correctly placed on the eye muscle.

Experimental results and quantitative analysis deal particularly with the perceptual and motor effects of eye muscle vibration in subjects with an upright standing posture. A preliminary report is given in connection with the other conditions.

Results

Postural effects of extraocular muscle vibrations

Application of low amplitude mechanical vibration to the extraocular muscles of a standing subject with his eyes closed elicited whole body shifts, the direction of which was found to depend strictly on which muscle was vibrated. The latency of these effects varied in the range of 1 to 2 seconds; this is illustrated by posturograms giving the combined time recordings of lateral (X) and anteroposterior (Y) components of postural forces. Figure 3 summarizes the postural vibration-induced effects of the simultaneous vibration of both eyes under all four different experimental conditions of stimulation.

Simultaneous vibrations of the two superior recti resulted in a whole body displacement forward (C), whereas a backward shift was observed during vibration of the two inferior recti (D). When the stimulation was applied simultaneously to the lateral rectus of the right eye and the medial rectus of the left eye (A), a leftward displacement was elicited, and vice-versa (B).

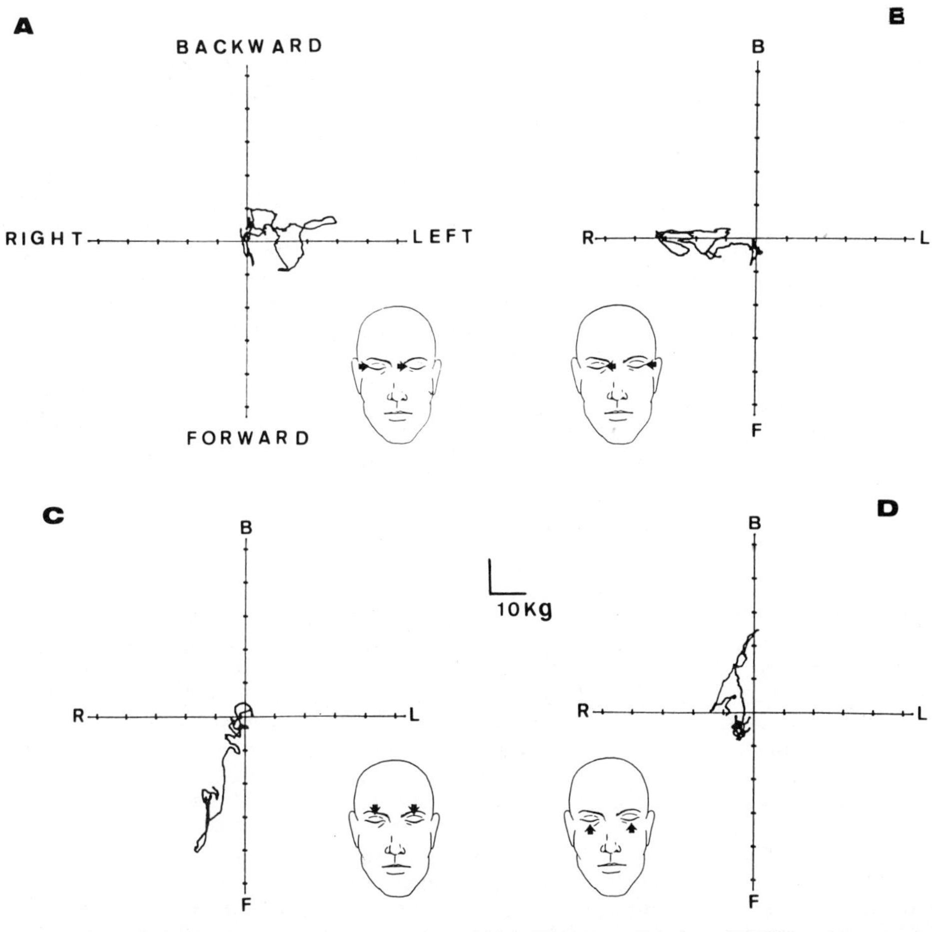

Fig. 3. Directional postural effects of application of high frequency vibrations (70 Hz) to homonymous extraocular muscles of both eyes. The arrows indicate the stimulated muscles. The corresponding stabilometric recordings show the postural force variations during the stimulation (5 sec.). Note the specific directionality of postural sways corresponding to each stimulated muscle.

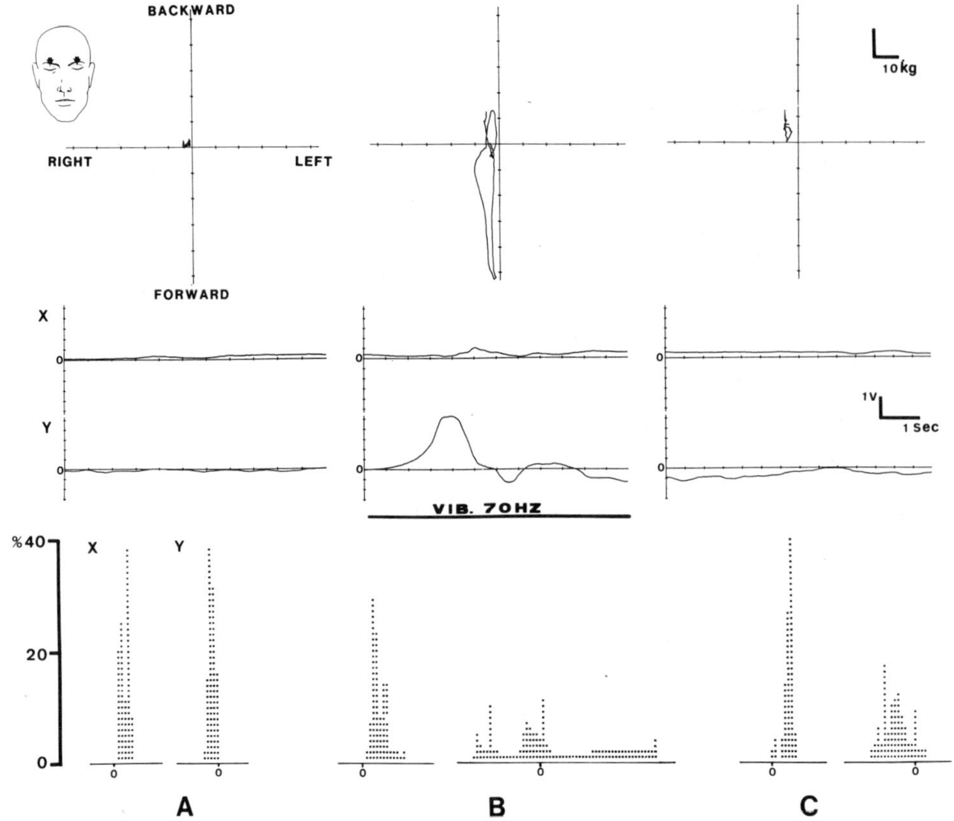

Fig. 4. Postural effects of vibratory stimulation of both superior recti in pre (A), per (B) and post (C) vibratory periods. From top to bottom: stabilometric recordings; time recordings of postural force variations on X lateral axis and Y anteroposterior axis; position error histograms on X and Y axis. The median horizotal line represents the onset and duration of the stimulation (vibration: 70Hz, 5 sec.).

These postural effects of extraocular muscle vibration can also be analysed through independent recordings of the force variations on the X and Y axis during pre, per and post vibratory periods.

The postural effects of the simultaneous vibration of the two superior recti in a standing subject for a duration of 5 seconds and at a frequency of 70 Hz, are described in Figure 4. The application of vibrations first resulted in a considerable increase in the equilibrium oscillation amplitude in the forward direction as compared to the control period, and was immediately followed by post vibratory oscillations in the opposite direction. The lower part of the diagram shows the position error histograms computed from time recordings on both axis. They illustrate the extent of postural equilibrium alterations in terms of mean position variations as well as their dispersion. Very few variations were observable on the X axis histograms in pre, per and post vibratory sequences, whereas an increasing spread of Y axis histograms due to the subject's imbalance could be observed during vibrations and, to a lesser extent, in the post stimulation sequence.

The results of individual trials (A) and mean results (B) for all subjects under all four conditions involving combined vibrations of both eyes are given in Figure 5. The vectors correspond to the polar coordinates of the maximum amplitude of displacement recorded during one trial. The modulus length of the vectors indicates the magnitude of the effects, and its angular position the direction of the postural shift. Vibration of the inferior rectus leads to backward body displacements, and vibration of the superior rectus to forward displacements. Likewise, vibration of one external and one medial muscle leads to lateral displacements. No significant difference was observed between the effects of single muscle stimulation and that of a pair of homonymous muscles.

As shown in Figure 6, the mean amplitude of postural shift induced by vibration depends on the frequency of vibratory stimulation. The posturograms show one subject's increasing forward displacement when a 5 second vibration, increasing in steps from 20 to 100 Hz, was applied to both inferior recti. This relationship between vibration frequency and the mean amplitude of the postural sway is illustrated in the lower part of the diagram. Both the path length and the maximum amplitude varied as a function of the vibration frequency.

Kinaesthetic effects of extraocular muscle vibrations

Besides these postural effects of extraocular muscle vibration, other perceptual effects, such as kinaesthetic illusions, can be elicited. These illusions of movement can affect the head, the trunk or the whole body, depending on the postural context (Fig. 7 A.B.C.).

Preliminary data show that the direction of these kinaesthetic illusions depends on which extraocular muscle is vibrated. The perceived sensations are those of a slow movement, that is a movement of some degrees per second. Their latency varies from 1 to 2 seconds and they stop when the stimulation stops. Their subjective content is that of a slow, continuous movement which is very similar to the vibration-induced kinaesthetic illusions previously described in connection with skeletal muscles (Roll et al., 1980).

When the subject was seated with his head free (A), the application of a vibration train to the lateral rectus of one eye, or to homonymous muscles of both eyes elicited an illusory sensation of head rotation the direction of which corresponded to a rightward horizontal rotation in the case of left lateral vibration, and a leftward horizontal rotation in that of right lateral vibration.

Likewise, vertical head-down sensation was experienced by subjects who underwent superior muscle vibrations and a head-up sensation by those who underwent inferior muscle vibrations. After a few seconds, the same vibratory stimulation generally elicited involuntary slow movements of the head in the same direction as that of the kinaesthetic sensation. When the subject's head was fixed by means of a mechanical device (B), vibration of the same lateral muscles induced kinaesthetic trunk illusions in the direction previously described for the head. Stimulation can also induce an involuntary trunk rotation in the same direction as that of the head in this situation.

Visual effects of extraocular muscle vibrations

Figure 7C shows the last experimental situation we studied. Vibrations were applied to extraocular muscles when the subject was looking monocularly at a small visual target in darkness. The vibratory stimulation resulted in a slow displacement of the visual target, the direction of which depended on the vibrated muscle. As with the postural and kinaesthetic effects, vibration of the inferior rectus induced an upward target shift and conversely, a downward shift occurred under superior rectus vibration. External rectus vibration elicited an illusory target displacement in the medial direction and vibration of the medial rectus a displacement in the external direction.

Likewise, an illusory shift of the visual target was elicited by vibration applied to neck muscles (Splenius, Sternocleïdomastoïdus or Trapezius) and ankle postural muscles (Soleus, Tibialis, Peroneus). The most noteworthy finding, which is summarized in Fig. 8, concerns the directional similarity of these effects. That is, vibration of either the inferior recti, the Sternocleïdomastoïdus or the Tibialis anterior muscles can induce, in darkness, identical illusory shifts of the target in the upward direction. In this case, since the only available cue is the position of the luminous target on the retina, the vibration-induced proprioceptive pattern simulating either an eye lift, a backward head or whole-body movement, is interpreted by the stationary subject in darkness as if it were an upward displacement of the target. These effects were found to add to each other when vibrations were applied simultaneously to the eye, neck and ankle postural muscles.

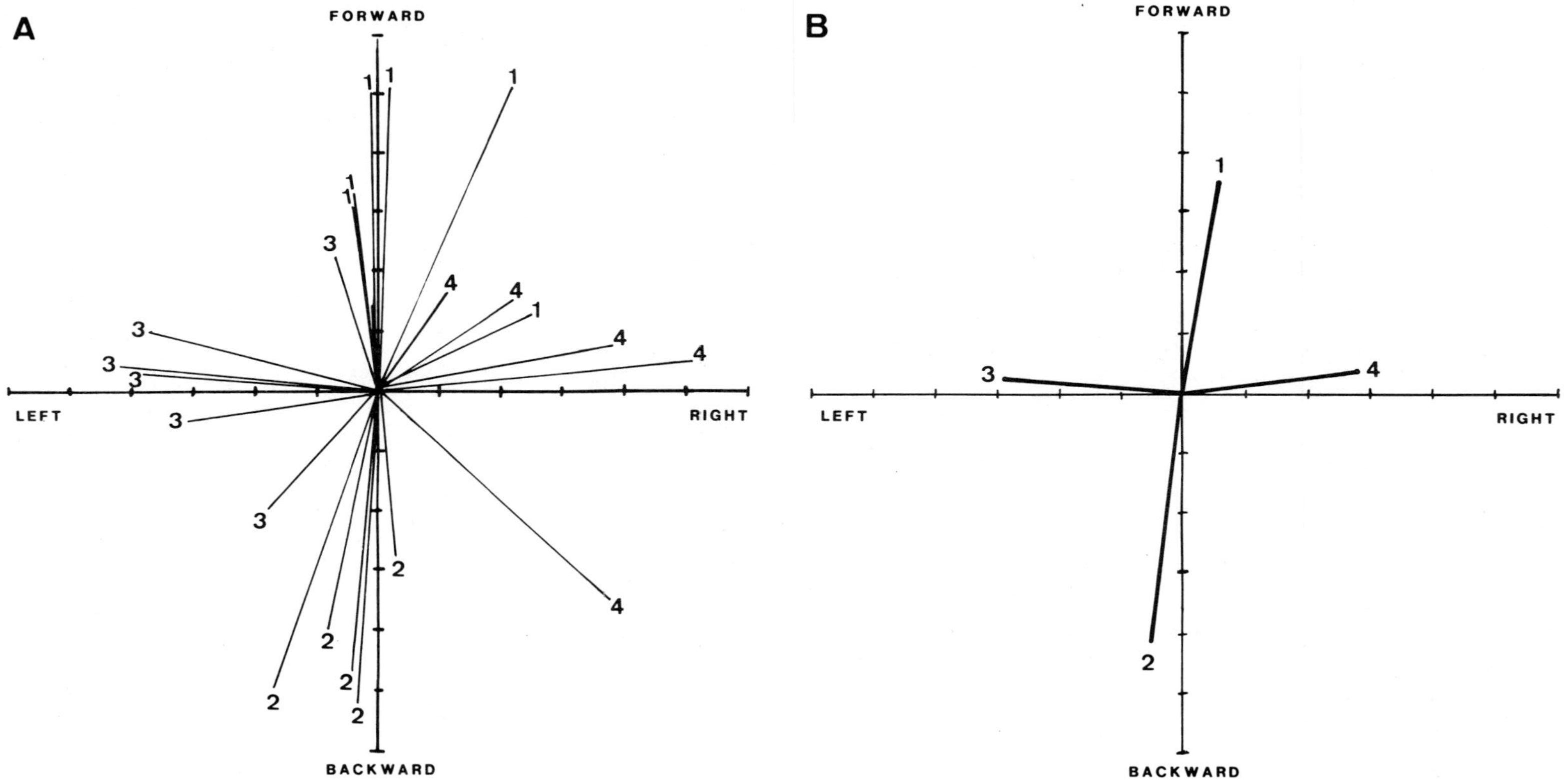

Fig. 5. Groups of individual (A) and mean (B) vectors corresponding to various combinations of extraocular muscle stimulation. 1) two superior recti; 2) two inferior recti; 3) lateral rectus of right eye together with medial rectus of left eye; 4) lateral rectus of left eye together with medial rectus of right eye. For each vector the modulus length signals the individual or mean maximum amplitude of the postural sways; its angular position indicates the direction of the shift.

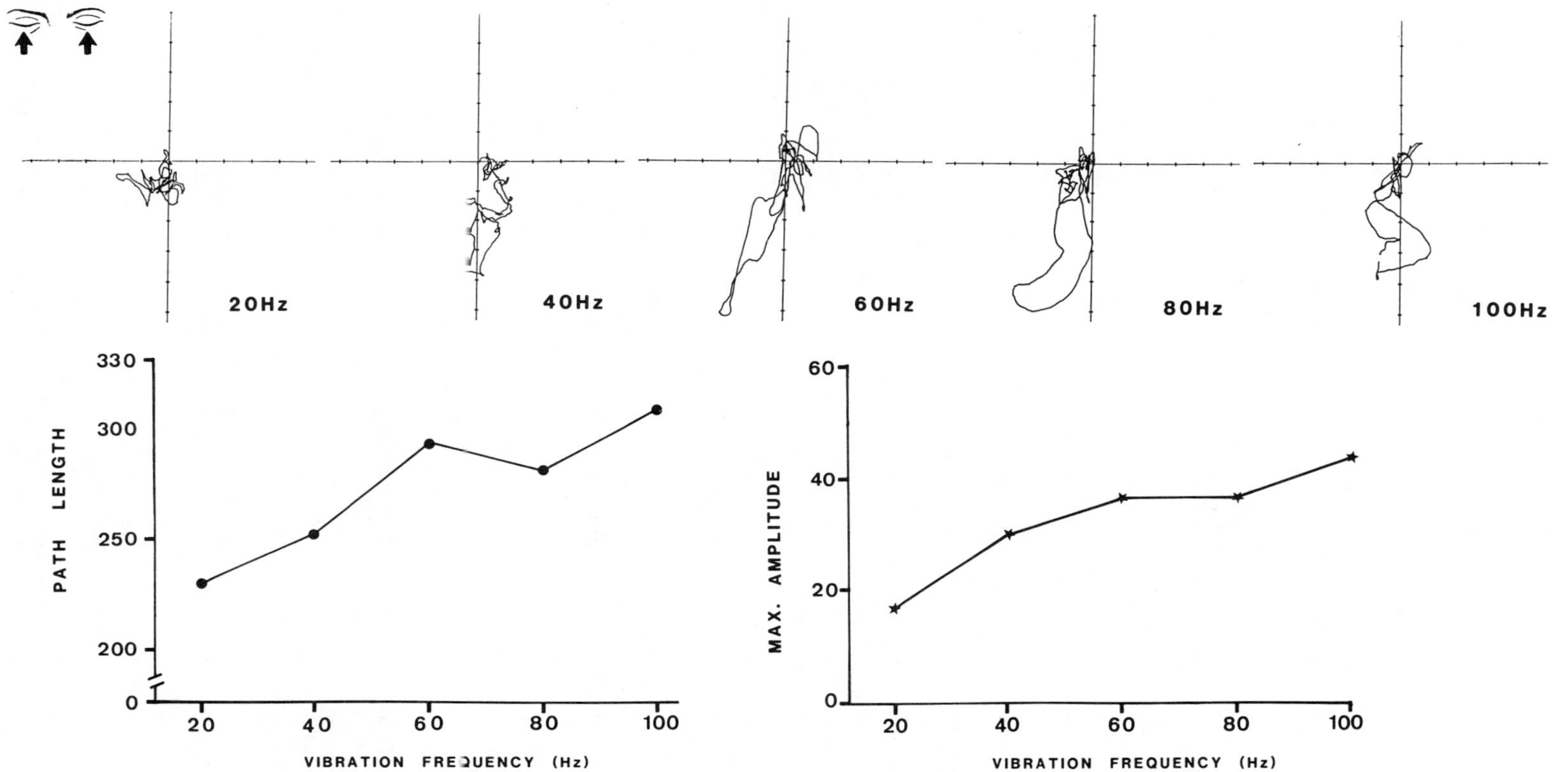

Fig. 6. Effect of the vibration frequency on the amplitude of the postural sways. Upper part: individual posturograms show the increasing amplitude of the backward shift induced by the inferior recti vibration of both eyes with an increasing frequency from 20 to 100 Hz. Lower part: vibration frequency with respect to the path length (left graph) and the maximum amplitude (right graph) of the postural sway.

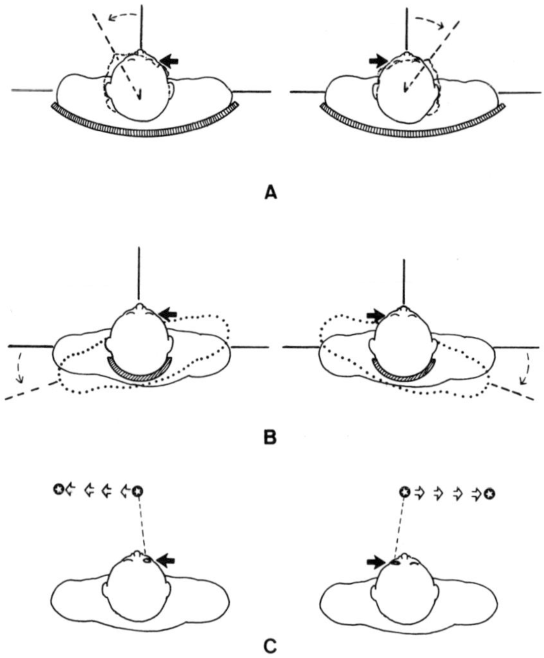

Fig. 7. Kinaesthetic and visual illusions induced by vibration on the right lateral rectus (on the left) and left lateral rectus (on the right). A) head free: illusory sensation of head rotation. B) head fixed: illusory sensation of trunk rotation. C) Monocular fixation of a luminous target: extraocular muscle vibration induced an illusory displacement of the target.

No clear eye movement sensation was reported by subjects in any of the situations studied. Moreover, preliminary recordings of eye movements by E.O.G. showed that no eyeball displacements took place during the vibratory sequences.

Discussion

Our assumption that extraocular proprioception exists in man and that it contributes to the regulation of spatially oriented behavior is based upon the following experimental data.

Extraocular muscle vibration entails perceptual and motor effects, in which each muscle has its own directional specificity.

This same directional specificity was found to exist in kinaesthetic, postural and visual extraocular muscle vibration-induced effects.

Under all our experimental conditions, the direction of both illusory and actual displacements was always that which corresponded to the stretch of the vibrated extraocular muscle.

At the postural level at least, the magnitude of the effects increased as a function of the stimulation frequency up to 80-100 Hz.

As with the skeletal muscles, painless vibration trains applied to the eye muscles might thus be thought to induce almost selective proprioceptive messages. However, under closed eye conditions, these messages do not seem to be decoded in terms of actual conscious eye movement, but more like conscious sensations of movement of the head, trunk and whole body depending on the postural context in which they occur. In this situation, the absence of eye movement perception may result from a sensory masking effect since the cornea and lid tactile receptors are also very sensitive to low amplitude high frequency vibration, and give rise to powerful cutaneous vibratory sensations (Vedel and Roll, 1983). In fact, extraocular muscle vibration induces low amplitude directional displacements of the eyeball, probably leading to selective high frequency stretching of the particular muscle stimulated. If the vibrator probe undergoes even a small displacement on the socket bone or the lids, the perceptual and motor effects reported above are no longer obtained. On the contrary, when the subject looks at the target, the proprioceptive input arising from extraocular muscles actually seems to be decoded like an eye movement since the target is perceived by the subject as moving in the expected direction in relation to the stimulation site. In this case, when the eyes are open and the target motionless, the proprioceptive signal elicited by vibratory stimulation of extraocular muscles informs the Central Nervous System that the eye is moving whereas the retinal image remains stable. This occurs for example during a smooth pursuit situation. Here the movement cue is interpreted as a target displacement. A similar effect has been reported in Lackner and Levine's experiment (1978), where the subject in darkness saw the luminous target attached to his index tip moving while his arm was being vibrated by the experimenter. This illusory arm movement induced by muscle tendon vibration has been interpreted in the same way as a target displacement in the same direction as the illusion. Likewise, Matin et al. (1982) have reported the illusory displacement of a visual target which was fixed in darkness by a subject whose eye was partially paralysed with curare. In addition, the direction of this illusion depended on the head position, and it disappeared on the illumination of the surroundings.

Although they do not reveal the exact neurophysiological process involved, these data suggest that the central processing of extraretinal signals depends to a great extent on the visual and/or postural context in which the experiments were carried out. The results described here could be interpreted along these lines. In the absence of retinal information, when the subject's eyes are closed, the extraretinal signal of eye muscle origin may not be assigned to the coding of the position of the eye itself, which would serve no useful purpose, but rather to the coding of the various body segment positions usually associated with eye rotation. Depending on the nature of the postural constraints involved, the head, the trunk, or the whole body might be concerned with such processing. The existence of a large amount of functional coupling between these various mobile segments is thus confirmed from perceptual and motor points of view by our experimental manipulation of eye proprioceptive feedback. It is possible that this kind of "proprioceptive coupling" might result from behavioral patterns which usually link eye movements to those of other mobile segments. This might occur, for instance, during the early eye-head directional linkage when a target is appearing on the periphery of the visual field; in particular, in the well-described "predictive mode" of eye-head coordination (Bizzi, 1974) when a conditioned monkey is able to turn its head in anticipation of the appearance of the target. In this case, contrary to the ordinary sequence (triggered mode), head turning precedes eye saccade. Thus the central command responsible for coordinated eye-head movements is liable to exhibit distinctive patterns depending on the experimental conditions. Both eye and head movements in the same direction could be linked centrally for functional purposes. Comparable observations in man have been described in the literature (Stroud and Burde, 1977). In view of the direction of the induced perceptual and motor effects, and the fact that the vibrated muscles behave as if they were stretched, extraocular proprioception can be said to be mainly involved in the directional transport program for both the eyes and head and, when appropriate, for the whole body.

In conclusion, our results show that it is possible to experimentally manipulate extraocular proprioception in man. Moreover, these data suggest that the extraretinal proprioceptive inflow may play an even more important part than that of the classically recognized outflow. The extraocular proprioceptive signals are able to give rise to concious sensations of head, trunk and whole body movement, and thus to contribute to the coding of egocentric references involved in spatial information processing.

By assuming that for a standing subject the ground is the reference providing the basic spatial information, the absolute eye-in-space position might be calculated by the

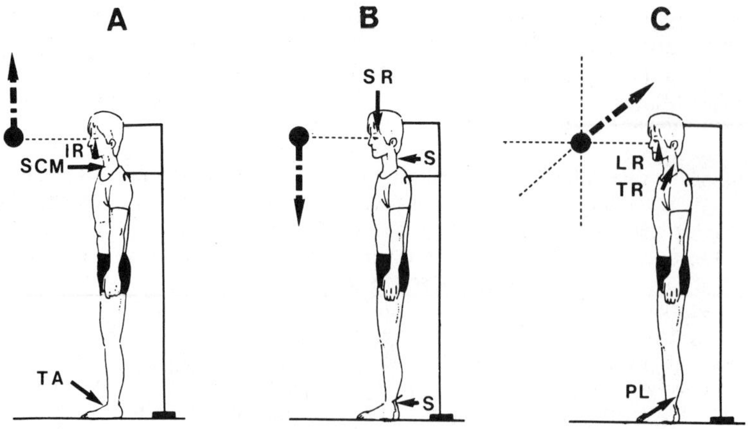

Extra-ocular Muscles	Inferior recti ↑	Superior recti ↓	Right Lateral Left Medial recti ←	Left Lateral Right Medial recti →
Neck Muscles	Sterno-cleido mastoïdeus ↑	Splenii ↓	Right Trapezius ←	Left Trapezius →
Leg Muscles	Tibialis anterior ↑	Soleus ↓	Right Peroneus longus Left Tibialis posterior ←	Left Peroneus longus Right Tibialis posterior →

Fig. 8. Direction of apparent shift of a monocularly fixed target when vibration (70 Hz) were applied to either eye, neck or/and ankle postural muscles. (For details see the text).

C.N.S. from the proprioceptive messages arising from various body segments which link eye with ground. The postural scheme elaborated centrally from this simultaneously collected information, might, moreover, be used in the spatial indexation of retinal information.

Finally, the muscle proprioceptive signals, in particular those associated with eye movements, might be involved in the interrelating, in terms of spatial coordinates, of body space with extrapersonal space in oriented behavior as a whole.

Acknowledgements

This work was supported by C.N.R.S. (U.A. 372), Fondation pour la Recherche Medicale and Ministère de l'environnement grants.

The authors thank F. Harlay and J.L. Vercher for helping with computer data analysis.

References

Ashton, J. A., Boddy, A., and Donaldson, I. M. L., 1984, Input from proprioceptors in the extrinsic ocular muscles to the vestibular nuclei in the giant toad, buffo mannus, Exp. Br. Res., 53: 409-419.

Bach y Rita, P., and Ito, F., 1966, Properties of stretch receptors in cat extraocular muscles, J. Physiol. (Lond.), 186: 663-688.

Barker, D., 1973, Muscle receptors, Handbook of sensory physiology, Springer-Verlag, ed. , Berlin, Heidelberg, New York, pp. 79-154.

Baron, J. B., 1955, Muscles moteurs oculaires, attitude et comportement des Vertébrés, Thèse Sciences, Paris.

Batini, C., and Buisseret, P., 1974, Sensory peripheral pathway from extrinsic eye muscles, Arch. ital. Biol., 112: 18-32.

Berthoz, A., 1974, Ocumomotricité et proprioception, Rev. EEG et Neurophysiol., 4: 569-586.

Berthoz, A., 1978, Rôle de la proprioception dans le contrôle de la posture et du geste, in: Hécaen H. and Jeannerod M., eds. , "Du contrôle moteur à l'organisation du geste", Massion, Paris, pp. 187-224.

Bizzi, E., 1974, The coordination of eye-head movement, Scientific American, 231: 100-106.

Brindley, G., and Merton, P., 1960, The absence of position sense in the human eye, J. Physiol., (Lond.), 153: 127-130.

Buisseret, P., 1978, Etude électrophysiologique chez le chat du développement des propriétés des champs récepteurs des neurones du cortex visuel. Rôle de la vision et de la proprioception extraoculaire, Doctorat ès Sciences, Paris, 89p.

Buisseret, P., and Gary-Bobo, E., 1979, Development of visual cortical orientation specificity after dark-rearing: role of extraocular proprioception, Neurosciences Letters, 13: 259-263.

Burke, D., Hagbarth, K. E., Lofstedt, L., and Wallin, G., 1976, The response of human muscle spindle endings to vibration of non contracting muscles, J. Physiol. (Lond.), 261: 673-693.

Cooper, S., and Daniel, P., 1949, Muscle spindles in human extrinsic eye muscles, Brain, 72: 1-24.

Cooper, S., Daniel, P. M., and Whittridge, D., 1955, Muscle spindles and other sensory endings in the extrinsic eye muscles, Brain, 78: 564-583.

De Cyon, E., 1911, "L'oreille", Alcan, ed. , Paris, 298p.

Dichgans, J. and Brandt, 1978, Visual-vestibular interaction: effects on self-motion perception and postural control, in: "Handbook of sensory physiology", Vol. VIII: Perception, R. Held, H.W. Leibowitz and H.L. Teuber , eds. , pp. 756-804.

Hein, A., and Diamond, R., 1982, Contribution of eye movement to representation of space, in: "Spatially oriented behavior", A. Hein and M. Jeannerod, eds. , Springer-Verlag, 983: 119-133.

Lacour, M., 1981, Contribution à l'étude de la restauration des fonctions posturocinétiques après labyrinthectomie chez le Singe et le Chat, Thèse de Doctorat ès Sciences, Université d'Aix-Marseille I, 152pp.

Lacour, M., Xerri, C., Barthelemy, J., et Borel, L., 1987, Neuronal coding of linear motion in the vestibular nuclei of the alert cat, II, Response characteristics to vertical visual motion cues, Exp. Brain Res, in press.

Lackner, J. H., and Levine, M. S., 1978, Visual direction depends upon the operation of spatial constancy mechanisms: the oculobrachial illusion, Neurosciences Letters, 7: 207-212.

Lee, D. N., and Aronson, E., 1974, Visual proprioceptive control of standing in human infants, Perception and Psychophysics, 15: 529-532.

Lee, D. N., and Lishman, J. R., 1975, Visual proprioceptive control of stance, Journal of human movement studies, 1: 87-95.

Matin, L., Picoult, E., Stevens, J., Edwards, M., and Mac Arthur, R., 1982, Oculoparalytic illusion: visual field dependent spatial mislocalization by human partially paralysed with curare, Science, 216: 198-201.

Matthews, P. B. C., 1981, Evolving views on the internal operation and functional role of the muscle spindle, J. Physiol., 320: 1-30.

Milleret, C., Gary-Bobo, E., and Buisseret, P., 1985, Responses of visual cortical cells (area 18) to extraocular muscles stretch or nerve stimulation in cats and normal or deprived kitten, Neurosciences Letters, 22 S 298.

Mukuno, K., 1986, Morphological demonstration of ocular muscle proprioception in the human extraocular muscles, Proceedings of International Workshop on proprioception of the ocular muscles, Hakone-Matsuyama, Japon, 1-1.

Nashner, L. M., 1970, Sensory feedback in human posture control, M.I.T. Report, MVT, 70-3.

Paillard, J., et Beaubaton, D., 1978, De la coordination visuo-motrice à l'organisation de la saisie manuelle, in: H. Hecaen et M. Jeannerod, eds. , "Du contôle moteur à l'organisation du geste", Paris, Masson, 225-260.

Roll, J. P., Gilhodes, J. C., and Tardy-Gervet, M. F., 1980, Effects perceptifs et moteurs des vibrations musculaires chez l'homme normal, Mise en evidence d'une reponse des muscles antagonistes, Archives italiennes de Biologie, 118: 51-71.

Roll, J. P., 1981, Contribution de la proprioception musculaire à la perception et au contrôle du mouvement chez l'Homme, Thèse de Doctorat ès Sciences, Marseille, 194pp.

Roll, J. P., Vedel, J. P., 1982, Kinaesthetic role of muscle afferents in man, studied by tendon vibration and microneurography, Exp. Brain Res., 47: 177-190.

Roll, J. P., and Roll, R., 1987, Kinaesthetic and motor effects of extraocular muscle vibration in man, in: "Eye movements, from Physiology to Cognition", J.K.O. Reagan and A. Levy-Schoen, eds. , Elsevier, North Holland, 57-68.

Stroud, M. H., and Burde, R. M., 1977, Eye-head coordination: an observation, Annals of Otolaryngology, 86: 94-96.

Tardy-Gervet, M. F., Gilhodes, J. C., Roll, J. P., 1984, Perceptual and motor effects elicited by a moving visual stimulus below the forearm: an example of segmentary vection, Behavioral Brain Res., 11: 171-184.

Vedel, J. P., and Roll, J. P., 1983, Muscle spindle contribution to the coding motor activity in man, Experimental Brain Research, suppl. 7: 253-255.

GRAVITATIONAL MECHANISMS IN THE MOTOR SYSTEM. STUDIES IN REAL AND SIMULATED WEIGHTLESSNESS.

I. Kozlovskaya, I. Dmitrieva, L. Grigorieva, A. Kirenskaya, and Yu. Kreidich

Institute of Biomedical Problems, Moscow, USSR

Introduction

The motor system of all terrestrial animals including man has evolved phylo-and-ontogenetically in the gravitational field that affects many (if not all) of the mechanisms, which provide reliability, accuracy and stability of motor responses on the Earth's surface. Thus, in the multicompenent motor system, zero-or-microgravity can dramatically affect the functioning of several individual mechanisms and the system as a whole. In addition to weight unloading that triggers muscle "disuse" disturbances, microgravity causes: i) redistribution of force along the body surface, followed by a reduction of support reactions, which play an important role in postural and static activity control; ii) alterations in the functioning of sensory inputs, such as otolithic, proprioceptive and others, which are deeply involved in motor regulation; iii) changes in the biomechanics of movements, due to any of the following environmental effects: altered relationships between the mass of an object and the effort required to move it and between resistance and inertia forces; absence of the gravitational force that on the Earth compensates for the static (postural) component of movements; facilitated use of cross synergy patterns (like swimming). All this should lead to an adaptive reorganization of motor coordination that will involve, for example, recalibration of effort scaling, accentuation of intermittent control mechanisms, and deprivation of all kinds of phasic-static linkage, possibly including α-γ-coactivation.

Since the structure in the motor system is strongly linked to its function, the above adaptive changes, if the exposure to a new environment is critically long, are likely to be followed by structural alterations in muscle, receptors, neuromuscular synapses and other parts of the system. Studies and observations performed during (or rather after) space flights (SF) of various durations have indeed revealed a wide spectrum of changes in almost every component of the musculoskeletal system including muscles (hypo- or atrophia, atonia - Berry, 1973; Gurovsky et al., 1975; Kakurin et al., 1971; Kozlovskaya et al., 1981a, Mitarai et al., 1978), sensory inputs and reflex mechanisms (hyper-reflexia of varying degrees - Baker et al., 1977; Kozlovskaya et al., 1981b; Reschke et al., 1979), and motor control mechanisms (posture, movements accuracy, etc. - Chekirda and Yeryomin, 1974; Chkhaidze, 1968; Homick et al., 1977; Kakurin et al., 1971; Kozlovskaya et al., 1982; Kubis et al., 1977).

Obviously, the weight of the above factors will vary depending on the durations of the flight. For instance, in short-term SF, support reaction activity seems to make a major contribution to the development of motor disfunction (Kozlovskaya et al., 1982; Kozlovskaya et al., 1983); on the contrary, in long duration SF, structural changes which are building up gradually in muscles (Berry, 1973; Gurovsky et al., 1975; Iljina-Kakujeva et al., 1979; Kozlovskaya et al., 1981a; Kozlovskaya et al., 1981b; Oganov and Potapov, 1979) neuromuscular junctions and in various areas of the CNS may be failure sites. A whole set of quantitative data concerning different compenents of the musculo-skeletal system during and after exposure to microgravity for various durations is required to reach better understanding of the mechanisms underlying weightlessness-induced motor abnormalities.

Materials and methods

Experiments were carried out to study the effects of real and simulated microgravity on several parts of the motor system. The data obtained were divided in two groups depending on the exposure time.

In group one, early, so-called acute effects of microgravity were studied in 18 members of Salyut-6 visiting crews and 22 participants of simulation experiments. In this case microgravity was simulated by dry immersion (DI), which according to Shulzenko and Vil-Viliams, 1976, entirely reproduced all the unloading effects without producing any uncomfortable sensations.The exposure to microgravity in this group lasted for 7 days (except for one 4- and one 14 day flight, each manned by 3 crew-members).

In group two, motor effects of long-term both real and simulated microgravity were studied in 15 crew-members of Salyut -6 and -7 that flew for 75-237 days and in 9 participants in a simulated 120-day head-down tilt study (-6°).

The experimental protocol, which was similar in all the groups under study, included procedures and tests designed to quantitatively evaluate specific parts of the motor system: sensory inputs, spinal mechanisms, muscles, mechanisms of postural regulation and voluntary movement control.

Sensory inputs. The activity of support and muscle-inputs was assessed in terms of the threshold of vibrosensitivity of the sole support zones (lateral pad of the foot, tuber calcaneaneus and the great toe pad and base), measured using the Vibrotester unit at 3 frequencies - 63, 125 and 250 Hz; and in terms of the recruitment functions of T- and H-reflexes of the triceps surae muscle (TS) showing the reflex amplitude as a function of the stimulation strength. To evoke T-Achilles tendon reflex, calibrated taps were applied over the tendon. The surface electromyographic responses were recorded. The reflex threshold, maximum amplitude and slope of the recruitment curve were analyzed. These parameters of the curve are independent (Artemjeva et al., 1977) and reflect various properties of the motoneuron pool: excitability of its low threshold (threshold) and high threshold (maximum amplitude) components and population homogeneity (the slope of the curve).

By comparing the recruitment curves recorded with both feet at rest and with the contralateral foot maintaining dorsal or plantar flexion, it was possible to determine the state of cross synergies, i.e., one of the major mechanisms of locomotor activity.

Muscular properties. The protocol used in the muscle studies involved isokinetic dynamometry, using a Cybex dynamometer. The tests were designed to measure the speed and strength of gastrocnemius and anterior tibial muscles with respect to the force moments and the EMG when moving the foot with high (180°/s), moderate (120°/s) and low (60°/s) speeds and during isometric contractions. Keeping in mind the purpose of the study, variations in the tone of leg muscle at rest and at maximum contraction were also carefully examined during DI and ABR and after space flights (SF) by measuring transverse stiffness. By using an electromyotonometer of a new design that can independently record the force with which the transducer is attached and the tissue responses to a wide range of deforming forces, as well as by ensuring a high standardization of measurement conditions (stability of the range of forces with which the transducer is attached, maintenance of the limb position, and control of the relaxation and tension of the muscle under study by means of an electromyographic signal used as a feedback), we significantly increased the accuracy and reproducibility of the data and the reliability of the method (Gevlich et al., 1983).

Motor control systems. Postural control systems were examined by means of stabilography and postural synergies. The stabilographic recordings were conducted according to the standard procedure (Kozlovskaya et al., 1981a) for 3 min: minute 1 - comfortable posture with eyes open, minute 2 - the same posture with eyes closed, minute 3 - Romberg's posture. In this situation, high frequency oscillations of the body mass centre were recorded and calculated as a percentage of the total number of oscillations. The postural stability was also tested in terms of the time within which balance recovered after being perturbed when pushes of a measured force were applied to the subject's chest. It is well known that postural imbalance is usually prevented (or restored) by the specific muscular activity, i.e. complex motor reactions precisely arranged in time and space and termed "postural synergies". When measuring the amplitude and time of EMG bursts of leg muscles during standing and perturbation tests, it was possible to qualitatively and quantitatively describe the state of the postural control system.

Two motor tasks were used to analyze the microgravity effects in preprogrammed and feedback movement control systems. When performing the former, subjects were required to make a series of gradually increasing isometric plantar foot flexions from minimal to submaximal (50-60 Kg) with minimal differences between consecutive efforts without feedback signals. The given level of integrated EMG (IEMG) of m.tibialis anterior (TA) or m.soleus (SOL.) and m. gastrocnemius lateralis (GL) was maintained with visual feedback (deflection of a galvanometer needle) and served as a model for feedback movements. In this case the level of IEMG was close to that recorded with efforts amounting to 7-10 % of the maximal value. Precise kinematic and electromyographic characteristics of movements were investigated.

Results

This study showed the existence of close similarities between the effects of real and simulated microgravity. However the extent of the changes obviously depended on the exposure time.

Sensory systems. Short-term exposure to microgravity was usually followed by a marked trend to develop proprioceptive hyper-reactivity. This included a pronounced decrease in vibrosensitivity threshold of the sole support zones in 8 out of 10 members of visiting crews and in all test subjects first exposed to immersion (Fig. 1), as well as a

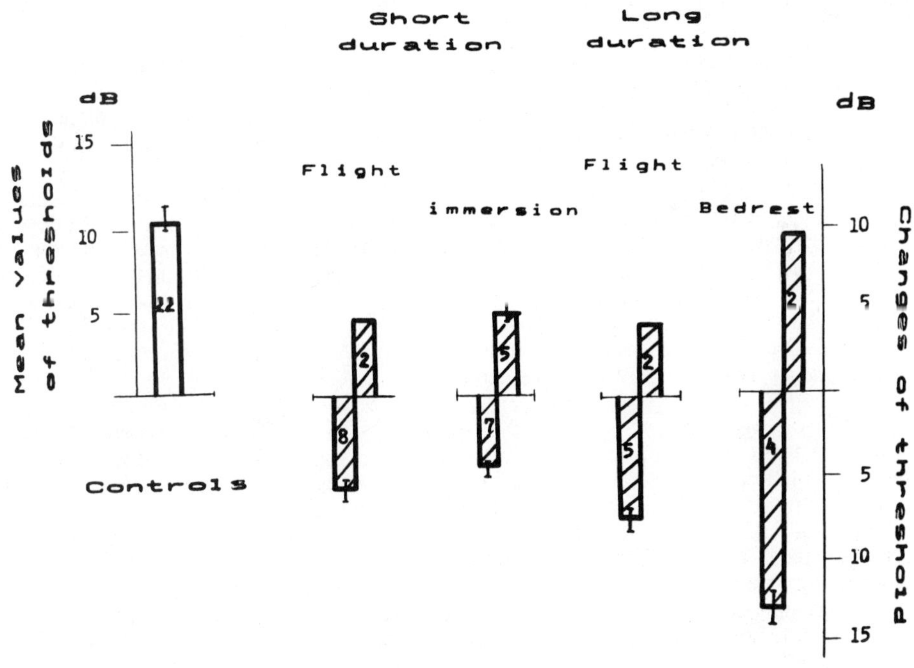

Fig. 1. Alterations of the vibrosensitivity thresholds after short and long exposures to micro- and hypogravity. The shaded bars show the mean value of changes in decibels: the standard errors are indicated by vertical lines. The figures inside the bars indicate the numbers of cases in the groups. The initial value of the vibrosensitivity threshold is shown by the empty bar.

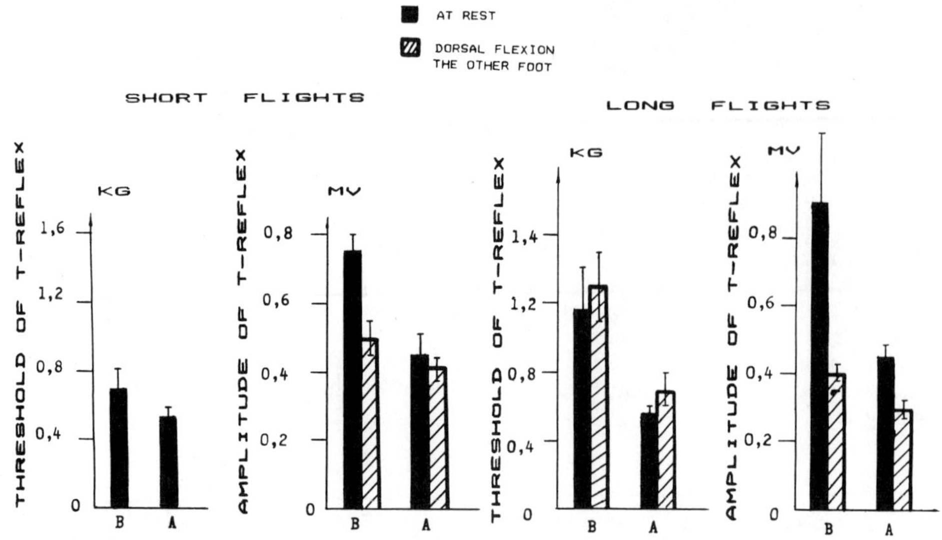

Fig. 2. Alterations of the amplitudes and the thresholds of the T-response after exposures to real microgravity (spaceflights). The black and shaded bars indicate the mean values of thresholds (in kG) and maximal amplitudes (in mV) of T-reflexes found before (B) and after (A) short term (on the left) and long term (on the right) space flights at rest and with the other foot dorsiflexed, respectively.

signifiant decrease in the T-reflex thresholds (Fig. 2). Analogous changes were recorded during immersion experiments in which on T-reflex threshold declined progressively beginning on the first day of exposure (Fig. 3).

Similar alterations in the H-reflex threshold (Fig. 3), which is generally thought to be a receptor-free counterpart of the T-reflex, and less distinct changes in the direct muscle (M) response thresholds suggest that the hyperactivity that accompanies support unloading is of a complex nature and involves both central (increased excitability of motoneurons) and peripheral mechanisms.

The suggestion concerning the involvement of peripheral mechanisms is supported by the finding that contrary to clinical hyper-reactivity, the threshold decline after exposure to microgravity was followed by a significant decrease in the maximal amplitude of the T-reflex and the slope of the reflex recruitment curve. As shown in Fig. 2 and 3, the maximal T-reflex amplitude decreased by more than half after a 7-day immersion (DI) and space flights (SF). However this decline was not accompanied by any analogous changes in the amplitude of H- and M-responses, which remained unchanged or even increased.

The changes in the T-, H- and M-responses as well as vibrosensitivity after long-term exposures to hypogravity were of the same sign and in five cosmonauts also included a pronounced decrease in vibrosensitivity thresholds of the support areas of the foot with every stimulation frequency (Fig. 1) as well as a drastic decrease in the tendon reflex thresholds. As can be seen from Fig. 2, the thresholds of the gastrocnemius muscle T-reflex, which averaged 1.2 Kg preflight, decreased by more than half postflight. After short- and long-term exposures the threshold decrease was followed by an enormous decline in the T-reflex maximal amplitude and a significant disturbance of cross synergy mechanisms: volontary dorsal flexion of the other leg, which induced reflex inhibition preflight, did not influence the reflex parameters postflight.

The ABR studies revealed two phases in changes of T-, H- and M-responses during prolonged exposure, which differed distinctly in magnitude and even in sign. The first phase was characterized by a marked hypersensitivity of the reflex and peripheral (muscle) mechanisms, a decrease in the thresholds and an increase in the maximal amplitudes of all responses under study (Fig. 3). During the second phase that began after the 30th day of ABR, hypersensitivity diminished gradually, pointing to the development of processes probably associated with modifications in muscular membrane properties: the thresholds of the responses increased, approximating on the 120th day of ABR the pretest value and the maximal amplitudes decreased accordingly. The maximal M-response amplitude on the 90th day of ABR decreased to 40-46 % of the control value (Fig. 3).

Muscular properties. The trend for a noticeable decline to occur in the strength of the leg muscles is a natural consequence of exposure to hypo- and microgravity of any duration.

After short-term exposure (7 days) to SF and DI similar changes in the strength and velocity properties were observed. As can be seen from Fig. 4, members of both groups exhibited a significant decrease (by 20-30 % after SF and 30-40 % after DI) of the maximum strength of leg extensors at all velocities including isometric contractions; at the same time they did not show any significant changes in the strength characteristics of the anterior tibial muscle (TA) that remained practically unaltered except during isometric contractions, which were altered in immersed subjects. Analysis of the ratio between EMG maximum amplitudes and muscles efforts show that it had increased in leg extensors after both space flight and immersion. This was another indication of decline in the functional capability of these muscles; no changes in the EMG effort ratio of the TA was observed after either type of exposure.

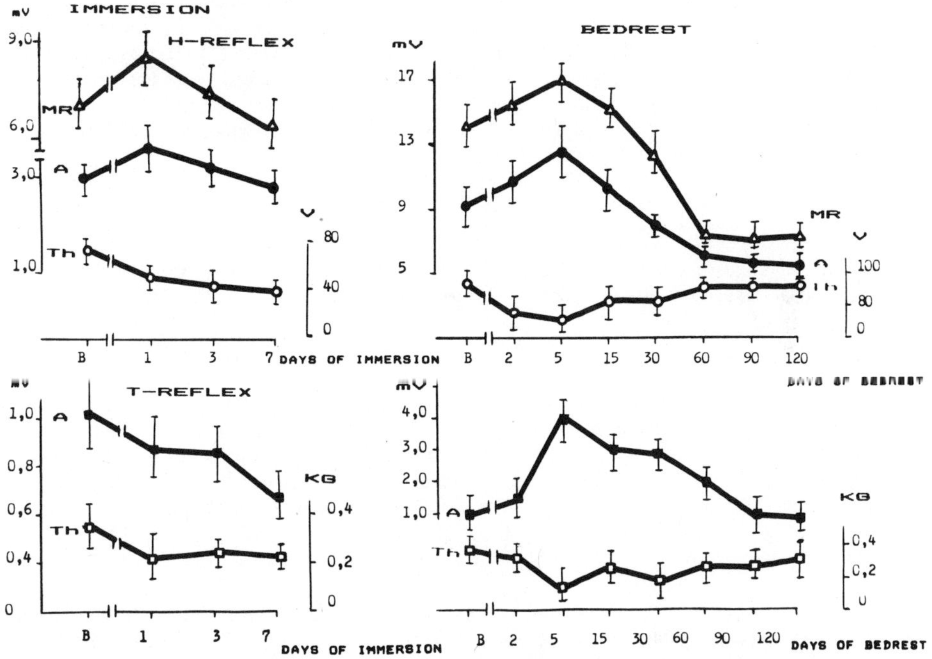

Fig. 3. Alterations of the amplitudes and the thresholds of the H-, T- and M-responses of triceps surae (TS) during exposures to simulated microgravity. The figures on the abscissae indicate days of exposure to hypogravity; the figures on the ordinates show the values of reflex amplitudes in mV (on the left) and the values of reflex thresholds in V (on the right, above) and kG (on the right, below). Each point on the curve indicates the mean value of parameters on a given day; the standard errors are shown by vertical lines. The amplitudes and the threshold value are represented by filled and open circles, correspondingly. The amplitudes of M-responses are shown by triangles.

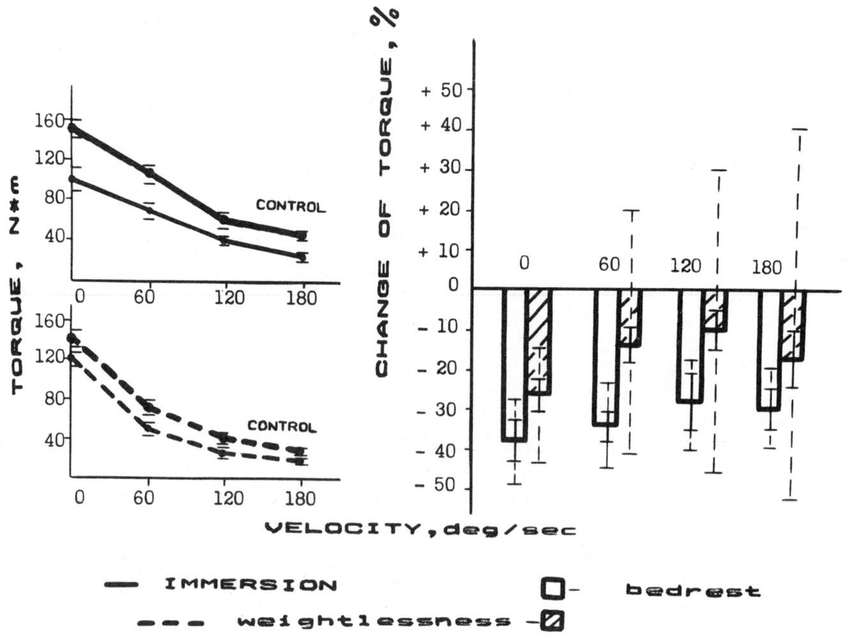

Fig. 4. Alterations of the force-velocity properties of the triceps surae muscle after short (on the left) and long (on the right) term exposures to the real (shown by dotted lines and shaded bars) and simulated (shown by open bars and continuous lines) hypogravity. The abscissae show the moment of force (the right part of the figure) and changes the moment of force (the left part of the figure). The ordinates give the velocities of movements. The control and 7-day exposure data are represented by thick and thin lines, respectively, in the right part of the figure. The open and shaded bars demonstrate the mean values of changes in the results obtained after long term exposures to simulated and real micrograyity, respectively. The standard errors are shown by the continuous vertical lines. The dotted vertical lines indicate the spread of value changes (from minimal to maximal) in the same group of data. 7-day immersion and 120-day antiorthostatic bedrest served as models of simulated microgravity in short and long term exposures, respectively.

Similar changes in leg muscle properties were observed in the ABR studies. However in this case they developed at a slower rate and changes in the TA muscle tested also occurred at all velocities.

The results of isokinetic dynamometry, which was performed after long-duration SF were different, as was to be expected. As shown in Fig. 4, postflight the cosmonauts exhibited only a slight decrease in the strength of the TS muscle and a noticeable decline in that of the TA muscle at every velocity except for 180 °/s. The decrease in muscle strength was manifested not only by a decrease in maximum efforts but also by a significant increase in the time they required. Analysis of individual results showed that the above changes differed greatly, since they ranged from -40 or -60 % to +10 or +15 %; this can be attributed to several countermeasures used by cosmonauts during SF on a preferential basis.

The myotonometric measurements performed by G.I. Gevlich (Gevlich, 1984) after all hypogravity exposures showed a significant decrease in TS muscle stiffness both at rest and during contraction. As shown in Fig. 5, the changes developed very rapidly during immersion, where the decrease reached maximum within the first few hours of exposure. The stiffness change in the three caputs of the muscle was different: it was highest in the soleus muscle and lowest in the caput mediale, which tented to recover its stiffness beginning with immersion day 2. The change in stiffness of the anterior tibial muscle at rest was insignificant. During maximum voluntary contractions the decrease in the extensors was greater than at rest, and the anterior tibial muscle also lost its stiffness.

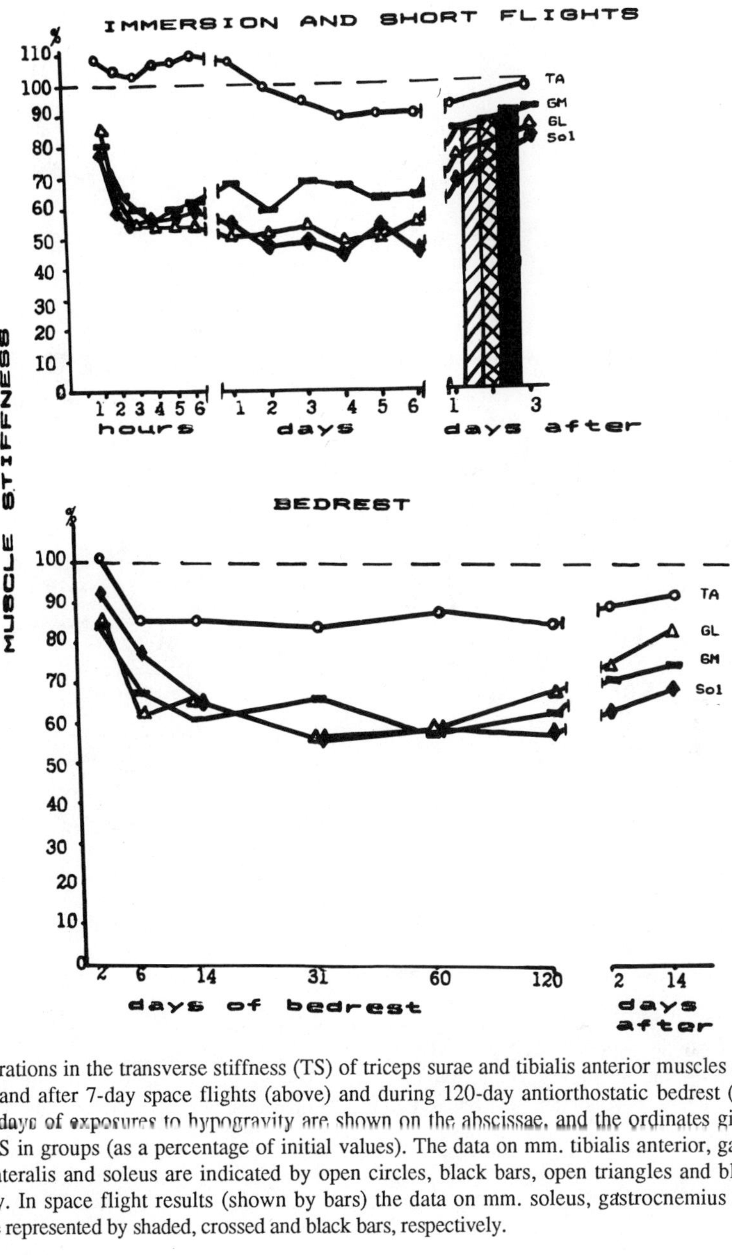

Fig. 5. Alterations in the transverse stiffness (TS) of triceps surae and tibialis anterior muscles during 7-day immersion and after 7-day space flights (above) and during 120-day antiorthostatic bedrest (below). The number of days of exposures to hypogravity are shown on the abscissae, and the ordinates give the mean values of TS in groups (as a percentage of initial values). The data on mm. tibialis anterior, gastrocnemius medialis, lateralis and soleus are indicated by open circles, black bars, open triangles and black squares, respectively. In space flight results (shown by bars) the data on mm. soleus, gastrocnemius lateralis and medialis are represented by shaded, crossed and black bars, respectively.

During ABR, similar changes in muscle stiffness were observed (Fig. 5), but their rate was different. After a slow decrease, which reached a maximum on days 14-30, the stiffness stabilized and began to grow on days 60-90, possibly due to atrophy of myofibers and substitution between the latter and the connective tissue.

Motor control systems. Micro- and hypogravity consistently modified the characteristics of preprogrammed and tracking movements, stabilograms and cyclograms of the gait, suggesting that disorders were affecting mechanisms controlling movements, posture and locomotion, as well as the fact that their efficiency and stability were lower.

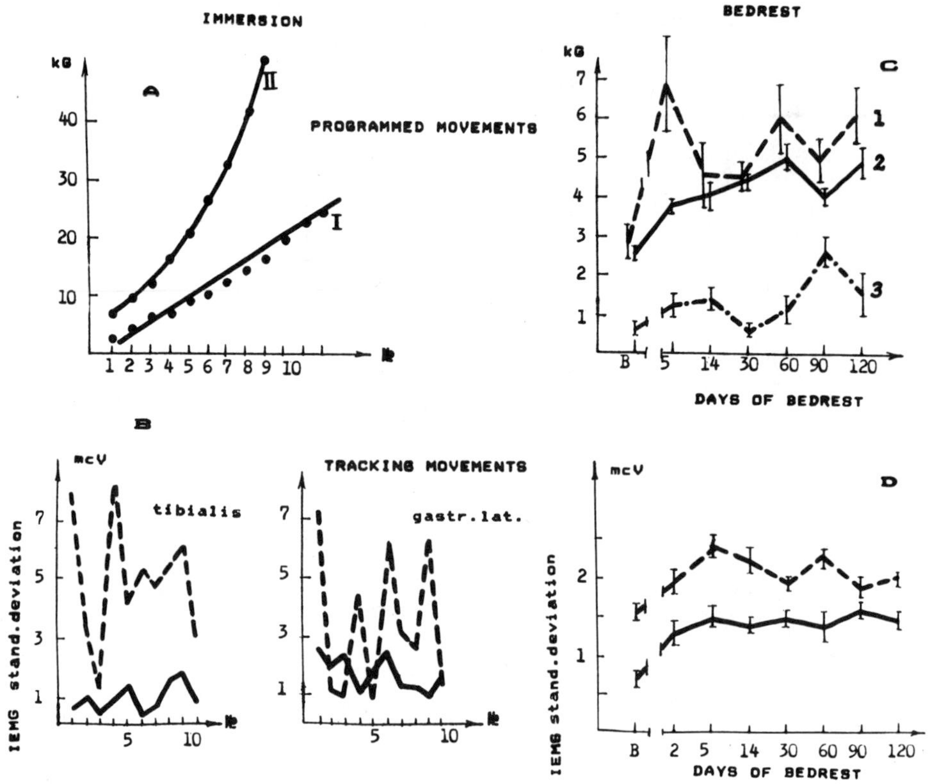

Fig. 6. Alterations in preprogrammed and feedback movement characteristics during 7-day immersion (on the left) and 120-day antiorthostatic bedrest (on the right). In the data given in A, the correlation between the number of attempts is shown on the abscissa, and the strength of effort (mean values) on the ordinate. Lines 1 and 2 represent the control and exposure data, respectively. B gives the standard deviation of the integrated EMG of m. tibialis anterior (on the left) and m. gastrocnemius lateralis (on the right) during 50s time intervals. The full and dotted lines represent the control and 3-day immersion exposure data, respectively. The ordinates give the IEMG standard deviations in mkV, and the abscissae the numbers of 5s time intervals. C give the changes of preprogrammed movement characteristics, namely absolute thresholds, differential thresholds and mean value of errors indicated by dashed, continuous and dotted lines, respectively. Days of bedrest are given on the abscissa, values of threshold efforts (in kG) and mean value of errors (in kG) on the ordinate. D gives the alterations in the integrated EMG of m. tibialis anterior (dotted line) and m. gastrocnemius lateralis (continuous line) measured in 50s time intervals. The day of bedrest and IEMG standard deviation (in mkV) are shown on the abscissa and ordinate, respectively.

When performing the feedback movement task, the subjects showed a greater variability in the TA and GL IEMG during both DI and ABR (Fig. 6). When performing the test in which muscular efforts were graded, they exhibited a failure of the precision control mechanisms: the number of differential efforts decreased and minimal and differential efforts grew in magnitude.

The time-course of the above alterations during DI and ABR was different: during the latter they developed slowly, going through two phases: the 1st phase lasted from ABR day 1 to 30 and the 2nd phase thereafter.

The stabilographic tests showed the existence of significant disorders in the mechanisms of postural regulation. This was manifested by changes in the stabilographic curve characterized, after exposure to hypo- and microgravity, by high frequency oscillations of the tremor type, a definite increase in calf muscle EMG when the subjects were standing in

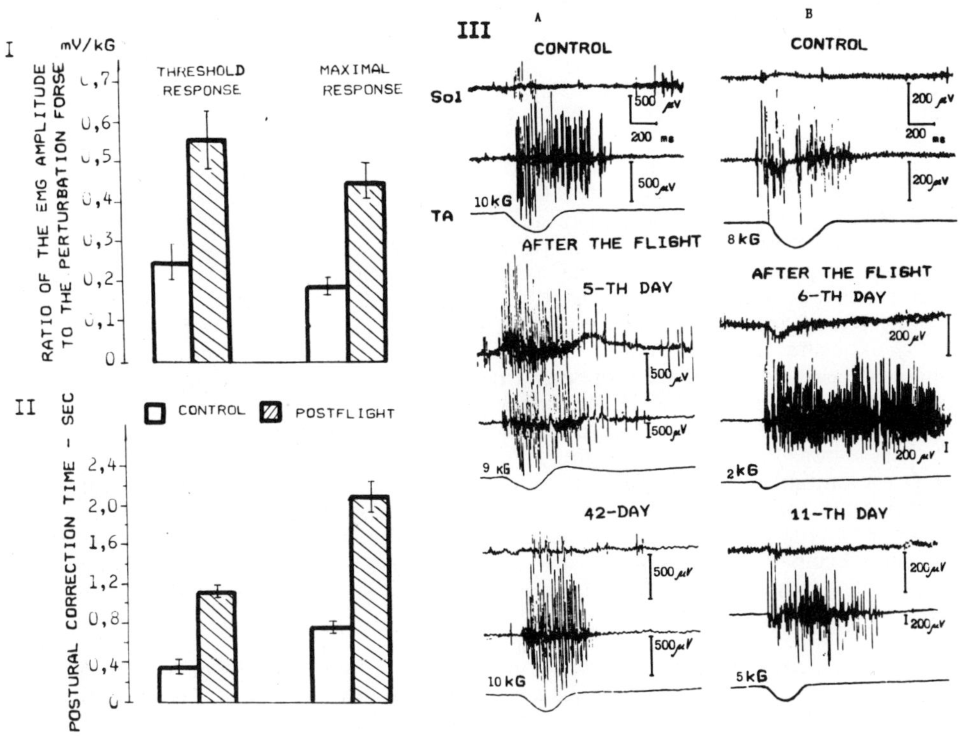

Fig. 7. Changes in characteristics of postural corrective responses after short (on the left) and long (on the right) space flights. The ratio of m. tibialis anterior EMG amplitude to perturbation force measured in threshold and maximal responses is shown in I; the correction time is presented in II. Open and shaded bars are used to indicate control and postflight data, respectively. Examples of postural EMG correction responses of two members (A and B) of long term (140- and 185-days) space flight teams are represented in III. In every sequence the first, second and third lines show EMG of mm. soleus, tibialis anterior and perturbation force, respectively.

the comfortable and Romberg poses, and an elongation of the postural imbalance when the subjects took the Romberg posture.

The perturbation test showed the existence of postural deficits. The time taken for balance recovery after external disturbances increased, and the thresholds of correction EMG responses decreased while their amplitudes and durations grew longer (Fig. 7).

The depth and duration of postural disorders, which were observed after short- and long-term exposures to hypo- or microgravity, varied appreciably, being shortest in DI and short-term SF crewmembers and increasing considerably with exposure time. In this case, corrective response amplitudes and durations increased greatly and the well organized postural synergy pattern disappeared: the extensors and flexors of the foot functioned simultaneously (Fig. 7).

Discussion

Experimental analysis, in which the quantitative spectrum of accuracy control disorders was specified, led us to the conclusion that these were atactic disorders and that the major pathogenic factors varied in the course of exposure. The study revealed the occurrence of similar changes in several compartments of the motor system after exposure to real (SF) and simulated (DI, ABR) microgravity; it also demonstrated the existence of differences due to the effects of secondary factors, such as the motor activity level (low during DI and ABR and high during SF) use of and the counter-measures during long-term SF. These may account for the fact that slightly greater motor disorders occurred after 7-day DI than 7-day SF, whereas less marked and extremely variable changes in muscle properties occurred after long-term SF and others.

Obviously, deterioration of motor capabilities observed during and after exposure to microgravity may be caused by disorders in any part of the motor system, e.g. changes in the feed-back activity, decrease in support reactions, reduction of the strength of gravitational muscles, etc. However, specific features of disorders, e.g. high correlation between the decrease in muscle tone and muscle strength, distinct relationship between the level of postural disorders and body harness rigidity (the disorders diminished when the freedom was decreased and were aggravated when it was increased), which were observed after 7-day exposure (Kozlovskaya et al., 1982), suggest that support unloading plays an important part in the genesis of ataxia developing in the first stage of exposure to microgravity - short-term SF, DI, the first 30 days of ABR (Kozlovskaya et al., 1983). The reduction of the afferent inflow from "support" areas of the body required for a high activity of motoneurons of extensors (Magnus, 1924.), and the subsequent decline in the tone of extensor (antigravity) muscles and muscle receptors may be responsible for other manifestations of hypogravitational ataxia, namely, disorders of postural synergies and hyper-sensitivity of spinal reflex mechanisms as well as proprioceptive sensory systems. Enhanced excitability of central nervous mechanisms and an inhibitory deficiency appear to be natural consequences of any kind of sensory deprivation (Drosdova, 1964; Kunstman and Orbeli, 1924; Magnus, 1924.).

The above suggestions were further supported by the results of the MU activity studies, which revealed, at an early stage of exposure to hypogravity, definite signs of motoneuronal hyperactivity: in ISI histograms low frequency MUs, which are usually not involved at low levels of muscle effort, and the obvious tendency to destabilization and synchronisation of MU rhythmic activity (Kozlovskaya et al., 1986).

At the second stage in hypogravitational exposure (long term SF, ABR) peripheral disorders, such as muscle hypo- and even atrophy, alterations of the neuromuscular transfer function, the muscular membrane properties, etc., can contribute considerably to the development of ataxia. It has been shown that 30-day or longer exposure to hypogravity usually causes a progressive development of morphologic and metabolic changes in muscles (mainly extensors), an obvious decline in their contractile properties, and structural disorders affecting neuromuscular synapses (Iljina-Kakujeva et al., 1979; Kakurin et al., 1971).

The alterations in MU rhythmic activity characteristic of 2nd phase of hypogravitational exposure (ABR) are consistent with this suggestion. The occurrence of a number of enormously long (two to three times longer than the mean duration) interspike intervals in ISI histograms points as a rule to disorders of the neuromuscular synaptic transmission (Kozlovskaya et al., 1986). The involvement of peripheral factors in the pathogenesis of the 2nd stage disorders also explains the high efficiency of muscle exercises used as a countermeasure by long-term SF crewmembers.

A large decrement in reflex and direct muscle electromyographic responses which was observed during and after long-term hypo- and microgravity even when muscle strength changed only slightly (185 day SF) led us to predict the presence of muscle membrane property alterations. This prediction was confirmed by the results of DI studies (Christova, 1986) that demonstrated substantial changes in the muscle conduction speed as well as the size of the muscle field potential.

The last point to be discussed concerns the nature of co-ordination disorders recorded after long-term SFs. The depth and recovery time (duration after SF completion) of these disorders were not closely correlated with other measurements, but depended mostly (but not fully) on the SF duration. It seems logical to attribute these disturbances to the fact that the processes reorganizing motor patterns in microgravity obey new biomechanics (see "Summary"). This would explain the disappearance of corrective postural responses or their substitution by other synergies which were observed in long-term SF crewmembers.

To sum up, the hypothesis that the genesis of hypogravitational ataxia is multifactorial allows for a variety of mechanisms that can contribute to the development of this pathological disorder. We have made a small step forward in studying some of these factors. However we are still far from knowing their exact respective contributions in real situations.

References

Artemjeva E.N., Kudinova M.P., Zalkind M.S., Kandel E.I., Koslovskaya I.B. 1977. Studies on mechanisms of supraspinal control of spinal activity in man. Physiol. man., 3, 913-923.

Baker J.T., Nicogossian A.E., Hoffler I.W., Johnson R.L., Hordinsky J. 1977. Changes in the Achilles tendon reflexes following Skylab mission. In: Biomedical results from Skylab. NASA, SP-377, 131-135.

Berry C.A. 1973. Weightlessness. In: Bioastronautics data book, 2nd ed., NASA, Washington, 349-416.

Chekirda J.F., Yeryomin A.V. 1974. Dynamics of cyclic and acyclic locomotions in Soyuz-18 crewmembers after 63 day Space Flight. Kosm. Biol. med., N 4, 9-13 (in russ.).

Chkhaidze L.V. 1968. Coordination of voluntary movements of man in spaceflight environment. Nauka, Moscow, p. 133.

Christova L.G., Gydikov A.A., Aslanova I.F., Kirenskaya A.V., Koslova V.G., Kozlovskaya I.B. 1986. Changes in parameters of muscle potentials in man induced by immersion. Kosm. Biol. a. Aviacosm. Med., N 6, 27-30.

Drosdova V.N. 1964. Consequences of full hind limb deafferentation in puppies and dogs. In: Mechanisms of compensatory adaptations. Nauka, Moscow, 99-103 (in russ.).

Gevlich G.I., Grigorieva L.S., Bojko M.G., Kozlovskaya I.B. 1983. Measurement of skeletal muscle tone by recording transverse stiffness. Kosm. Biol. , N 5, 86-89.

Gevlich G.I. 1984. Mechanisms of muscle tone disturbances under conditions of hypogravity. Thesis of dissert., Moscow, p. 24.

Gurovsky N.N., Eryomin A.V., Gazenko O.G., Egorov A.D., Bryanov I.I., Genin A.M. 1975. Medical investigations during Soyuz-13, Soyuz-14 and Salyut-3 space flights. Kosm. Biol., N 2, 48-53 (in russ.).

Homick J., Reschke M., Miller E. 1977. The effects of prolonged exposure to weightlessness on postural equilibrium. In: Biomedical results from Skylab, NASA, SP-377, 104-112.

Iljina-Kakujeva E.I., Petrova N.V., Portugalov V.V. 1979. The influence of space flight on skeletal muscles and neuronal apparatus of muscles. In: The influence of space flight factors on animals. Nauka, Moscow (in russ.).

Kakurin L.I., Tcherepahin M.A., Pervushin V.I. 1971. The influence of space flight factors on the muscle tone in man. Kosm. Biol., N 2, 63-68 (in russ.).

Kozlovskaya I.B., Kreydich Yu.V., Oganov V.S., Koserenko O.P. 1981a Pathophysiology of motor functions in prolonged manned space flights. Acta astronautica, 8, 1059-1072.

Kozlovskaya I.B., Kreydich Yu.V., Rachmanov A.S. 1981b. Mechanisms of the effects of weightlessness on the motor system of man. The Physiologist, 24, 59-63.

Kozlovskaya I.B., Aslanova I.F., Grigorieva L.S., Kreydich Yu. v. 1982. Experimental analysis of motor effects of weightlessness. The Physiologist, 25, 49-52.

Kozlovskaya I.B., Aslanova I.F., Barmin V.A., Grigorieva L.S., Gevlich G.I., Kirenskaya A.V., Sirota M.G. 1983. The nature and characteristics of a gravitational ataxia.

Kozlovskaya I.B., Aslanova I.F., Kirenskaya A.V. 1986. The effect of support unloading in characteristics of motor control systems activity. In: "Motor control", New-York, Pergamon Press, p. 149-153.

Kubis J.F., Mc Laughlin E.L., Jackson J.M., Rusnak R., Mc Bride G., Saxon S.V. 1977. Task and work performance on skylab missions 2,3, and 4. In: "Biomedical results from Skylab". NASA, SP-377, 136-154.

Kunstman K.I., Orbeli L.A. 1924. Consequences of the hindlimb deafferentation in dogs. News of Petrogr. inst. by A.E. Lesgaft, 9, 187-194 (in russ.).

Magnus R. 1924. Korpersfellung. Springer Verlag, Berlin, p. 540.

Mitarai G., Muno T., Mori H., Jamasaki J., 1978. Compensatory leg muscle function shift during adaptation to stimulated weightlessness. XXVI Inter. Congr. Aerosp. Med., London, 4-8.

Oganov V.S., Potapov A.N. 1979. Study of skeletal muscle functions in experiments carried out on biosatellites. In: "The influences of space flight factors on animals".

Reschke M., Homick J., Baker J. 1979. Vestibulospinal reflexes in man as a function of linear acceleration. Ann. Sci. Meeting Aerosp. Med. Ass., Washington, 298-299.

Shulzenko E.B., Vil-Viliams I.F. 1976. Simulation of prolonged water immersion effects by "dry" immersion. Kosm. Biol., N 2, 82-84 (in russ.).

SYNAPTIC INFLUENCES ON MOTONEURONES OF THE SPINAL CORD IN VERTEBRATES

A. I. Shapovalov and N. P. Vesselkin

I.M. Sechenov Institute of Evolutionary Physiology and Biochemistry
U.S.S.R. Academy of Sciences, Leningrad, U.S.S.R.

Summary

The mode of synaptic transmission between coupled spinal motoneurones and between primary somatic afferent fibres and motoneurones was studied in various vertebrate species. In frogs and carp the majority of interacting motoneurones were found to be coupled electrically. In lampreys, only electrotonic motoneuronal interactions were observed. In rats the interactions between motoneurones were found to be mostly mixed electrical and chemical ones. Direct monosynaptic control of motoneurones by the somatic primary afferents has been shown to occur in two aquatic species (lamprey, carp). Consequently, their mechanism does not depend on the presence of muscle spindles, which are absent in both species.

In the lamprey, carp and frog the transmission in these synapses is a bimodal electrical and chemical one.

Comparison between transmission modes in homologous rows of synapses reveals an evolutionary trend towards a decrease in the number of synapses exhibiting the electric mode of transmission.

Introduction

It is known that the neurones of the vertebrate CNS are connected by synapses which differ in their structure, mode of transmission and mediatory specificity. But up to now it has not been established why this or that particular type of synaptic interaction occurs rather than another. Comparative investigations on the structural and functional organization of junctions between homologous neurones in different vertebrates can help to elucidate the general principles underlying the functional or evolutionary predominance of a given type of synapse. At our laboratory the motoneurone of the spinal cord was chosen as a model for investigation and the characteristics of its inputs, consisting of primary afferents, propriospinal and supraspinal fibres, and neighbouring motoneurones were studied. In the present paper, the results of investigations on intermotoneural interactions and the direct segmentary sensori motor control involved will be reported.

I. Intermotoneuronal interactions

It has been known since the 1960s that activation of the ventral root (VR) by stimuli which are subthreshold for some of the fibres can produce a gradual depolarization in some motoneurones. It is believed that this depolarization is the p.s.p. reflecting direct interaction between motoneurones. This type of direct intermotoneuronal interaction caused by VR stimulation - VR-e.p.s.p. - has been described in Cyclostomes (Tamarova et al., 1978), Teleosts (Auerbach and Bennett, 1969; Bando, 1975; Suderevskaya, 1983), Amphibians (Grinnell, 1966; Magherini et al., 1976; Shapovalov and Shiriaev, 1978) and Mammals (Fulton et al., 1980; Arasaki et al., 1984).

Analysis of the parameters of VR-e.p.s.p.s. in various animals strongly suggests the existence of electrotonic interactions between the motoneurones. It is generally believed that these interactions involve dendro-dendritic synapses. The recurrent axonal collaterals are also assumed to make contact with neighbouring motoneurones and evoke e.p.s.p.s. (See for ref. Shapovalov and Shiriaev, 1984).

However, the methods which have been used to stimulate the whole VR have some limitations as far as the analysis of motoneuronal interactions is concerned. Numerous VR fibres are stimulated and many inputs to one cell can be activated. It is known that the VR contains the afferent somatic and sympathetic fibres which may also be activated (Shapovalov and Shiriaev, 1984; Chung et al., 1985). The most precise data on the mode of transmission is that obtained by parallel intracellular recordings on two inter-connected cells.

Synaptic interaction between single motoneurones in the frog spinal cord

Shapovalov and Shiriaev (1984) have recorded 89 connected pairs of motoneurones in isolated preparations of frog spinal cord. When the intracellular stimulation of one motoneurone by the depolarizing pulse evoked an action potential, the second cell of the coupled pair generated the e.p.s.p. In the majority of cell pairs the latency of these e.p.s.p.s. was 0.5-2.3 ms after the stimulus artefact and 0-1.3 ms after the onset of the action potential. Thus, the synaptic delay between the spike in the "triggering" cell and the e.p.s.p. in the coupled neuron indicated the existence of an electrical connection. The replacement of Ca^{2+} by Mn^{2+} in the perfusing solution failed to influence the e.p.s.p.s and demonstrated their electrotonic nature (Fig. 1). The interaction between the coupled cells in the cases described was symmetrical: the spike in the first cell produced the e.p.s.p. in the second, and vice versa. Therefore the majority of connected motoneurones in the frog spinal cord are coupled via electrotonic synapses.

Only about 4% of the cell pairs were found to be coupled purely by chemical synapses. In these cases the latency of the e.p.s.p.s was longer (2.6-4.3 ms after the artefact

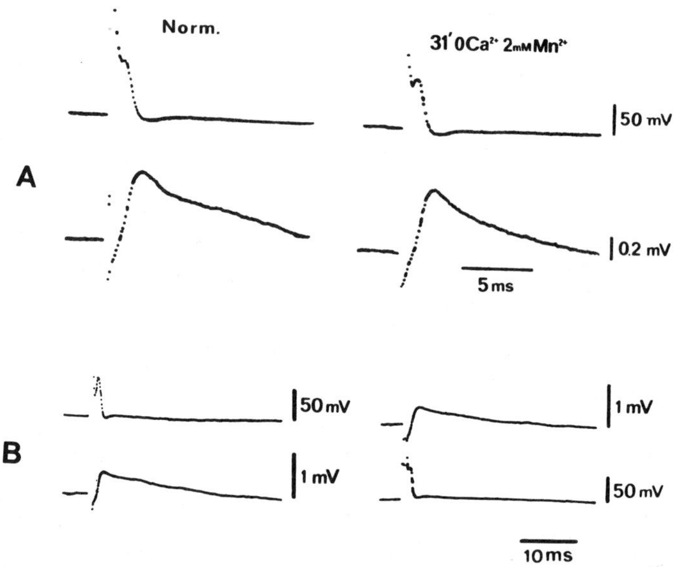

Fig. 1. Electrotonic synaptic interaction between motoneurones of the frog. Averaged elementary e.p.s.p.s. A. The replacement of Ca^{2+} by Mn^{2+} did not affect the e.p.s.p. Upper traces - the action potential in the first motoneurone evoked by an intracellular depolarizing pulse. Lower traces - e.p.s.p.s in the second motoneurone. Left - with the preparation in normal Ringer solution. Right - in the modified perfusion solution. B. Bidirectional transmission in the electrotonic contact. Left - the action potential in the first motoneurone (upper trace) and the e.p.s.p. in the second cell (lower trace). Right - the e.p.s.p. in the first cell (upper trace) evoked by the action potential in the second neurone (lower trace).

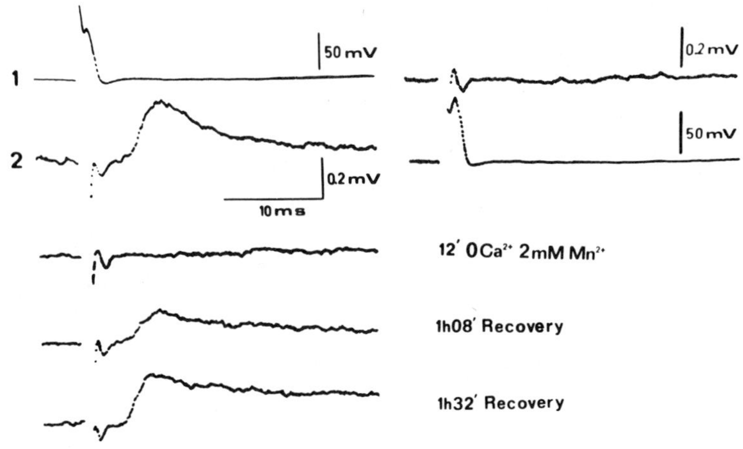

Fig. 2. Chemical transmission in the intermotoneuronal synapses of the frog. Averaged elementary e.p.s.p.s. Left row - effects of replacement of Ca^{2+} by Mn^{2+} in the perfusion Ringer solution. 1: action potential in the first motoneurone. 2: e.p.s.p. in the second motoneurone in normal Ringer solution; 12 min. after removal of Ca^{2+} and addition of Mn^{2+}; 1h08 and 1h32 min after renewing the perfusion with normal Ringer solution. Right - the action potential in the second motoneurone (lower trace) did not produce the e.p.s.p. in the first cell (upper trace).

and 1.3-3.3 ms after the initial component of the action potential). (Fig. 2). These e.p.s.p.s displayed a slow rising phase. Repetitive stimulation with interstimulus intervals of 10 to 20 ms produced a potentiation of the test-response, and its amplitude increased up to 40%. The transmission in these synapses was solely unidirectional. It was reversibly blocked in Ca^{2+} - free Mn^{2+} - containing Ringer solution. The e.p.s.p.s fluctuated in amplitude and were completely absent from some recordings.

Lastly, in about 3% of the cases, the stimulation of one cell produced a two-component e.p.s.p. in the coupled motoneurone (Fig. 3). The start of the first component almost coincided with the beginning of the action potential in the "triggering" neuron. The latency of the second component was about 1.4 - 1.8 ms longer. The initial component was insensitive to the replacement of Ca^{2+} by Mn^{2+} in the Ringer solution and it was transmitted bidirectionally. The second component, which was observed in only one of the two coupled cells, was potentiated during paired stimulations (increased by up to 60%), and its amplitude fluctuated and was reversibly blocked by the ion substitution in the perfusing solution. This indicated that the first component was mediated electrically and the second chemically.

The data presented show that the inter-motoneuronal interaction in frogs is mainly electrotonic. Most probably, this interaction is mediated by electrotonic dendro-dendritic contacts. Sotelo and Taxi (1970) have found gap junctions in the frog spinal cord, in the intermediary gray matter, where the motoneuronal dendrites are located. Sotelo and Grofova (1976) and Motorina (1978) have described gap junctions in the lateral motor column. The stability of the amplitudes of e.p.s.p.s recorded in either direction as well as the decrease in the e.p.s.p.s when the soma-dendritic component was blocked but spared the initial segment component and therefore did not alter the propagation of action potential in the collaterals, argue against axonal recurrent collaterals being involved in this mechanism (Shapovalov and Shiriaev, 1984).

Furthermore, these collaterals very probably play a role in mediating the chemical components in the case of mixed and pure chemical interactions. It should be mentioned that the existence of these collaterals in frogs has been confirmed morphologically in preparations processed with the Golgi method and in experiments employing the intracellular injection of HRP.

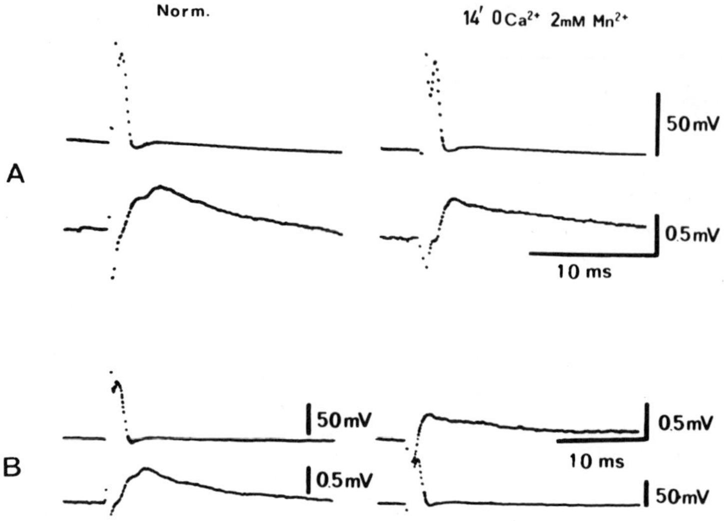

Fig. 3. Bimodal electrical and chemical interaction between frog motoneurones. Averaged elementary e.p.s.p.s. A. Replacement of Ca^{2+} by Mn^{2+} in the perfusing Ringer solution selectively blocked the second component of the e.p.s.p. Upper traces - action potentials in the first cell, lower trace - e.p.s.p.s in the second one. B. Bidirectional transmission of first component. Left - the action potential in the first motoneurone (upper trace) produces mixed e.p.s.p. in the second one (lower trace). Right - spike in the second cell (lower trace) evoked only electrotonic e.p.s.p. in the first one.

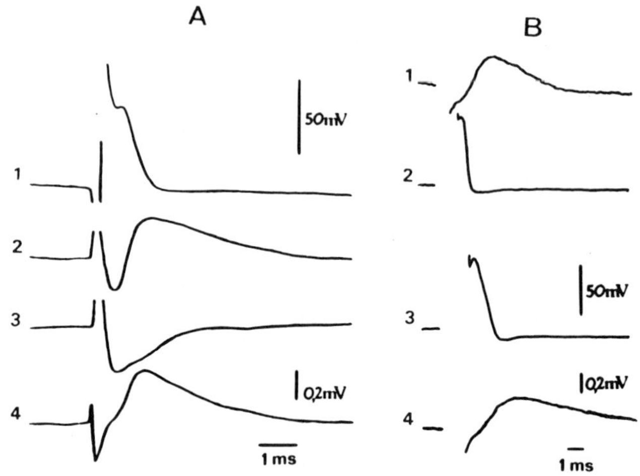

Fig. 4. Electrotonic interaction between two pairs of motoneurones (A, B) in the spinal cord of the carp. Averaged responses. A: 1 - action potential in the first motoneurone evoked by intracellular depolarizing pulse; 2 - intracellular response in the second motoneurone; 3 - field potential, recorded extracellularly; 4 - elementary e.p.s.p. resulting from subtraction of field potential (3) from the intracellular response (2). B. elementary e.p.s.p. in the first motoneurone (1) evoked by the action potential in the second one (2); 3, 4 - transmission in the opposite direction.

Synaptic interactions between single motoneurones in the carp spinal cord

The results obtained at our laboratory on teleosts had much in common with those reported on frogs. Karamian and Suderevskaya (1986) investigated 9 pairs of interacting motoneurons in the carp using the same method. In 6 of the pairs, the intracellular stimulation of one motoneurone produced an e.p.s.p. in the second cell which was probably electrotonically mediated. More specifically, these responses exhibited a very short latency, a stable amplitude and bidirectional conduction (Fig. 4). A similar coupling probably also occurs with the gap junctions described in the corresponding spinal cord regions of fish (Schnitzlein and Brown, 1975).

In 3 coupled pairs, the stimulation of one cell evoked an e.p.s.p. in the second neurone with two components (Fig. 5). The latency of the second component was O.7-1.5 ms longer than that of the first. On the basis of the criteria described above, the first component was electrotonic, whereas the second was chemically mediated. Statistical analysis of the amplitude fluctuations of the second component showed that the distribution of the amplitudes is well approximated by a Poisson or bimodal distribution. This is in agreement with the model of quantal release of transmitter in these synapses.

There exists a difference between the frog and fish intermotoneuronal contacts. In the frog, the electrical component of the mixed e.p.s.p. could be recorded in both directions, while in the carp both components of the e.p.s.p.s were transmitted in one direction only. This is normal with chemical transmission, but not typical of electrotonic synapses. These characteristics of bimodal e.p.s.p.s can be explained by the fact that both components are evoked by the contacts made upon the motoneurones studied by the axonal collaterals of neighbouring motor cells. In this case, even if the electrical transmission in these synapses is bidirectional, the passive depolarization of the collateral in the opposite direction is limited by decrement and does not reach the soma.

The reported data show that in the carp spinal cord the motoneurons are mainly coupled through dendro-dendritic contacts with electrical transmission. Mixed contacts of axon collaterals with the soma are more infrequent. This conclusion is supported by the fact that, in several investigations, mixed synapses have been found to exist on the motoneurons of fish (Charlton and Gray, 1966; Schnitzlein and Brown, 1975; Bennett et al., 1978). This

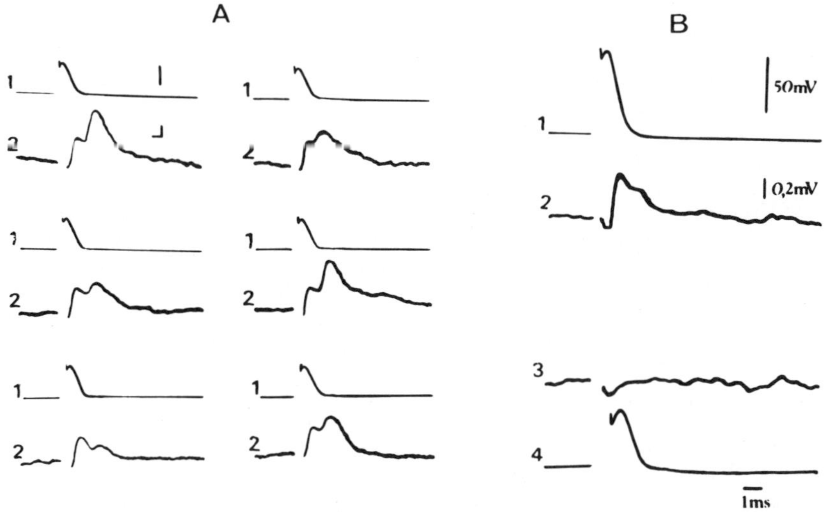

Fig. 5. Two-component moto-motoneuronal e.p.s.p.s in the carp. A: spontaneous amplitude fluctuation of the second (chemical) component and stability of the first (electric) one. 1 - action potential in the first cell; 2 - e.p.s.p. in the second cell. B: unidirectional transmission of both components. 1 - action potential in the first cell; 2 - e.p.s.p. in the second cell; 3, 4 - absence of transmission in opposite direction.

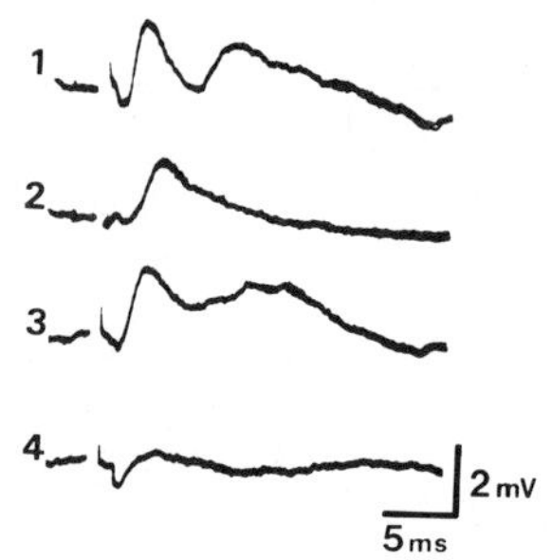

Fig. 6. Mixed electrical and chemical interaction between motoneurones in 12-day old rat. VR e.p.s.p.s. 1 - responses in normal Ringer solution; 2 - the early (electric) component was unaffected, the late (chemical), was blocked in the Ringer solution with Ca^{2+} replaced by Mn^{2+} (17th min); 3 - recording in the normal Ringer solution (25th min); 4 - field potential, extracellular registration.

also suggests that in fish the contact of collateral with motoneurone is organized differently from in the frog, where this contact is purely chemical.

The characteristics of the interaction between motoneurones when recorded from two coupled cells were therefore investigated in two vertebrates: in fish and the frog. It has been mentioned that in cyclostomes, interactions between motor cells have been studied only on the basis of VR-e.p.s.p.s (Tamarova et al., 1978). It may be concluded that motoneurons in the lamprey are coupled electrotonically both by the dendro-dendritic synapses and by contacts of axonal collaterals with motoneurones. In mammals, coupled pairs of motoneurones have also been investigated. The existence of electrotonic coupling between motoneurons has been demonstrated in two studies (Fulton et al., 1980; Arasaki et al., 1984).

The VR-e.p.s.p.s in the motoneurones of the spinal cord of the young rat

At our laboratory the VR-e.p.s.p.s in motoneurones of isolated preparation of the spinal cord of young rats (9-12 d.) were studied by Karamian and Kozhanov. It was observed that in these animals motoneurones were coupled with motoneurones of both the same and adjacent segments. The motoneuronal interaction found to exist in their experiments was mostly mixed. The VR-e.p.s.p.s when recorded displayed two components: the early and the late components. The late component was blocked by the replacement of Ca^{2+} by Mn^{2+} in the Ringer solution, wheras the early potential was not influenced by substitution (Fig. 6), which is characteristic of electric transmission. In some cases unimodal chemical or electrical e.p.s.p.s were also recorded.

The analysis of intermotoneuronal connections in mammals is now at the stage where it is premature to make any precise conclusions concerning the mode of interneuronal contacts. However there is evidence suggesting that motoneurones in mammals are connected by three types of contacts: chemical, electrical and mixed. This suggestion is also supported by morphological data published by Motorina (1986). In her ultrastructural studies on the motor nuclei of the mouse and young rat she described not only chemical axosomatic and axodendritic synapses, but also morphologically mixed axodendritic, axosomatic and purely electrotonic axodendritic, dendrodendritic and dendrosomatic synapses. Although these findings are not conclusive, they provide some indication as to the possible structural bases for motoneuronal interactions in mammals.

Comparative studies on motoneuronal interaction therefore support the view that direct intermotoneuronal coupling is an important general functional characteristic of vertebrates, regardless of their evolutionary level and type of locomotion. The second, principal conclusion is that electric transmission plays an important role in motoneuronal coupling in all the vertebrates investigated, though it would seem that the amount of chemical intermotoneuronal contacts is greater in mammals.

II. The synaptic transmission from the primary sensory afferents to the spinal cord motoneurones

The second class of homologous synapses studied is constituted by the contacts formed by the terminals of primary sensory afferents and motoneurones. This problem has two aspects. First, it is generally considered that the direct monosynaptic control of motoneurones by the primary afferents is a feed-back channel from stretch receptors and exists only in the more developed animals, which have muscle spindles. The verification of this point of view is of primary importance. Secondly, it is interesting to compare the mode of synaptic transmission in homologous synapses in different species which have monosynaptic sensorimotor control.

In the lamprey (Cyclostoma), the spinal ganglion cell counterparts consist mainly of the dorsal cells located in the spinal cord. It has been shown (Batueva, 1983; Shapovalov and Batueva, 1984) that about 20% of the dorsal cells make contact directly upon motoneurones and that the transmission is bimodal. This conclusion is based on the analysis of elementary e.p.s.p.s recorded in motor cells after stimulation of single dorsal root fibres and the subsequent intracellular labeling of coupled units.

Using the HRP labeling of dorsal root fibres, Batueva has found fine terminals on the surface of muscle fibres (Fig.7). These data confirm those concerning the sensory nerve endings in the muscles and skin of the lamprey (Whiting, 1948). Thus it is now clear that in lampreys, there is a direct segmentary control of motor cells and that the transmission in the sensorimotor synapse in these animals is mixed (electrical and chemical). It is likely that the free nerve endings in the muscle are the source of sensory fibres of dorsal cells directly contacting the motoneurones.

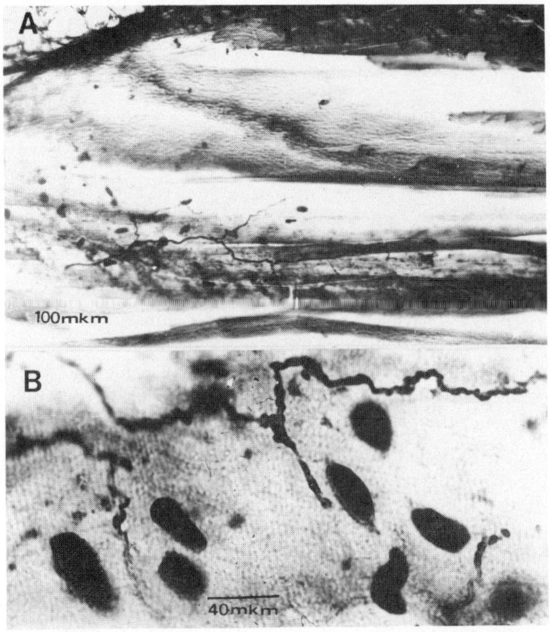

Fig. 7. Sensory nerve branching in the trunk somatic muscle of the lamprey revealed by retrograde axonal HRP transport. The enzyme was applied to the spinal cord after dorsal root transsection. Carmine counterstaining. A: the dichotomic branching of the nerve following two parietal muscle fibres. B: the terminals of the sensory nerve.

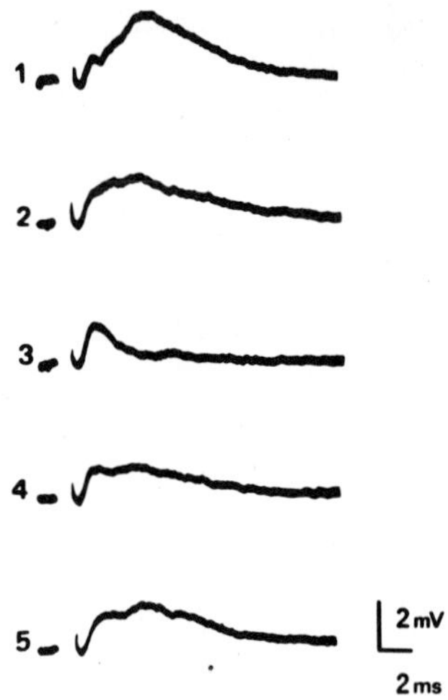

Fig. 8. Mixed electrical and chemical transmission from primary afferent fibers of the motoneurones in the carp spinal cord. E.p.s.p.s in the motoneurones of isolated spinal cord after dorsal root stimulation. Normal Ringer solution (1); 20th min (2) and 60th min of perfusion with solution with Ca^{2+} replaced by Mn^{2+}; 45 min (4) and 70 min (5) after the renewal of perfusion with the normal Ringer solution. Recovery of chemical transmission.

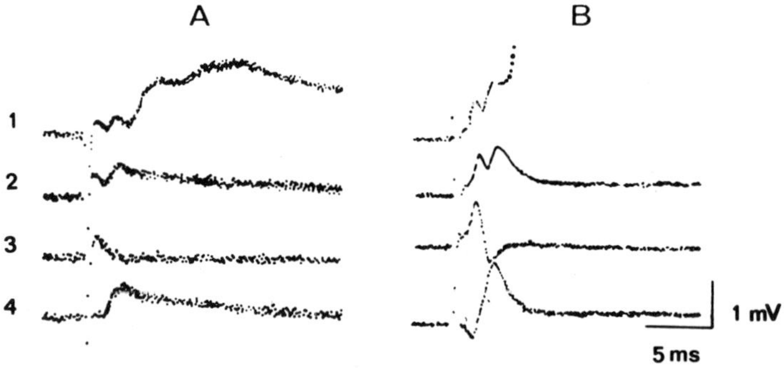

Fig. 9. Mixed electrical and chemical transmission from the primary somatic afferent to the motoneurones of the isolated spinal cord of Rana ridibunda tadpoles in stages XIV and XXV of individual development. Averaged responses. e.p.s.p. in the normal Ringer solution; 2 - 20th min of perfusion with Ringer solution containing Mn^{2+} and no Ca^{2+}; 3 - field potential, extracellular recording; 4 - result of subtraction of extracellular field potential (3) from intracellular response (2). Ionic substitution blocked mono- and polysynaptic chemical components of e.p.s.p., but did not influence the electrical one.

In the carp, the HRP and cobalt labeling injected into the muscle (after the VR was cut) or applied to the dorsal root, was localized within the zone of motor nuclei of the corresponding and neighbouring segments. It is therefore reasonable to assume that in this fish direct contacts exist between the primary afferents and motoneurones (Motorina and Suderevskaya, 1983). Direct stimulation of the carp dorsal root produced, in 70% of the motoneurones of either the corresponding or adjacent segments, short latency e.p.s.p.s, the temporal parameters of which reflected monosynaptic responses. The majority of these e.p.s.p.s comprised two components (Fig. 8). The replacement of Ca^{2+} by Mn^{2+} in the perfusion solution reversibly blocked the second component whereas the first was not affected. The latter indicates the electrotonic nature of the first component and the chemical nature of the second. Since morphologically mixed synapses have been described in the case of motoneurones in fish (Charlton and Gray, 1966; Schnitzlein and Brown, 1975; Bennett et al., 1978), it would seem probable that in our experiment, the transmission occurs in the bimodal (electrical and chemical) synapse as has been shown recently to be the case in lampreys (see below) and frogs (Shapovalov and Shiriaev, 1984). Direct monosynaptic control of motoneurones by the primary afferents has already been shown to take place in two aquatic species (lamprey, carp). Consequently, this mechanism exists in animals of quite different phylogenetic levels and does not depend on the presence of muscle spindles, which are absent in both species.

Various authors have described the particularities of the monosynaptic primary afferent control of motoneurones in various vertebrates. In cyclostomes, the transmission in these synapses is bimodal: electrical and chemical. The majority of recordings have shown the same mode of transmission to occur in teleosts and amphibians. But judging from the literature and our own data, this transmission appears to be chemical in mammals (Brock et al., 1952; Tamarova et al., 1978; Shapovalov, 1980).

Since the data indicate that even at the lowest evolutionary level studied, the sensorimotor synapses are mixed, we thought it might be interesting to retrace the formation of these contacts during the course of ontogenesis, beginning from the first moment at which

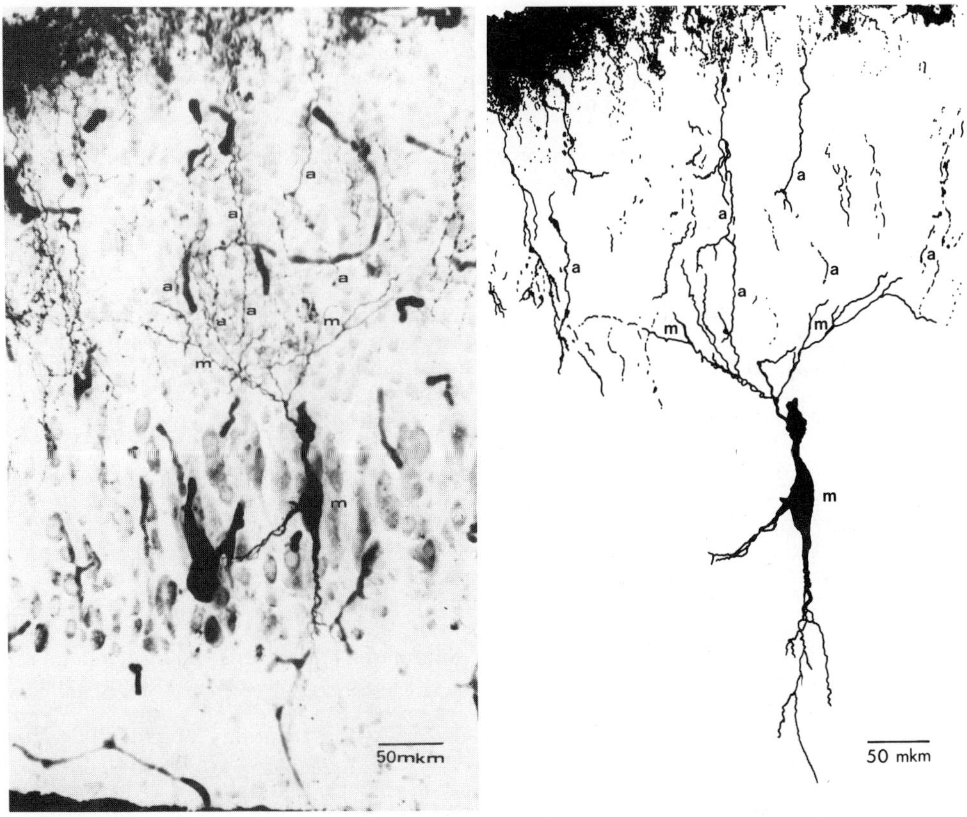

Fig. 10. Camera lucida drawing of primary somatic afferents contacting an electrophysiologically identified motoneurone. Rana ridibunda, metamorphosis stage. The HRP was applied simultaneously at the dorsal root and into the motoneurone. Sagittal cut of the spinal cord at the X segment level: m - motoneurone and its dendrites, a - primary afferent.

they appeared. This has been investigated at our laboratory by Shupljakov in two anurans: Rana ridibunda and Xenopus laevis. All the experiments were performed on isolated, perfused preparations of the spinal cord. It was demonstrated electrophysiologically that the first direct contact of primary afferents with motoneurones in Rana ridibunda occurred at the period of prometamorphosis (Shupljakov, 1986). In Xenopus laevis these contacts appeared at the same stage. At the same time, dorsal root stimulation evoked e.p.s.p.s in single motoneurones which appeared to be monosynaptic (Fig. 9).

The primary afferent fibres approaching the dendritic and cellular zones of the motoneurones are clearly visible at this stage when the afferents and the motoneuron are labeled with HRP. The fibres reach the motoneurone and very probably can make contact with it (Fig. 10). This suggestion was confirmed under the electron microscope in the case of junctions identified with HRP. It was found that the primary afferents contact the dendrites of motoneurones. These contacts have the typical specializations of chemical synapses and gap junctions. In other respects they are morphologically mixed synapses.

These results demonstrate that the synapses between the primary afferents and motoneurones acquire the properties of mixed contacts (morphologically and functionally) as of the first moment of their formation.

All the present data support the idea put forward by Shapovalov that in the CNS, the purely electrotonic synapses interconnect predominantly with homotypic cells, whereas the mixed synapses connect with heterotypic cells (Shapovalov, 1980); in our study, we have been dealing with synapses between the motor cells and between the primary afferents and motor cells.

An evolutionary trend is known to exist towards a decrease in the number of synapses with the electronic mode of transmission. This can be seen to be the case, for example, in sensori-motoneuronal and reticulo-motoneuronal synapses (Shapovalov, 1983). The latter data have not been reported here.

Moreover, the fact that electrotonic transmission is preserved up to mammalian level suggests that chemical transmission probably cannot replace it satisfactorily in several functional mechanisms.

References

Arasaki, K., Kudo, M., and Nakanischi, T., 1984, Firing of spinal motoneurones due to electrical interaction in the rat: an in vitro study, Exp. Brain Res., 54: 437-445.

Auerbach, A.A., and Bennett, M.V.L., 1969, A rectifying electrotonic synapse in the central nervous system of a vertebrate, J. gen. Physiol., 53, 211-237.

Bando, T., 1975, Synaptic organization in teleost spinal motoneurons, Jap. J. Physiol., 25: 317-331.

Batueva, I. V., 1983, Properties of excitatory postsynaptic potentials elicited by stimulation of single segmental afferents in spinal motoneurones of the lamprey Lampetra fluviatilis, Zh. Evol. Biokhim. Fiziol., 19: 54-61.

Batueva, I.V., Shapovalov, A.I. and Shiriaev, B.I., 1985, Combined morphological and electrophysiological description of connections between single primary afferent fibres and individual motoneurons in the frog and fish spinal cord, 5th International Symposium "Motor Control 85", Abstracts, Warna, p. 15.

Bennett, M.V.L., Sandri, C., Akert, K., 1978, Neuronal gap junctions and morphologically mixed synapses of the spinal cord of a teleost, Sternarchus albifrons (Gymnotoidei), Brain Res., 143: 43-60.

Brock, L.G., Coombs, J.C., and Eccles, J.C., 1952, The recording of potentials from motoneurones with an intracellular electrode, J. Physiology, 117: 431-460.

Charlton, B.T., and Gray, E.G., 1966, Comparative electron microscopy of synapses in the vertebrate spinal cord, J. Cell Sci., 1: 67-80.

Chung, J.M., Lee, K.H., Kin, J., and Coggeshall, R.E., 1985, Activation of dorsal horn cell by ventral root stimulation in the cat, J. Neurophysiol., 54: 261-272.

Fulton, B.P., Miledi, R., and Takahashi, T., 1980, Electrical synapses between motoneurons in the spinal cord of the newborn rat, Proc. Roy. Soc. Lond., B208: 115-120.

Grinnell, A.D., 1966, A study of the interaction between motoneurones in the frog spinal cord, J. Physiol., 182: 612-648.

Karamian, O.A., and Suderevskaya, E.I., 1986, Synaptic interactions between single spinal motoneurons, Neurophysiology (Neirofiziologiya), 18: 262-266.

Magherini, P.C., Precht, W., Schwindt, P.C., 1976, Electrical properties of frog motoneurons in the in situ spinal cord. J. Neurophysiol., 39: 459-473.

Motorina, M.V., 1978, Ultrastructure of synapses of gap junction type in the motor nuclei of the frog spinal cord, Arch. Anat., Histol & Embryol., 75: 27-34.

Motorina, M.V., 1986, Electrotonic synapses in the spinal cord of mammalia, Arch. Anat., Histol. & Embryol., 91: 13-20.

Motorina, M.V., and Suderevskaya, E.I., 1983, Distribution of synaptic projections of the dorsal root in the spinal cord of the carp Cyprinus carpio, Zh Evol. Biokhim. Fiziol., 19: 524-528.

Schnitzlein, H.N., and Brown, H.K., 1975, Spinal motoneurons of the goldfish (Carassius auratus), Brain Behav., Evol., 12: 207-228.

Shapovalov, A.I., 1980, Interneuronal synapses with electrical, dual and chemical mode of transmission in vertebrates, Neuroscience, 5: 1113-1124.

Shapovalov, A.I., 1983, Types of synaptic connections between nervous cells and their evolution, Zh. Evol. Biokhim. Fiziol., 19: 341-349.

Shapovalov, A.I., and Batueva, I.V., 1984, Interactions between primary afferent fibres and motoneurones in the spinal cord of the river lamprey, Fiziol. Zh. SSSR, 70: 1178-1188.

Shapovalov, A.I., and Shiriaev, B.I., 1978; Two types of electrotonic EPSP evoked in amphibian motoneurons by ventral root stimulation, Exp. Brain Res., 33: 313-323.

Shapovalov, A.I., and Shiriaev, B.I., 1984, Synaptic interaction between single motoneurons in isolated spinal cord of frog, Neurophysiology (Neirofiziologiya), 16: 619-629.

Shupljakov, O.V., 1986, Transmission in sensory-motor synapses in the lumbar spinal cord of the tadpoles Rana ridibunda, Zh. Evol. Biokhim. Fiziol., 22: 174-180.

Sotelo, C., and Taxi, J., 1970, Ultrastructural aspects of electrotonic junctions in the spinal cord of the frog, Brain Res., 17: 137-141.

Sotelo, C., and Grofova, J., 1976, Ultrastructural features of neurons and synapses of the spinal cord, in: "Frog Neurobiology" , R. Llinas and W. Precht, eds, Springer, Berlin: pp. 707-727.

Suderevskaya, E.I., 1983, The study of motoneurons synaptic inputs in the isolated spinal cord of fish, in: "Neuron and interneuronal integration", Nauka, Leningrad: pp. 54-55.

Tamarova, Z.A., Shapovalov, A.I., and Shiriaev, B.I., 1978, Action of magnesium and manganese ions and calcium deficiency on the synaptic transmission in the isolated perfused rat spinal cord, Neurophysiology (Neirofiziologiya), 10: 530-533.

Whiting, H.P., 1948, Nervous structure of the spinal cord of the young larval brook-lamprey, Quart. J. Micr. Sci., 89: 359-383.

NEURONAL ORGANIZATION OF THE CORTICORUBRAL SYSTEM

V. V. Fanardjian and V. L. Gorodnov

L. A. Orbeli Institute of Physiology
Academy of Sciences of Armenian SSR
22 Orbeli Bros. Str.
375028, Yerevan, USSR

Summary

EPSPs evoked in rubrospinal neurons by stimulation of the sensorimotor cortex were studied in pentobarbitonized cats with the intracellular recording technique. Participation of corticospinal inputs in initiating the above-mentioned EPSPs was determined by selectively activating of corticospinal fibres at the level of the medullary pyramids and by studying peculiarities of their interaction with effects of cortical stimulation. Convergence of excitatory influences from the sensorimotor cortex and medullary pyramids was observed in 94.2% of the rubrospinal neurons studied. In some neurons (8.1%), the summation of cortical and pyramidal inputs was 100%, which suggests that the corticorubral neurons played a predominant role in the genesis of cortical EPSPs. In the majority of cells, the summation of inputs tested was less than 100%, proving that both corticospinal and corticorubral neurons participated in initiating the EPSPs evoked by cortical stimulation. A collision component consisting of the difference between the algebraic and experimental sum of the EPSPs and reflecting that part of the depolarization which "dropped out" of the summed EPSP as a result of impulse collision in pyramidal fibres carrying stimulation effects from both the sensorimotor cortex and the medullary pyramids, was distinguished when studying interactions between the inputs mentioned above. From this analysis, it was possible to separately assess pyramidal and extrapyramidal participation in corticofugal control. It was found that slow-conducting corticospinal and corticorubral neurons participated equally in initiating the first two components of complex EPSPs. Cell composition and the mechanisms underlying corticofugal influences on the red nucleus neurons are discussed.

Introduction

In previous experiments we showed (Fanardjian and Gorodnov, 1985) that stimulation of the sensorimotor cerebral cortex evoked EPSPs in the red nucleus rubrospinal neurons, which consisted of several components. It was also found that the complex character of these EPSPs persisted after removal of the gray matter, which suggested that these EPSP components did not result from transsynaptic activation of cortical neurons elicited by stimulation of the cortex. Some peculiarities of these EPSPs suggest that the first component may be of monosynaptic origin, as well as the second component in the case of some EPSPs (Fanardjian and Gorodnov, 1985). It was assumed that the EPSP components described resulted from the activation of slow-conducting pyramidal and corticorubral neurons. It is not yet clear, however, to what extent the latter participate in the genesis of corticofugal EPSPs in the red nucleus neurons.

This problem may possibly be solved by physiological experiments in which corticospinal fibres in the lower part of the brain are selectively activated and their interaction with effects of cortical stimulation subsequently studied (Kitai et al., 1969; Tsukahara et al., 1969). In the present study corticospinal fibres were activated separately by stimulating them at the level of the medullary pyramids.

Methods

Experiments were carried out on pentobarbitonized cats (30 mg/kg, administered intraperitoneally). After surgical preparation (Tsukahara et al., 1964), recording glass micropipettes with bevelled tips (average tip resistance, 10 MΩ) were inserted into the caudal part of the red nucleus so as to intracellularly record the activity of rubrospinal neurons. These neurons were identified by the antidromic action potential response to stimulation of the rubrospinal tract in the region of the second cervical segment of the spinal cord. Bipolar metal electrodes with an interpole distance of 1.0-1.5 mm were used to stimulate the sensorimotor cortex. The resistance between their tips was 20-70 KΩ. Electrodes were inserted to a depth of 1.5-2.0 mm from cortical surface. Similar bipolar electrodes were implanted according to Horsley-Clark's coordinates (P 7, L 1, H-1), (Reinoso-Suarez, 1961), into the medullary to stimulate the pyramidal tract. Single rectangular pulses with a duration of 0.01-0.1 ms and a frequency of 10 imp/s were used to evoke EPSPs in rubrospinal neurons.

The electric potentials evoked in the red nucleus were amplified with a DC amplifier. The PSPs were averaged and processed in a real time computer system consisting of a multicanal analyser and a joint minicomputer. EPSPs were averaged by the analyser (50 superimposed sweeps). The averaged EPSPs recorded from one cell in response to various

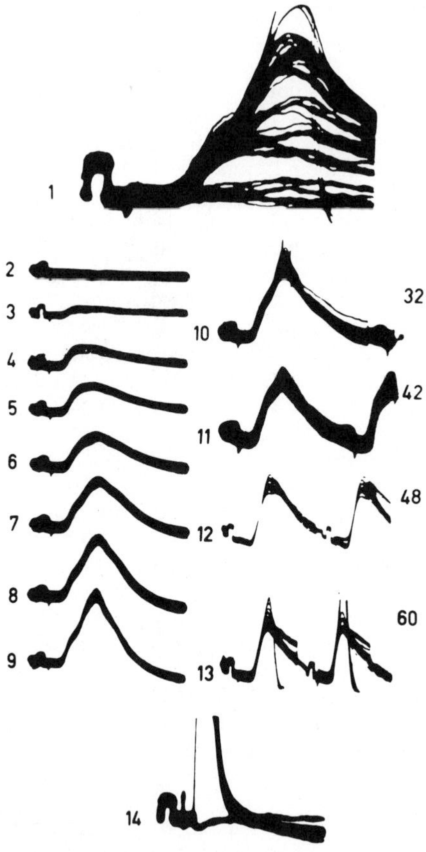

Fig. 1. Complex EPSPs evoked in rubrospinal neurons by stimulation of the sensorimotor cortex. 1- EPSPs at various stimulation intensities. 70 sweeps superimposed. 2-9, EPSPs at increasing (from top to bottom) stimulation intensities. Each curve represents 30 sweeps superimposed. 10-13, stimulation with various frequencies. Stimulation frequency is indicated by figures given on the right of each wave. 14, antidromic action potential at the threshold intensity of spinal cord stimulation. Calibration : 0.5 mV; 1ms.

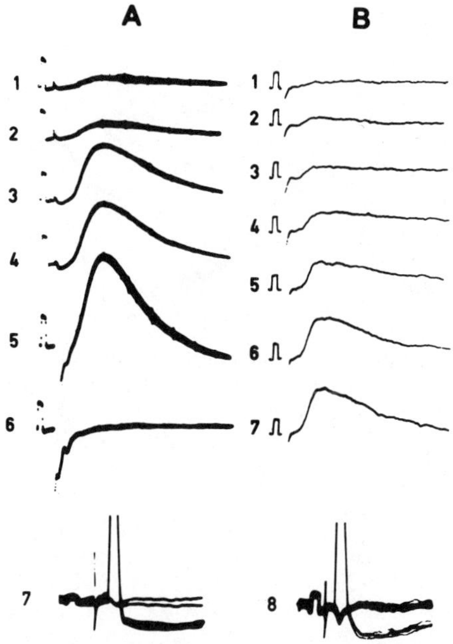

Fig. 2. EPSPs evoked in rubrospinal neurons by stimulating the medullary pyramidal tract. A, 1-5, averaged EPSPs at increasing stimulation intensities (from top to bottom). A, 6, extracellular field potential, stimulation intensity was the same as in A, 5. Curves 1-6 were obtained by computer by summing 50 sweeps. A, 7, antidromic action potential at the threshold intensity of spinal cord stimulation. B, 1-7, another neuron. EPSPs at increasing stimulation intensities (from top to bottom). B, 8, antidromic action potential at the threshold intensity of spinal cord stimulation. Calibration : 0.5 mV; 1 ms.

inputs were stored in the memory of the system to be used for evaluating of the collision effect and separating of the corticorubral component. The sum of the EPSPs, $U_\Sigma = U_c + U_p$, where U_c and U_p were averaged EPSPs elicited by cortical and pyramidal inputs, was calculated by the computer. The summed curve U_Σ was compared with the averaged EPSP U_{cp} obtained with simultaneous stimulation by both inputs. The difference $\Delta U = U_\Sigma - U_{cp}$ obtained by the computer was used to compare functions U_Σ and U_{cp}. Averaged EPSP curves as well as U_Σ and ΔU functions were displayed on the analyser.

After the experiments, the tissue was destroyed electrolytically and the localization of the stimulation area was checked by histological examination.

Results

The intracellular activity of 52 cat rubrospinal neurons was studied. Fig. 1 shows multicomponent EPSPs elicited in a rubrospinal neuron by stimulation of the sensorimotor cortex at various intensities (1-9) and frequencies (10-13). Increased stimulation evoked a second EPSP component which could be reproduced at frequencies not exceeding 45 imp/s. The conditioning technique was applied between the responses evoked by stimulation of the sensorimotor cortex and medullary pyramids to determine the participation of the corticospinal and corticorubral fibres in the initiation of various components of the corticofugal EPSPs evoked in the red nucleus neurons. The contribution of the corticospinal input to the genesis of the EPSPs described was determined by studying the interaction (collision) of presynaptic impulses arriving at the red nucleus via the same pathway (axon collaterals of the pyramidal tract neurons).

Fig. 2 shows the effects of stimulation of corticospinal fibres at the medullary pyramid level. A gradual rise in depolarization as a result of increased stimulation intensity

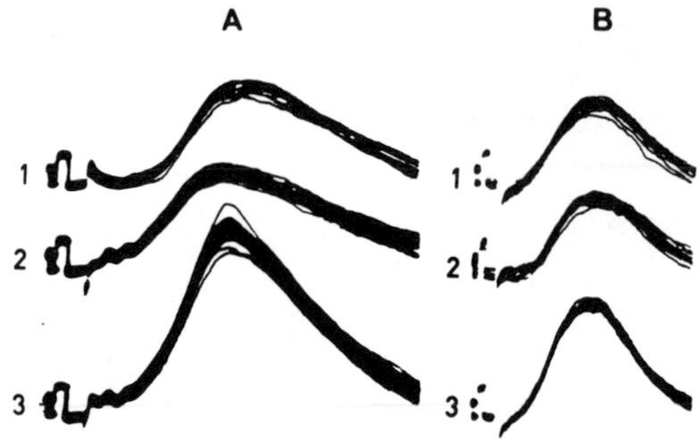

Fig. 3. Summation of EPSPs evoked by stimulation of the sensorimotor cortex and medullary pyramidal tract. A, B, EPSPs of two rubrospinal neurons. A, 1, B, 1, EPSPs to the sensorimotor cortex stimulation. A, 2, B, 2, EPSPs to stimulation of medullary pyramids. A, 3, B, 3, summed EPSP obtained under simultaneous stimulation of the sensorimotor cortex and medullary pyramids. Calibration : 0.5 mV; 1 ms.

could clearly be seen to occur. EPSPs evoked by stimulatión of medullary pyramid were characterized by latencies of 1.4-3.9 ms (mean 2.12 ± 0.51 ms; n = 38) and a rising phase of 1.9-8.0 ms (mean 4.55 ± 1.48; n =38). They were complicated by additional components (Fig. 2, B) in less than 10% of the cases studied. The majority of rubrospinal neurons responsible for the sensorimotor cortex stimulation also responded with EPSPs to stimulation of the medullary pyramid (49 from 52 neurons, 94.2%). Fig. 3 shows the interaction between EPSPs evoked in two neurons by simultaneous stimulation of sensorimotor cortex and medullary pyramid. In both cells, summation of depolarization was observed (Fig. 3, A; B,3). With the neuron shown in Fig. 3, A, the summation effect was 100% : the total value of the summed EPSPs was equal to the sum of amplitudes of EPSPs evoked separately by stimulation of the sensorimotor cortex (Fig. 3, A, 1) and medullary pyramid (Fig. 3, A, 2). This result, obtained at various sensorimotor cortex stimulation intensities, suggests that the corticorubal neurons played a predominant role in initiating the EPSPs evoked by cortical stimulation. However, the type of interaction we have described was observed in only a few cases (4 out of 49 neurons, i. e. 8.1%). Fig. 3, B shows another neuron in which summation of EPSPs from two inputs was 75%. This type of summation was observed in the majority of neurons studied and suggests that the activated pyramidal tract also participated in the genesis of corticofugal EPSPs. Reduction of the amplitude of the summed response to 75% depended on the decrease in the number of excited presynaptic axons resulting from the collision of impulses from the sensorimotor cortex and medullary pyramid. To obtain this kind of interaction, the intensity of the stimuli applied seems to be rather important, as it determines the number of pyramidal tract fibres involved.

Another explanation for the decrease in the sum of two inputs as compared with the algebraic sum could possibly be that a process of non-linear summation of post-synaptic potentials takes place in neurons of the red nucleus. This seems rather unlikely, however, since in the neurons of the red nucleus where the interaction between cortical and cerebellar EPSPs took place, the summation was linear. Besides, differences in the efficiency of summation were found to exist when the interstimulus intervals were varied (Fig. 4).

Percentage summation was studied with the two inputs in relation to the intervals between stimuli. This relationship plotted in Fig. 4, A, which shows that maximal reduction of the summed EPSP was observed when the two stimuli were applied simultaneously. Further analysis of interactions between the responses to stimulation of the sensorimotor cortex and medullary pyramid was conducted using averaged EPSPs obtained by computer. Fig. 5 gives the results of one experiment. Collision between two inputs was tested at various stimulation intensities (A, B). The neuron responded with a complex EPSP to stimulation of the sensorimotor cortex (A,1; B, 1) while a simple EPSP not complicated by additional components was observed in response to stimulation of the medullary pyramid components (A, 1; B, 2). The algebraic sum of these EPSPs obtained by computer is given in oscillograms A, 3; B, 3; the experimental sum of the same potentials is given in A, 4; B, 4. Oscillograms A, 5, B, 5 show the difference between the algebraic and experimental sums of the EPSPs. The latter is taken to represent the part of depolarization which dropped out of the summed EPSP as a result of impulse collision in the pyramidal fibres carrying impulses from the both sensorimotor cortex and the medullary pyramid. In the discussion below this difference will be called by convention "a collision component". With the neuron studied in Fig. 5, the collision component coincides in time with the first component of the corticofugal EPSP, which suggests that the corticospinal fibres may play some role in initiating that component.

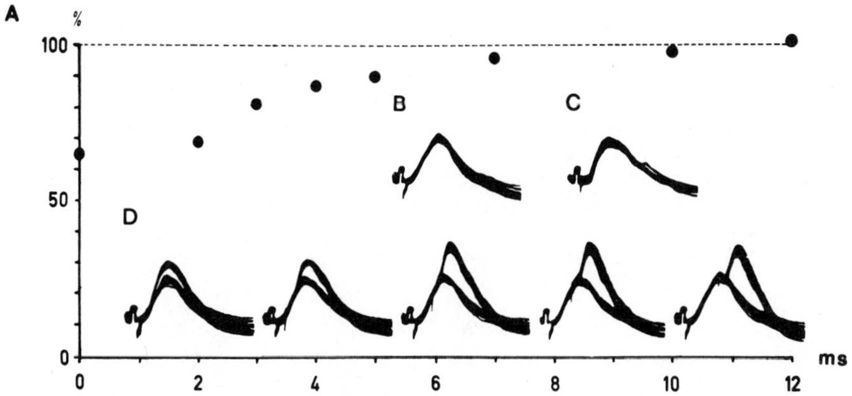

Fig. 4. Summation of postsynaptic potentials dependent on the intervals between stimulation of the sensorimotor cortex and medullary pyramids. A, diagram expressing dependence of the EPSP amplitude on the interval between stimulation of the sensorimotor cortex and medullary pyramids. The peak amplitude of the summed EPSP are plotted as a percentage of the algebraic sum of the two inputs as a function of the time interval between stimuli. The algebraic sum of he two inputs is indicated by the dotted line. B, EPSPs evoked by stimulation of the sensorimotor cortex. C, EPSPs evoked by stimulation of the medullary pyramids. D, summation of EPSPs at various intervals between stimulation. Lower traces, EPSPs recorded in response to the sensorimotor cortex stimulation. Upper traces, summed EPSPs recorded in response to stimulation of both inputs. Calibration : 0.5 mV; 1 ms.

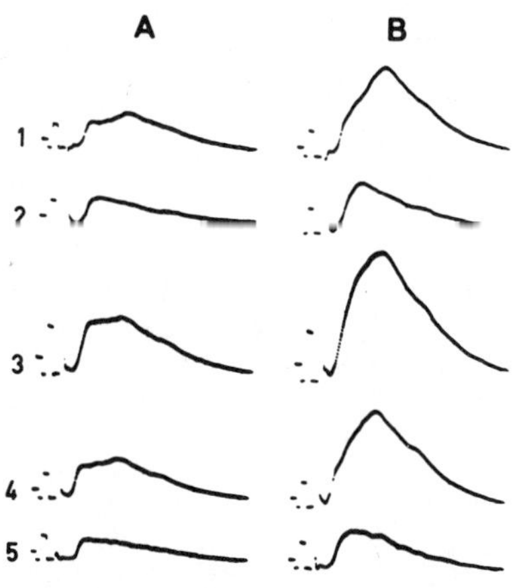

Fig. 5.Interaction of averaged EPSPs evoked by stimulation of the sensorimotor cortex and medullary pyramids in a rubrospinal neuron. A, 1, B, 1, EPSPs evoked by stimulation of the sensorimotor cortex with various intensities. A, 2, B, 2, EPSPs evoked by stimulation of the medullary pyramids with various intensities. A, 3, B, 3, algebraic sums of the EPSPs given in A, 1, A, 2, and B, 1, B, 2, respectively, obtained by computer. A, 4, B, 4, experimentally obtained sums of the EPSPs given in A, 1, A, 2, and B, 1, B, 2, respectively. A, 5, B, 5, difference between A, 3 and A, 4, and between B, 3 and B, 4 respectively. Collision component. Calibration : 1 mV (for A, 3, B, 3) and 0.5 mV (for the remaining recordings); 1 ms.

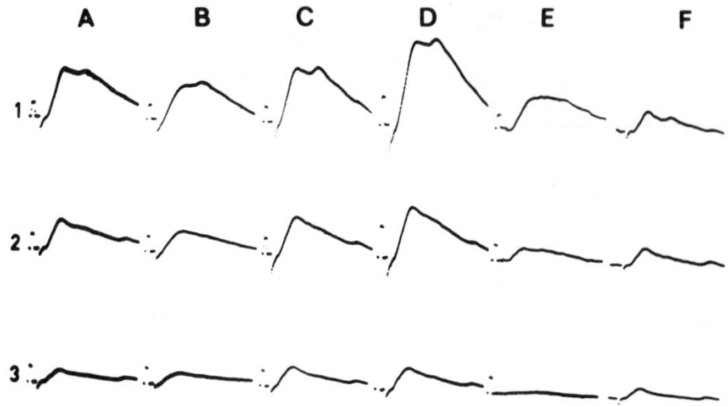

Fig. 6. Separation of the pyramidal and extrapyramidal components of the corticofugal EPSPs in a rubrospinal neuron. A, EPSPs evoked by stimulation of the sensorimotor cortex with various intensities (1-3). B, EPSPs evoked by stimulation of the medullary pyramids with various intensities (1-3). C, summed EPSP obtained with simultaneous stimulation of both inputs. D, algebraic sum of the EPSPs given in A and B, obtained by computer. E, difference between D and C. Collision component. F, difference between A and E. Calibration : 1 mV (for D, 1-3) and 0.5 mV (for the remaining of recordings); 1 ms.

Fig. 6 illustrates the response of a rubrospinal neuron to three different stimulation intensities (1-3). The following reactions are shown: effects of sensorimotor cortex stimulation (A), effects of medullary pyramid stimulation (B), experimentally obtained sum of EPSPs (C), algebraic sum of EPSPs (D), collision component (E) and difference between cortically-evoked EPSP and collision component (F). The latter is assumed to reflect the result of activation of the corticorubral fibres as well as that of the corticospinal fibres not influenced by collision interaction. It is obvious from Fig. 6 that the magnitude of the collision component is directly dependent on the intensity of the medullary pyramid stimulation (E, 1-3). To obtain a more complete involvement of all the corticospinal fibres participating in the initiation of the corticofugal EPSP in question, it was necessary to increase the intensity of medullary pyramid stimulation to a point where further intensification of stimulation was ineffective and led to no further increase in the amplitude of the collision component. It was assumed that under these conditions, the corticospinal fibres activated at the described intensity of sensorimotor cortex stimulation and participating in the genesis of the corticofugal EPSPs, were probably affected by the collision with impulses evoked by the medullary pyramid stimulation. Under these conditions, this collision component would most fully reflect the corticospinal component, hence, the difference between the corticofugal EPSP and collision component would provide the most accurate means of determining the value of the corticorubral component. The results of this experiment are given in Fig. 7.

The sensorimotor cortex was stimulated with a constant intensity (A) whereas the intensity of the medullary pyramid stimulation (B) was increased and was accompanied by an increase in the amplitude of the collision component (C). Fig. 7, D gives the difference between the corticofugal EPSP and the collision component, i. e. the presumed corticorubral component. It can be seen from the figure that the corticorubral and collision components were inversely proportional; this relationship was confirmed by changing the stimulation intensity. In Fig. 7, D, 1, 2 an increase in the intensity of the medullary pyramid stimulation (B, 1-3) led to a rise in amplitude of the collision component (C, 1, 2) and a simultaneous reduction in the corticorubral EPSP. In the corticorubral EPSPs, changes took place predominantly in the early phase, corresponding to the first component of the complex EPSP (A).

A further increase in the intensity of the medullary pyramid stimulation did not cause any subsequent change in the amplitude of the collision component (C, 3 and oscillograms in D, 3), thus showing quite definitely that corticospinal and corticorubral components of the complex EPSP were separate.

Participation of corticospinal and corticorubral fibres in the initiation of various components of the corticofugal EPSPs was analysed additionally in five further rubrospinal neurons. The results of this study are given in Fig. 7, F : the collision percentage is given in the columns for the first (light columns) and second (shaded columns) components of a complex EPSP interacting with an EPSP evoked by stimulation of the medullary pyramid. The mean values of the collision component of EPSPs of the neurons mentioned were 58.3% with the first component and 53.3% with the second. The figures show clearly that the corticospinal and corticorubral neurons play approximately equal parts in the genesis of the first and second components of the EPSPs evoked in rubrospinal neurons by corticofugal impulses.

Discussion

Electrophysiological (Padel et al., 1973; Tsukahara and Kosaka, 1969) and morphological (Kuypers and Lawrence, 1967; Mabuchi and Kusama, 1966; Rinvik and

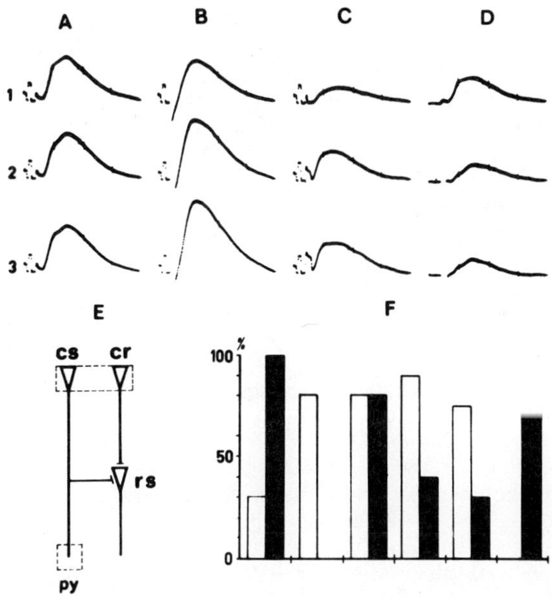

Fig. 7. Characteristics of components of the complex EPSPs evoked by the sensorimotor cortex stimulation. A, EPSPs evoked by stimulation of the sensorimotor cortex with one intensity (1-3). B, EPSPs evoked by increasing intensities (1-3) of medullary pyramid stimulation. C, collision component with increasing intensities (1-3) of medullary pyramid stimulation. D, difference between A and C with increasing intensities of (1-3) of medullary pyramid stimulation. Calibration : 0.5 mV; 1 ms. E, diagram showing neuronal interaction of the red nucleus with corticospinal and rubrospinal systems. Stimulation regions are circled with dotted lines. cs, corticospinal neuron with an axon collateral to the red nucleus; cr, corticorubral neuron; rs, rubrospinal neuron; py, medullary pyramidal tract. F, diagram expressing the percentage of collision of the first (light columns) and second (shaded columns) components of the complex EPSPs of rubrospinal neurons in response to corticofugal stimulation. Abscissa, number of neurons. Ordinate, percentage of EPSP suppressed by collision.

Walberg, 1963)) studies have revealed the somatotopic character of projections from the sensorimotor cortex to the red nucleus. The rubrospinal tract has been found to have a similar organization (Niberg-Hansen and Brodal, 1964; Pompeiano and Brodal, 1957). There is evidence suggesting that in many ways the corticorubral system duplicates the functions of the corticospinal system in the control of motor reactions of the limbs (Hongo et al., 1969). This can be observed most clearly in pyramidotomized animals, in with the compensatory rehabilitation of motor deficiency is arrested after additional lesion of the rubrospinal tract (Lawrence and Kuypers, 1968). It has been shown that the above-mentioned groups of efferent neurons projecting to the red nucleus are localized predominantly in the fifth layer of the cerebral cortex (Coulter et al., 1976; Humphrey and Rietz, 1976; Palmer et al., 1981). The fact that these cells have physiological effects on the red nucleus neurons is manifested in the disynaptic inhibitory influences from the fast-conducting pyramidal neurons and monosynaptic excitatory influences from the slow-conducting pyramidal as well as corticorubral neurons (Tsukahara et al., 1969; Tsukahara and Kosaka, 1968). However, as already mentioned above, it has not yet been established how this contribution to cortically evoked EPSPs in the rubrospinal neurons is divided up between these two populations of neurons.

The cell composition of corticofugal neurons in the monkey's precentral hand area projecting to the red nucleus has been studied by Humphrey and Rietz (1976). It has been shown that the overwhelming majority of these cells were slow-conducting neurons (81%), a very small proportion of which (8%) belonged to the pyramidal tract. This finding was confirmed in another study by the same authors carried out on a larger number of neurons (Humphrey et al., 1978). There it was found that the contribution of the corticorubral neurons to the total excitatory influence of the cerebral cortex on the red nucleus neurons was greater than that of the corticospinal neurons (Humphrey and Rietz, 1976). A different technique was used in the present study. Collision between presynaptic impulses to the red nucleus was used to study the participation of various populations of neurons in corticofugal influences on the red nucleus. This technique was successfully used earlier to study descending inputs to brain stem structures (Kitai et al., 1969; Pilyavsky et al, 1977; Tsukahara et al., 1969). The data obtained show that corticospinal and corticorubral neurons participate equally in the genesis of the first two components of complex EPSPs. These corticofugal neurons were found from their axon-conduction velocity to be predominantly slow-conducting (5.4-10.7 m/s; 13.6-26.1 m/s). This agrees well with data available in the literature (Humphrey and Rietz, 1976; Kitai et al., 1969) and once more proves the modulatory character of the cortical impulses reaching the rubrospinal neurons (Fanardjian, 1975; Tsukahara and Kosaka, 1968).

References

Coulter, J. D., Ewing, L., and Carter, C., 1976, Origin of primary sensorimotor cortical projections to lumbar spinal cord of cat and monkey, Brain Research, 103: 366-372.

Fanardjian, V. V., 1975, On neuronal organization of cerebellar efferent systems, "Nauka", Leningrad, 74pp, in Russian.

Fanardjian, V. V., and Gorodnov, V. L., 1985, Multicomponental synaptic potentials in rubro-spinal neurons of cat brain evoked by corticofugal impulsation, Neurofisiologia, 17: 665-673, in Russian.

Hongo, T., Jankowska, E., and Lundberg, A., 1969, The rubrospinal tract, I. Effects on alpha-motoneurons innervating hindlimb muscles in cats, Exp. Brain Res., 7: 344-364.

Humphrey, D. R., Corrie, W. S., and Rietz, R. R., 1978, Properties of the pyramidal tract neuron system within the precentral wrist and hand area of primate motor cortex, J. Physiol., 74: 215-226.

Humphrey, D. R., and Rietz, R. R., 1976, Cells origin of corticorubral projections from the arm area of primate motor cortex and their synaptic actions in the red nucleus, Brain Research, 110: 162-169.

Kitai, S. T., Oshima, T., Provini, L., and Tsukahara, N., 1969, Cerebrocerebellar connections mediated by fast and slow conducting pyramidal tract fibres of the cat, Brain Research, 15: 267-271.

Kuypers, H. G. J. M., and Lawrence, D. G., 1967, Cortical projections to the red nucleus and the brain stem in the rhesus monkey, Brain Research, 4: 151-188.

Lawrence, D. G., and Kuypers, H. G. J. M., 1968, The functional organization of the motor system in the monkey, II. The effects of lesions of the descending brain stem pathways, Brain, 81: 15-36.

Mabuchi, M., and Kusama, T., 1966, The cortico-rubral projections in the cat, Brain Research, 2: 254-273.

Nyberg-Hansen, R., and Brodal, A., 1964, Sites and mode of termination of rubrospinal fibres in the cat: An experimental study with silver impregnation methods, J. Anat., 98: 235-253.

Padel, Y., Smith, A. M., and Armand, J., 1973, Topography of projections from motor cortex to rubrospinal units in the cat, Exp. Brain Res., 17:315-332.

Palmer, C., Schmidt, E. M., and McIntosh, J. S., 1981, Corticospinal and corticorubral projections from the supplementary motor area in the monkey, Brain Research, 209: 305-314.

Pilyavsky, A. I., Gribov, S. N., and Gokin, A. P., 1977, Peculiarities of activation of medullar reticulospinal neurons of collaterals of pyramidal fibres, Neurofisiologia, 9: 495-503, in Russian.

Pompeiano, O., and Brodal, A., 1957, Experimental demonstration of a somatotopical origin of rubrospinal fibres in the cat, J. Comp. Neurol., 108: 225-252.

Reinoso-Suarez von F., 1961, Topographischer Hirnatlaz der Katze für experimental-physiologische Untersuchungen, E. Merck, Darmstadt, 67s.

Rinvik, E., and Walberg, F., 1963, Demonstration of a somatotopically arranged cortico-rubral projection in the cat, J. Comp. Neurol., 120: 393-407.

Tsukahara, N., Fuller, D. R. C., and Brooks, V. P., 1969, Collateral pyramidal influences on corticorubrospinal system, J. Neurophysiol., 31: 467-484.

Tsukahara, N., and Kosaka, K., 1968, The mode of cerebral excitation of red nucleus neurons, Exp. Brain Res., 5: 102-117.

Tsukahara, N., Toyama, K., and Kosaka, K., 1964, Intracellularly recorded responses of red nucleus neurons during antidromic and orthodromic activation, Experientia, 20: 632-633.

ROLE OF THE MONKEY SUBSTANTIA NIGRA PARS RETICULATA IN SENSORIMOTOR CONTROL OF EYE-HEAD-ARM COORDINATION

F. Lestienne, P. Cailler, and P. Liverneaux

Laboratoire de Physiologie Neurosensorielle du C.N.R.S.
15 rue de l'Ecole de Médecine
75270 Paris Cedex 06, France

Summary

The role of the substantia nigra pars reticulata (SNpr) has been studied in the head-free monkey during orienting behaviour in response to visual instruction signals triggering head positioning and conditioned arm movement. Electomyographic activities of neck muscles and triceps brachii, head movement, horizontal electro-oculogram and single unit activity of SNpr neurons have been recorded. Neurons located in the medial part of SNpr showed a moderate decrease in tonic firing rate during postural preparation preceding the orientation toward eccentric visual signal. This decrease, unrelated with saccadic eye movements per se, was followed by a marked pause observed when the conditioned arm movement was executed to get the reward. These data suggest that the pause in discharge of these SNpr neurons are time locked with behaviourally relevant visual stimuli and/or appropriate motor responses.

Introduction

The pars reticulata of the mammalian substantia nigra is a major output of the basal ganglia. SNpr neurons receive mostly inhibitory afferents from the caudate nucleus (Yoshida and Precht, 1971), the putamen (Parent et al., 1984) and the subthalamic nuclei (Rouzaire-Dubois et al., 1983). The SNpr projects heavily to the intermediate layers of the superior colliculus (Graybiel, 1978), the pars compacta of the nucleus tegmenti pedunculopontis (Parent et al., 1983) and the medial parts of the VA and VL thalamic nuclei (Carpenter et al., 1976).

Although the activity of SNpr neurons was thought to be predominantly related to oral-facial movements (Mora et al., 1977), these anatomical connections suggest that the SNpr belongs to complex neuronal circuits subserving sensory and motor processes involved in the preparation and execution of a large spectrum of behavioral acts such as orienting responses to visual and auditory stimuli.

In the last five years, changes in neuronal discharge activity in the SNpr in relation to several aspects of the behavioral situation have been investigated in cats and monkeys. It has been found that the activity of monkey and cat SNpr neurons decreased during saccadic eye movements in response to visual and acoustic stimuli (Hikosaka and Wurtz, 1983; Joseph and Boussaoud, 1985). Furthermore a large group of SNpr neurons in monkeys increased or decreased their activity during forelimb movements together with or after the onset of EMG activity in prime mover muscles (Schultz, 1986). In studies dealing with the orienting behavior of cats, only a few cells showed a change in activity related to active head movement (Joseph and Boussaoud, 1985). Similar results have been obtained in monkeys (Lestienne and Cailler, 1985). In addition, some SNpr cells decreased or increased their activity in relation to instructive light and acoustic stimuli indicating the behavioral situation (Schultz, 1986).

On the basis of these previous results, we used a well-defined sequence of motor events involved in the preparation and execution of goal-directed eye-head-arm movements in response to visual stimuli (Lestienne et al., 1981, a and b). We studied the role of the monkey SNpr in orienting behavior, in which animals responded to visual instruction signals triggering head positioning and associated arm movements (Lestienne and Cailler, 1986; Cailler, 1986). By precisely analysing the various motor events underlying the visuo-motor task, we attempted to evaluate the importance of behavioral contingency for the occurrence of responses in neurons located in the medial part of the SNpr.

Methods

The experiments were performed on 3 adult rhesus monkeys.The animals were trained to visually discriminate between horizontal and vertical hairlines superimposed on a luminous target spot. Target spots were placed on a horizontal perimeter arc. By sequentially presenting the target spots we induced the monkeys to make coordinated eye-head movements to direct their gaze toward the targets. A sequence of targets began with the presentation of a blank target spot (B target) situated in the center of the gaze. Then, at other positions in the arc, a spot came on that contained combined horizontal (H target) and vertical hairlines (V target). The monkey was rewarded with a drop of water by pressing a bar by means of an extension of the forelimb to the left when the V target was on. Inappropriate bar pressing when the B target or H target were on brought the trial to an end.

After training, monkeys were anesthetized with pentobarbital and operated on aseptic conditions. Vertical and Horizontal EOG electrodes were implanted. EMG electrodes were implanted. EMG electrodes were thrust into the left biceps and triceps brachii. A head connector and a recording well were attached to the skull. The activity of SNpr neurons was recorded using flexible electrodes designed for recording from neurons when the head is

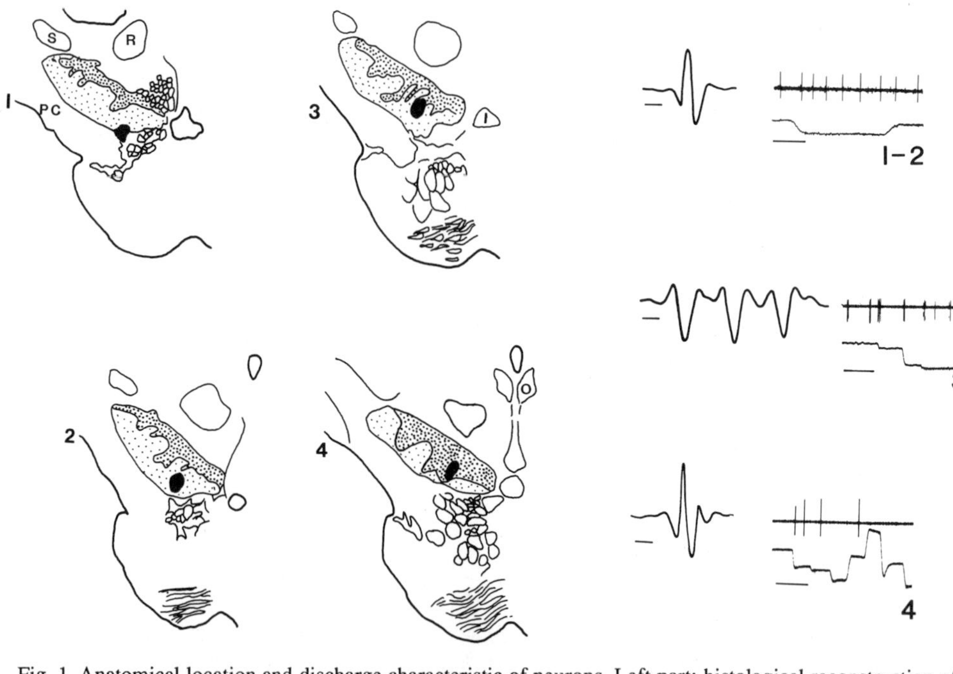

Fig. 1. Anatomical location and discharge characteristic of neurons. Left part: histological reconstruction of one lesion. S: Nucleus subthalamicus, R: Nucleus ruber, PC: Cerebral peduncule, O: Oculomotor nuclei, I: interpeduncular Nucleus. Small and large dots represent SNpr and SNpc respectively. Lesion is drawn in black. Right side: Impulse characteristics of SNpr neurons (1-2) and pars compacta neurons (3-4). Time scale: 1 ms for the single potential and 100 ms for pattern of discharge and horizontal EOG.

moving (Whittington et al, 1984). The location of recording tracks and of small lesions placed at the end of selected tracks were determined histologically.

Results

Anatomical location and discharge characteristics of neurons

In this study, electrode penetrations were aimed towards the medial part of the SNpr. In this area the activity of a total of 73 single neurons in 3 monkeys was recorded: it was established histologically that these neurons were located in the pars compacta (N=26) or pars reticulata (N=38). An example of the histological reconstruction of lesions placed at the end of two neighbouring tracks (spaced by O.5 mm) is shown in figure 1. The left part of this figure shows the location of a lesion on 4 serial coronal sections through the basal midbrain of a rhesus monkey between stereotaxic planes A8 and A7. Impulse characteristics of the neurons recorded along these two tracks (N=4) are displayed on the right part of the figure.

SNpr neurons (serial chartings 1 and 2) discharged spontaneously with rates of 20-40 spikes/s, and impulses were mostly less than 1 ms in duration. The rate of discharge was not modulated during spontaneous eye and head movements as long as the monkey was sitting in darkness without performing the task or any detectable behavioral activity. On the contrary, during performance of the orienting task these neurons showed a pronounced decrease in the activity related to the visual stimuli and behavioral events. Pars compacta

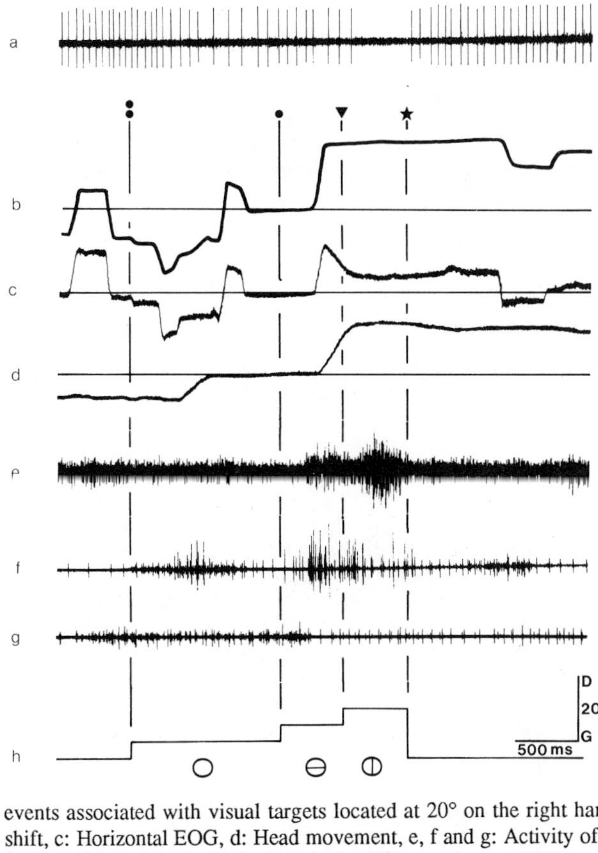

Fig. 2. Behavioral events associated with visual targets located at 20° on the right hand side. a: SNpr cell discharge, b: Gaze shift, c: Horizontal EOG, d: Head movement, e, f and g: Activity of triceps brachii, right and left splenius capitis, h: Sequence of visual targets. The different phases of this paradigm and the mean value of the onset of the pressure on the bar are represented by the following symbols: 2 black dots onset of B target, single black dot onset of H target, triangle onset of V target, star onset of bar pressing. Time scale: 500 ms. Calibration vertical line: 20° right.

neurons (SNpc) discharged sporadically (serial chartings 3 and 4), showing either polyphasic impulses 3-6 ms in duration or single impulses of more than 1 ms at a rate at less than 7/s. None of the SNpc neurons tested showed any discharge modulation that was related to the sensory signals or the motor responses of the paradigm.

We report here the responses of histologically verified SNpr neurons (N=19) the activity of which was related to specific sensory signals associated with the task and the subsequent orienting behavior to those sensory signals of the task and the subsequent orienting behavior to those sensory stimuli. Each of the neurons with a signal to noise ratio of 5:1 was extensively studied with different target locations (10°, 20°, 30° left and right) and with each target a minimum of 20 trials was performed. Furthermore, each of these neurons was tested for somato-sensory inputs by touching the face, the trunk and the hand and for vestibular and neck proprioceptive inputs.

Behavioral events

The sequence of behavioral events during performance of the visual-motor task in the free head situation is shown in Figure 2 together with the activity of an SNpr cell. This figure is a representative example of triggered eye-head coordination (Bizzi et al., 1971) accompanied by a conditioned arm movement to the left side.

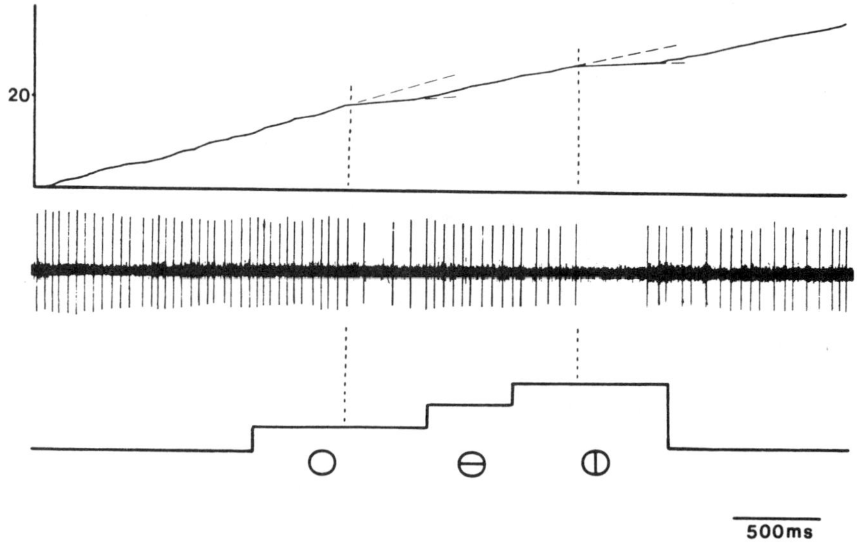

Fig. 3. Activity of the SNpr neuron during the sequential presentation of the B, H and V targets. Above: Cumulative frequency distribution. Calibration on the left indicates 20 spikes/s. Symbols as in Fig. 2.

During the presentation of the B target located in the center, the monkey aligned the direction of his gaze with the angular position of this first visual event. The gaze remained stable until the next visual event occurred. The gaze shift and its stabilization were accomplished by eye-head coordination. The onset of a 20° right HV target was followed by a saccadic eye movement and after a short interval (20-30 ms) by a movement of the head. This movement was associated with a pause in the EMG of the antagonist neck muscle and a synchronous burst of EMG activity in agonist muscles. Concerning the EMG activity of the triceps which is the prime mover in forearm extension, the onset of activity was time locked with that of the neck agonist muscles. We always observed sustained tonic EMG activity lasting for the entire duration of the fixation of gaze in front of the V target (between 0.5 and 1.5 s). This activity was accompanied by tonic contraction of the biceps. On the basis of previous data, this co-contraction may be related to an obligatory synergy between arm flexors and extensors (Lestienne, 1979) and also the neck muscles which serves to stabilize the shoulder and neck joints during fixation of the gaze (Lestienne et al., 1984). Following the tonic activity, the triceps shows a distinct burst of activity, specific to the extension of the arm and the bar press.

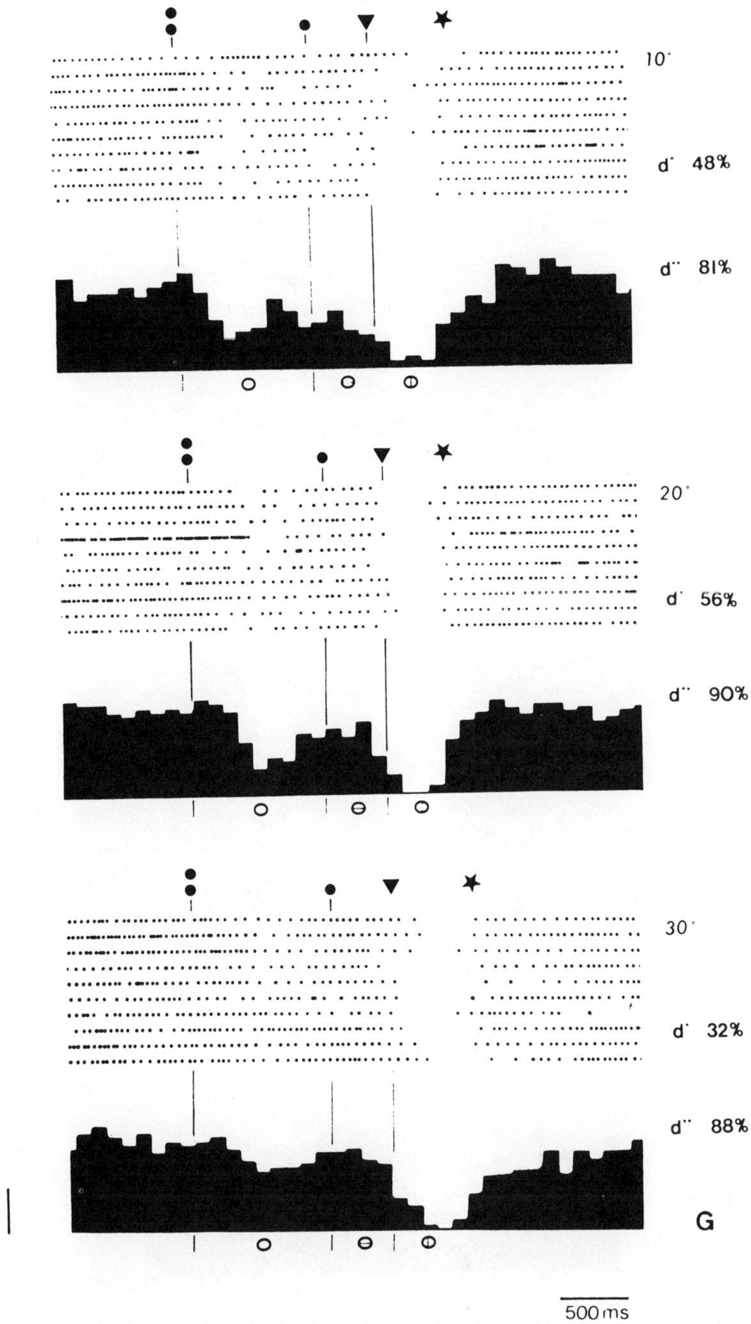

Fig. 4. Activity of a SNpr neuron during the presentation of the targets located at different positions on the left (10, 20 and 30°). The site of recording was on right. Single dots in the raster indicate the time of occurrence of single impulses. Ten consecutive trials are shown as successive lines in the rasters. Lines are aligned on the onset of the B target. The ten successive lines are summated to produce peristimulus-time-histograms. Calibration of histograms indicates 10 spikes/100 ms. Bin width is 100 ms. Symbols as in Fig. 2.

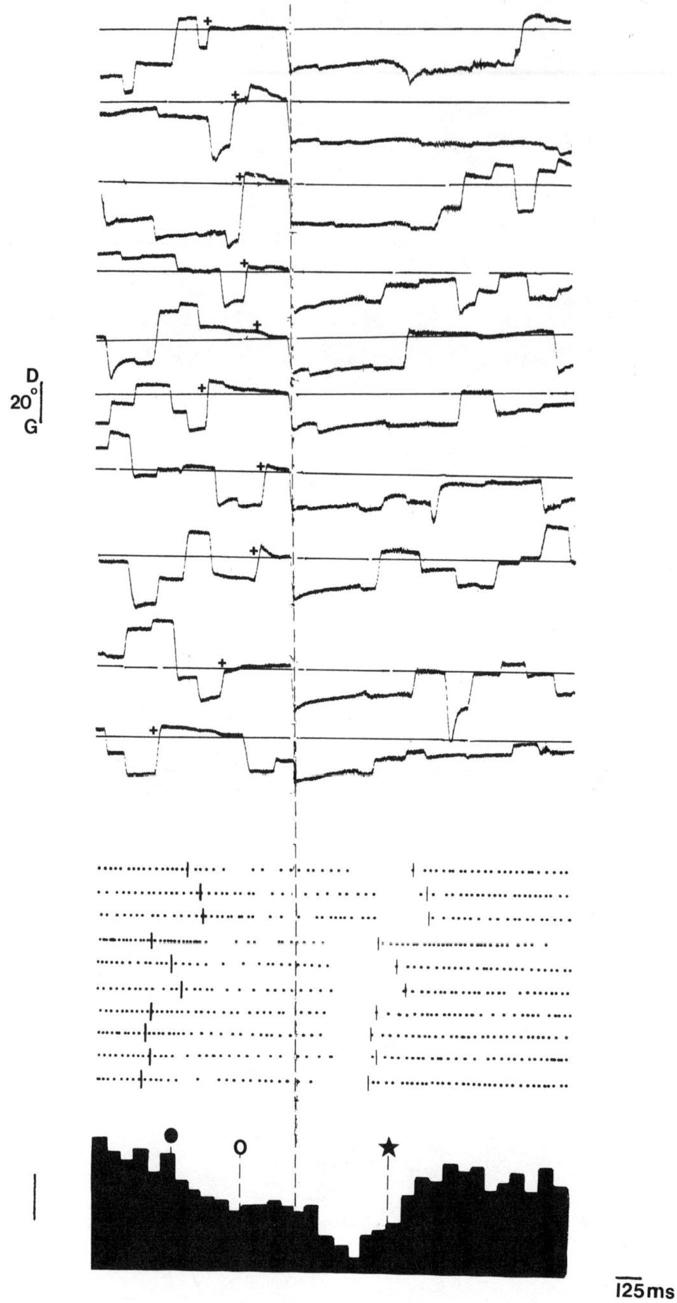

Fig. 5. Activity of a SNpr neuron associated with oculomotor activity (A) and EMG activity of the triceps (B). A) Above: 10 successive horizontal eye movements aligned on the onset of gaze fixation on H targets located at 20° left. Small crosses (+) indicate onset of the fixation of the B target. Below: raster display of unit activity and peristimulus-time-histogram (bin width 75ms; 8 spikes/125ms). Thick vertical bars on the raster indicate onset of the presentation of the B target. Thin vertical bars indicate offset of V target. Mean timing of: the onset of B target (●), the onset of gaze fixation on B target (O) and the offset of V target (★).

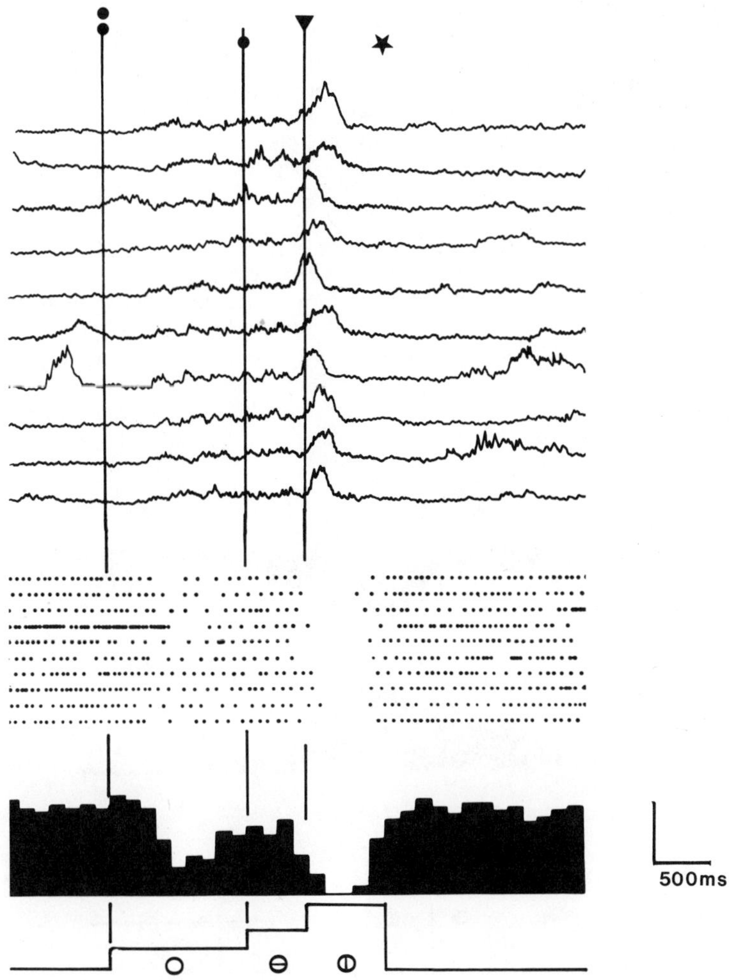

Fig. 5: B) Above: rectified and integrated EMG activity of the triceps. Below: raster display and peristimulus-time-histogram of unit activity, (bin width 125 ms; 13 spikes/125 ms). The H-V target is located at 20° left. Symbols as in fig. 2.

Neurons in this class (N=14) belong to the 19 neurons characterized by systematic and significant changes in activity during the visuo-motor task, but not during spontaneous saccades executed in the dark. A common property of these 14 neurons was that their changes in activity paralleled the various sequences of the visual signals (Figure 3).

The responses of this class of neurons are illustrated on the raster display of ten successive trials performed with 3 different target locations, i.e. 10, 20 and 30° left (Figure 4). This cell was responsive to the onset of the B target. Latencies ranged from 80 to 300 ms (mean latency: 190 ms). In 85% of the trials the onset of the gaze shift lagged less than 160 ms behind onset of the visual response. After this first response the firing rate increased but never resumed the level of spontaneous activity. A second and more pronounced decrease was observed in response to the HV target. Latencies were comparable to those of the first response. The duration of the second response covaried with the duration of the gaze fixation on the target.

Visually contingent arm movement related response

Figure 5A shows responses of SNpr neurons occurring when the monkey made saccades toward the visual target. All the neurons of this class showed a decrease in the firing rate before or at the onset of the saccade. This response was absent when saccades were not related to the target stimulus. This was also true during spontaneous saccades performed in darkness. Similar neurons have also been described in the superior colliculus (Mohler and Wurtz, 1976).

During the gaze fixation on HV target we noticed a clear cut off in the discharge of the cells while no signs of any large eye and head movements were observable (Figures 2, 5A). It should be stressed that during the fixation period, a burst in triceps activity time-locked to the extension of the arm toward the bar occurred concomitantly with the activity of agonist neck muscles.

Fig. 6. Relationship between the depression of the activity of an SNpr neuron and the EMG activity of the left triceps brachii during the performance of behavioral task. Targets were located 20° to the left (A) and 30° to the right (B). With each target position the figure shows above: Depression of activity of the SNpr cell obtained from the peristimulus-time-histogram of 10 trials (bin width 125 ms). Below: peristimulus-time-histogram of integrated EMG activity of the triceps (bin width 125 ms). Calibration: vertical lines: 16 spikes and 200 picks of EMG activity/125 ms. Symbols as in Fig. 2.

Figure 6 shows the relationship between the depression of activity of an SNpr neuron and the increase in triceps EMG activity (Cailler, 1986). Every unit belonging to this class of SNpr neurons was found to have a similar profile of activity depression that was clearly related to the EMG activity of the triceps. In order to assess the extent to which the depression of neuronal activity might be associated with the EMG activity of either neck or arm muscles, we undertook, in addition to the behavioral experiments, a series of tests during which the animal performed a) free left forelimb extension directed to the bar, b) voluntary head turning while the animal was scanning the surroundings. Only a small percentage of cells responded to the active extension of the arm. In these few cells the depression of neuronal activity was considerably reduced (70%) and was observed only during large bursts of EMG activity. In the same context, we failed to observe any significant modulation of neuronal activity during active head movements recorded during free scanning (Lestienne and Cailler, 1986).

Discussion

A substantial number of neurons were observed in the pars reticulata of the monkey substantia nigra that modulated their activity in response to visual instruction signals triggering head positioning associated with conditioned forelimb movements. This finding helps to elucitade several aspects of the behavioral involvement of SNpr neurons in orienting behavior and arm movement. This population of neurons lying in the medial part of the pars reticulata is characterized by a low background discharge rate (20-40 spikes/s) compared to the high rate of tonic activity of SNpr cells located in the lateral part (80-100 spikes/s) (Hikosaka and Wurtz, 1983). The other characteristic of these cells was the fact that their responses took the form of decreases in discharge rate.

One of the most striking results was the enhancement of the activity depression observed when the visual stimulus was associated with the occurrence of arm muscle activity and not with the onset of the saccade as has been shown in the lateral part of the SNpr by Hikosaka and Wurtz (1983). Recent studies on the activity of SNpr neurons in monkeys in relation with sensory and motor processes (Schultz, 1986) demonstrate the existence of task-related neurons showing temporal relationships with forelimb reaching movements. The activity of these neurons was enhanced or depressed in relation to forelimb movement. However most changes in these arm movement-related cell activities were independent of the temporal pattern of activity of individual muscles. Furthermore the changes in the SNpr activity temporally related to forelimb movements were quantitatively minor when compared with those related to eye movements. These neurons were found in intermediate and lateral parts of SNpr, in which no selective enhancement of visual responses related to forelimb movement has been reported. On the basis of anatomical studies, it is known that the targets of the lateral part of the SNpr are mainly the thalamic nucleus ventralis anterior and the intermediate layers of the superior colliculus, while the target of the medial part of the SNpr is the thalamic nucleus ventralis lateralis (Chevalier, 1985). In the light of these anatomical connections and in conformity with the suggested segregated pathway from the frontal association cortex and the primary sensory-motor cortex through the basal ganglia and the thalamus (Delong et al., 1983) it is possible that our results may contribute to revealing the role of the medial part of the SNpr in controlling the arm movements and postural adjustments involved in visual orienting behavior.

Another important question concerns the importance of behavioral contingencies for the occurrence of responses in SNpr neurons. Indeed we have shown (Lestienne and Cailler, 1986) that these responses were not necessarily linked to the sensory inputs or motor outputs themselves. In the same behavioral situation as reported here, when the HV target was randomly suppressed during the conventional paradigm, correct fixation on the "remembered" location of the target was commonly observed and rewarded. In this paradigm, we observed a reduction of 40-50% in the second depression of the neuron in response to the fixation of the "remembered" target and the arm movement. When the HV target was not rewarded, the response after a few repetitions of the unrewarded paradigm rapidly vanished. In other words, the visual stimulus may evoke a response in SNpr neurons when this stimulus is behaviorally significant but not when the same stimulus does not carry this significance (Evarts et al., 1984). These complex relationships with the task might involve neuronal mechanisms dealing with certain aspect of preparation for the correct response depending on the indicative stimuli (Schultz, 1986) and/or selective attention and neglect (Joseph and Boussaoud, 1985) including mnemonic processes. In light of the close anatomical connections between basal ganglia, superior colliculus, thalamus, supplementary motor area and prefrontal cortex (Graybiel and Ragsdale, 1979; Beckstead et al., 1982; Parent et al., 1983; Schell and Strick, 1984; Kievit and Kuypers, 1977) and on the basis of neurophysiological data including our findings and results on the lateral part of the SNpr, the putamen and the globus pallidus (Hikosaka and Wurtz, 1983; Schultz, 1986; Evarts et al., 1984; Liles, 1985; Delong et al., 1985) we suggest that responses of many basal ganglia neurons are mainly involved in cognition.

References

Beckstead, R. M., and Frankfurter, A., 1982, The distribution and some morphological features of substantia nigra neurons that project to the thalamus, superior colliculus and pedunculopontine nucleus in the monkey, Neuroscience, 7: 2377-2388.

Bizzi, E., Kalil, R. E., and Tagliasco, V., 1971, Eye-head coordination in monkeys: evidence for centrally patterned organization, Science, 173, 452-454.

Cailler, P., 1986, Rôle de la substance noire pars reticulata dans les comportements d'orientation visuellement déclenchés chez le singe, Thèse de Doctorat, 109 p.

Carpenter, M.B., Nakano, K., and Kim, R., 1976, Nigrothalamic projections in the monkey demonstrated by autoradiographic technics, J. Comp. Neurol., 165, 401-416.

Chevalier, G., 1985, Analyse des relations entre ganglions de la base et centres moteurs: Etude anatomo-physiologique des projections nigro-tectales et nigro-thalamiques, Thèse de Doctorat d'Etat, 62 p.

Delong, M.R., Georgopoulos, A.P., and Crutcher, M.D., 1983, Cortico-basal ganglia relations and coding of motor performance, Exp. brain Res., Suppl.7, 30-40.

Delong, M.R., Crutcher, M.D., and Georgopoulos, P., 1985, Primate globus pallidus and subthalamic nucleus: functionnal organization, J. Neurophysiol., 53: 530-543.

Evarts, E.V., Kimura, M., Wurtz, R.H., and Hikosaka, O., 1984, Behavioral correlates of activity in basal ganglia neurons, TINS, 7: 447-453.

Graybiel, A.M., 1978, Organization of the nigrotectal connection: an experimental tracer study in the cat, Brain Research, 143: 339-348.

Graybiel, A., and Ragsdale, C.W., 1979, Fiber connections of the basal ganglia, Progress in Brain Res., 51: 239-283.

Hikosaka, O., and Wurtz, R.H., 1983, Visual and oculomotor functions of monkey substantia nigra pars reticulata. I. Relation of visual and auditory responses to saccades, J. Neurophysiol., 49: 1230-1253.

Joseph, J.P., and Boussaoud, D., 1985, Role of the cat substantia nigra pars reticulata in eye and head movements. I. Neural activity, Exp. Brain Res., 57: 286-296.

Kievit, J., and Kuypers, H.G.J.M., 1977, Organization of the thalamo-cortical connections of the frontal lobe in the Rhesus monkey, Exp. brain Res., 29: 299-322.

Lestienne, F., 1979, Effects of inertial load and velocity on the braking process of voluntary limb movements, Exp. Brain Res., 35: 407-418.

Lestienne, F., Whittington, D.A., and Bizzi, E., 1981a, Single cell recording form the pontine reticular formation in monkey: behavior of preoculomotor neurons during eye-head coordination, In: Fuch and Becker, eds, "Progress in oculomotor research", Elsevier, North Holland, pp. 325-333.

Lestienne, F., Polit, A., and Bizzi, E., 1981b, Functional organization of the motor process underlying the transition form movement to posture, Brain Res., 230: 121-131.

Lestienne, F., Vidal, P.P., and Berthoz, A., 1984, Gaze changing behavior in head restrained monkey, Exp. Brain Res., 53: 349-356.

Lestienne, F., and Cailler, P., 1985, Role of the monkey substantia nigra pars reticulata in behavioral responses to visual stimuli. Neurosci. Lett., Suppl. 22: S484.

Lestienne, F., and Cailler, P., 1986, Role of the monkey substantia nigra pars reticulata in orienting behavior and visually triggered arm movements, Neurosci. Lett., 64: 109-115.

Liles, S.L., 1985, Activity of neurons in putamen during active and passive movements of wrist, J. Neurophysiol., 53: 217-236.

Mohler, C.W., and Wurtz, R.H., 1976, Organization of monkey superior colliculus: intermediate layer cells discharging before eye movements. J. Neurophysiol., 39: 722-744.

Mora, F., Mogenson, G.J. and, Rolls, E.T., 1977, Activity of neurons in the region of the substantia nigra during feeding in the monkey, Brain Res., 133: 267-276.

Parent, A., Mackey, A., Smith, A., and Boucher, R., 1983, The output organization of the substantia nigra in primate as revealed by a retrograde double labeling method, Brain Res., Bull. 10: 529-537.

Parent, A., Bouchard, C., and Smith, Y., 1984, The striatopallidal and striatonigral projections: two distinct fiber systems in primate, Brain Res., 303: 385-390.

Rouzaire-Dubois, B., Scarnail, E., Hammond, C., Crossman, A.R., and Shibazaki, T., 1983, Microiontophoretic studies on the nature of the neurotransmitter in the subthalamo-entopeduncular pathway of the rat, Brain Res., 271: 11-20.

Schell, G.R., and Strick, P.L., 1984, The origin of thalamic inputs to the actuate premotor and supplementary motor areas, J. Neurosci., 4: 539-560.

Schultz, W., 1986, Activity of pars reticulata neurons of monkeys substantia nigra in relation to motor, sensory, and complex events, J. Neurophysiol., 55: 660-677.

Whittington, D.A., Lesticnnc, F., and Bizzi, E., 1984, Behavior of preoculomotor burst neurons during eye-head coordination, Exp. Brain Res., 55: 215-222.

Yoshida, M., and Precht, W., 1971, Monosynaptic inhibition of the substantia nigra by caudo-nigral fibers, Brain Res., 32: 225-228.

THE PARTICIPATION OF THE CEREBRAL CORTEX AND BASAL GANGLIA IN MOTOR ACT PROGRAMMING

A.S. Batuev, A.A. Orlov and V.I. Shaefer

Department of Higher Nervous Activity, Leningrad State University
Leningrad, 199034, USSR

Summary

Long-lasting unit activity of cortical areas (prefrontal, parietal, motor) and subcortical structures (neostriatum, pallidum, n. ventralis anterior) was studied in conscious monkeys during the performance of a delayed alternation task. The behavioural task was as follows: on presentation of the right or left cue the monkey had to press the right or left lever, respectively. A separate group of spatial selective neurons was found to exist. The properties of these units, which were found in prefrontal and posterior parietal areas, neostriatum and n. ventralis anterior, suggest that these structures are involved in the programming of a forthcoming goal-directed motor act. Motor and anterior parietal cortices and pallidum participate rather in the execution of the motor act.

*
* *

The comparative analysis of the unit activities of different brain structures in awake animals during preparation and performance of complex behavioural acts appears to be a useful approach to motor programming (Fuster, 1980; Pirogov and Batuev, 1981; Hyvärinen, 1982; Batuev, 1982, 1987; Evarts et al., 1984). The consistency of the behavioural tasks and the presence of a temporal interval between the cue onset and the trigger stimulus are important prerequisites for this aim to be achieved. A suitable paradigm for primates is the indirect two-choice delayed alternation response (Kubota, 1975; Niki and Watanabe, 1976; Batuev et al., 1980).

In the present study a monkey was seated in the primate chair (1) (Fig. 1,B) in front of which was a panel permitting free operation of the levers. To the right and left of the midline were two opaque windows with lamps (2). Below each was a lever (4). A shutter (3) screened the levers from the monkey during the learning stage of the programme. A flash lamp (5) was fixed above the panel. Electromagnetic relays switched on an automatic reward system whenever the correct lever was pressed. The food-reward was delivered to the feeding-spoon (6).

The experimental sequence was as follows (Fig. 1, C). 6 seconds after the onset of the programme an anticipatory signal (1) - a series of flashes that occurred within 1 s - was given, while the shutter was closed. 4 or 9 s later a cue was lit for 3 or 5 s on the right or left (2) at random. After an interval of 5, 10 or 16 s the shutter was lifted to show the levers (3). Our behavioural task thus incorporates the probabilistic component, i.e. the location of the cue. The shutter was raised for 7 s. Correct trials were rewarded. After the lever was pressed by the appropriate forelimb (4), the automatic reinforcement system was switched off.

Neuron activity from the left hemisphere was recorded by several microelectrode bundles (Fig. 1, A). The electrodes were assembled in a plastic plug which was mounted upon the micromanipulator (2,3). Plugs were then implanted into the skull for fixing micromanipulators and a metal holder onto which a micro-4-channel source follower was

Table 1. The distribution of neurons sampled from various brain structures.

The brain structure	The number of recorded units
Prefrontal area	61
Parietal anterior (area 5)	55
Parietal posterior (area 7)	98
Motor area	100
Neostriatum	30
Pallidum	21
N. ventralis anterior	24

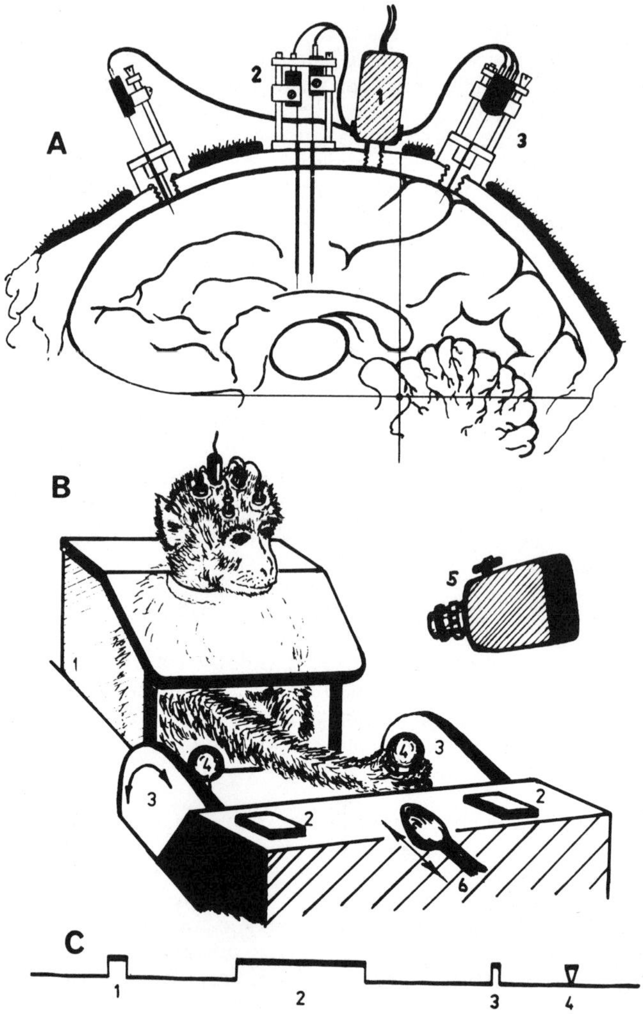

Fig. 1. An experimental apparatus for testing motor act programming in monkeys. A - disposition of the preampliphier (1) and micromanipulators (2,3) in the skull. B - experimental situation. 1 - primate chair, 2 - opaque window with lamp, 3 - shutter opened, 4 - lever, 5 - flash lamp, 6 - spoon for the food reward. C - experimental sequence. 1 - anticipatory signal, 2 - cue, 3 - trigger signal, 4 - lever pressing.

mounted. The holder served as an indifferent electrode. The source follower transmitted unit activity into a 4-channel ampliphier (1), then the signal went to a 6-channel magnetograph. The computer processing then took place.

Long-term single unit activities were recorded simultaneously from different cortical areas: prefrontal (periprincipal sulcus), anterior and posterior parietal (around s. intraparietalis, areas 5 and 7) and motor (g. precentralis, forelimb projectional zone), basal ganglia (neostriatum and pallidum) and thalamic nucleus (n. ventralis anterior) (see Table 1).

The details of the microelectrode system, apparatus, and data processing have been described previously (Batuev et al., 1980, 1985).

First spontaneous activity of the sampled units was evaluated. Prefrontal units were characterized as a rule by low background activity (lower than 10 imp/s) while in the parietal area units with both high (area 5) and low (area 7) firing rates could be found. The firing rate of motor cortex units was predominantly high (more than 10 imp/s).

Units with both low and high background activities were analyzed. Prefrontal and parietal units with low firing rates appear to reflect the spatial location of the cue (see also Mountcastle, 1976; Niki and Watanabe, 1976; Kojima and Goldman-Rakic, 1984).

For example, the activity of the parietal cortex neuron recorded from area 7 in response to cue displays on the right or left is illustrated in Fig. 2. The discharge rates varied significantly between the cue onset and the shutter closure. With the cue lit on the right the discharge frequency was several times lower than with the cue on the left. Moreover, the pattern of discharge of the same neuron in response to the cue display on the right was opposite to that with the left cue location. This type of differential activity persisted until the end of the cue onset-shutter closure interval (table interval) but disappeared completely once the lever was pressed. As the monkeys had been trained for a long time before the neuronal

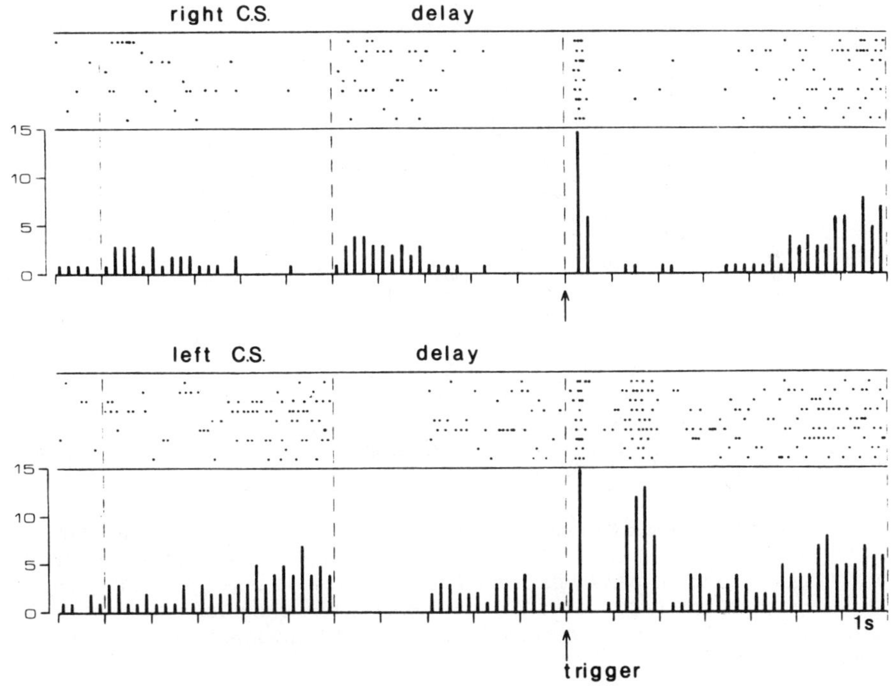

Fig. 2. Activity of a parietal area unit during the cue display on the right and left and during the task interval. Histogram and raster displays for 10 programme runs during the correct choice. Vertical lines (from left to right), the onset and offset of the cue and the trigger onset. The abscissa, time in s, the ordinate, number of unit discharges for Δ t = 200 ms. Calibration 1 s.

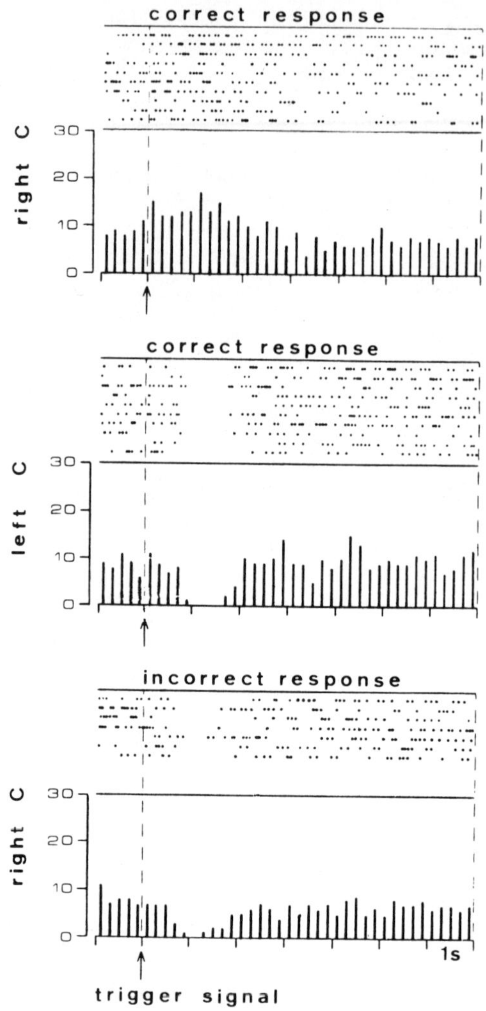

Fig. 3. Activity of a prefrontal area unit after the trigger signal in programmes with different cue displays during correct and error choice. Designations are the same as in Fig. 2.

recording started, we did not present here the variations in neuron discharges from trial to trial during the early stages of learning. Therefore, all the data presented refer to the period of learning when the task performance and the response pattern was stable. Histograms and rasters in Fig. 2 correspond to this period of learning and are based on 10 programme runs during correct choice.

The activity of these neurons was studied with various task interval durations (5, 10 and 16 s). Rasters and histograms were built for neurons which changed their discharge rate significantly when the cue was on the right, but were unresponsive or were characterized by other discharge patterns to the cue display on the left. Differential responses to the cues on the right and the left were also observed during the task interval with the three interval durations tested.

In the posterior parietal area this type of unit accounted for 21% (n=19), and in the prefrontal area for 28% (n=17) or the units sampled. This type of spatio-selective unit is not found in the motor cortex or among the units with a highfiring rate in the anterior parietal cortex.

The existence of spatio-selective neurons in the posterior parietal cortex which remain differentially active during intervals of different durations may suggest the participation of this structure in a short-term trace storage. The monopoly of the prefrontal cortex (periprincipal are, so called "focus" of spatial memory) in these processes (Butters et al., 1971) seems doubtful.

One of the most appropriate ways of studying the central motor programme is to analyse the impulse activity of spatio-selective units depending on correct or error performance after the trigger signal. Therefore, we looked for some correlation between cue displays, unit activities and motor act success. The activities of spatio-selective neurons in correct and error trials were then compared. These units in the prefrontal and posterior parietal areas fall into two groups. One group consisted of 7 prefrontal units and 10 parietal units. The activity patterns of these units were related only to the cue location irrespective of the trial success. Various types of neuron activity both in prefrontal and parietal cortices were described within this group. But in all cases, activity of these neurons was found to be independent of the motor act efficiency. In our opinion these units are therefore not involved in motor act programming.

The other group of neurons was characterized by the following property: in error trials the activity of these neurons during cue display and the subsequent interval differed considerably from that recorded in correct trials. This group included 8 prefrontal and 8 parietal units. Out of these a small population of neurons was singled out which revealed their spatial selectivity only after shutter opening, i.e. during the trigger period when the animal could perform the corresponding lever-press. Fig. 3 shows the activity of one spatio-selective prefrontal neuron after the trigger signal onset, when the monkey was able to carry out the previously constructed motor programme. Upper and middle parts of Fig. 3 are the fragments of rasters and histograms representing the neuron activity during the correct choice. In these cases the motor acts corresponded to the cue spatial location. A fragment of the same neuron activity during the error choice, i.e. during the incorrect response, is shown at the bottom of Fig. 3. It can be clearly seen that during correct choices, when the monkey performed similar movements to the right (upper) or left (middle) lever, opposite patterns of neuron activity occurred: the activity increased when the right lever was pressed and decreased when the left lever was pressed. The above patterns of firing activity may reflect the differences between two alternative motor programmes. It is also possible that this type of neuronal responses may be related rather to the programme of a forthcoming motor act than to the side of the cue display.

Unlike the spatial location of the cue, the temporal parameters of the behavioural task were stable during a long session. We usually employed programmes with the fixed 5 s task interval. But with some neurons (7 in the prefrontal and 12 in the parietal area) the interval was soon prolonged. The initial temporal interval (5 s) can be clearly seen from the pattern of activity. There may exist a neuronal set which stores the temporal sequence of the behavioural programme. This neuronal system is quite inert and numerous trials are required to reconstruct the neuronal pattern according to the new temporal sequence. In addition, there exists another population of neurons which rearrange their pattern of activity very promptly. These units are included in a very plastic neuronal system the activity of which parallels changes in the temporal sequence of a behavioural task (Batuev et al., 1985).

We have previously shown that parietal units with a high firing rate are not characterized by spatial selectivity (Batuev, 1987). Their activity was correlated with the performance of the motor act (pressing a lever and taking a food-reward).

The activity of neurons in the neostriatum and globus pallidum was recorded during the same behavioural sequence. The striatal neurons were characterized by low background activity: 97 % of the sampled units generated no more than 10 imp/s. On the contrary, pallidal neurons fired with a high frequency: 74 % of the sampled units had a background activity of more than 10 imp/s. Using the above procedure we observed spatio-selective units in the striatum (24 %) but not in the pallidum.

Fig. 4 shows the activity of two neurons (1, 2) which were recorded simultaneously from the neostriatum (ventromedial part of the head) using two microelectrodes of one and the same bundle. The histograms demonstrate their activity in correct trials when the cue was displayed on the right (A) and on the left (B). The spatial selectivity of these units was marked during the cue and subsequent interval. Their discharge rates changed significantly starting from the cue presentation until shutter closure. Differential activity persisted until the end of the interval and continued after the trigger (see cell 1).

It is interesting to compare A and C histograms. Histogram A corresponds to correct choice during the cue on the right. Histogram C averaged 10 error trials when the cue display was on the right but the monkey pressed the left lever. Note that histograms show similar activity patterns after triggering when the monkey pressed the left lever with the cue display on the left, i.e. correct choice (Fig. 4, B, 1) and the posttrigger histogram when the monkey pressed the left lever with cue display on the right, i.e. error choice (Fig. 4, C, 1). When comparing histograms A and C, one should note the absence of the specific pattern of activity during the cue presentation and subsequent interval when the monkey performed error choice.

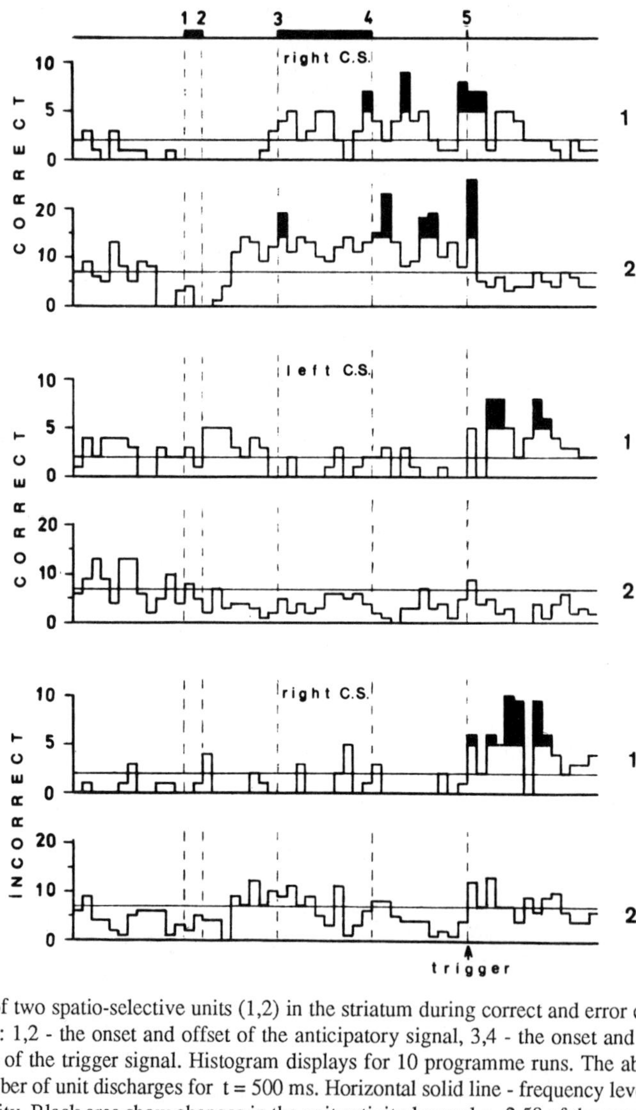

Fig. 4. Activity of two spatio-selective units (1,2) in the striatum during correct and error choice. Dash-lines from left to right: 1,2 - the onset and offset of the anticipatory signal, 3,4 - the onset and offset of the cue, and 5 - the onset of the trigger signal. Histogram displays for 10 programme runs. The abscissa, time in s, the ordinate, number of unit discharges for t = 500 ms. Horizontal solid line - frequency level (average) in the background activity. Black area show changes in the unit activity beyond +- 2.58 of the average level.

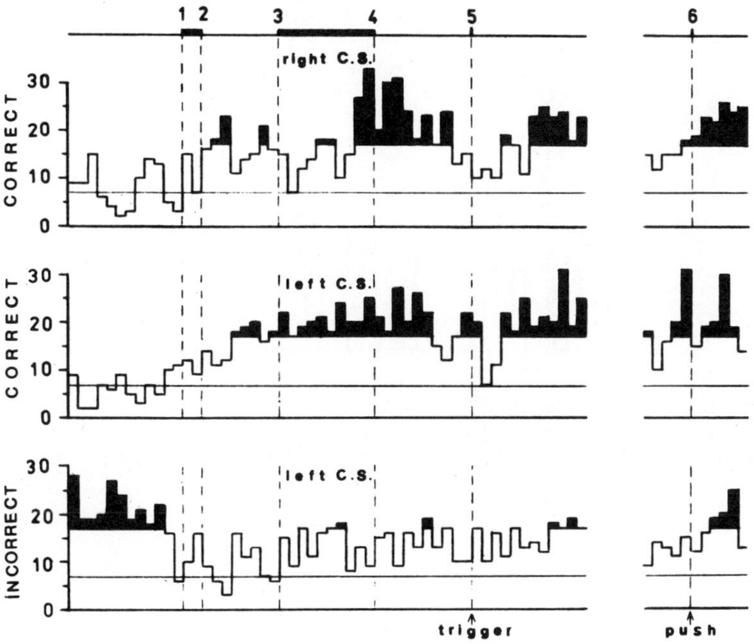

Fig. 5. Histograms of low frequency unit activity of pallidum during correct and error choice. Designations are the same as in Fig. 4.

It can be suggested therefore that the type of neuronal response produced is related to the side of the cue display and to the programme of the forthcoming motor act.

Fig. 5 shows histograms of a pallidal neuron response during the correct choice of the right lever (1) and the left lever (2) and during the error choice of the left lever (3). The cue location appears to have had no effect on the activity of this unit. In correct trials, a significant increase in firing began after the anticipatory signal and lasted until the lever was pressed. In error trials (3) the histogram of the unit differs from that mentioned above. The spontaneous high firing rate accompanied the subsequent instructive period of the programme. The histograms after a dash-line (6) were synchronized according to the lever pressing. Histograms for the correct and error choices do not differ significantly.

The activity of the sampled pallidal neurons was connected with various movements of the animal and predominantly with voluntary saccades toward the relevant object. No such correlation between the activity of striate or prefrontal neurons and that of simultaneously recorded pallidal units was observed (Orlov et al., 1984). We therefore assume that pallidal units are active rather during motor act performance than during motor programming. The same finding was made with high frequency neurons in the motor and parietal cortex (Batuev, 1987).

It is well known that n. ventralis anterior (VA) is a nonsensory thalamic relay nucleus via which neostriatum and pallidum are projected onto the cerebral cortex (Kuo and Carpenter, 1973; Strick, 1976; De Long et al., 1981; Alexander et al., 1986). It is also a relay-station for the thalamo-cortical inflow to the frontal lobes from unspecific thalamic nuclei and medio-dorsal nucleus. Therefore VA plays an important role in passing information from different cerebral sources to the prefrontal cortex. Thus, the investigation of VA neuronal activity under the above described experimental conditions is of special interest. The main features of VA units were low frequency firing rate and their spatial selectivity, which is more pronounced during the executive period, i.e. after the shutter opening (Fig. 6).

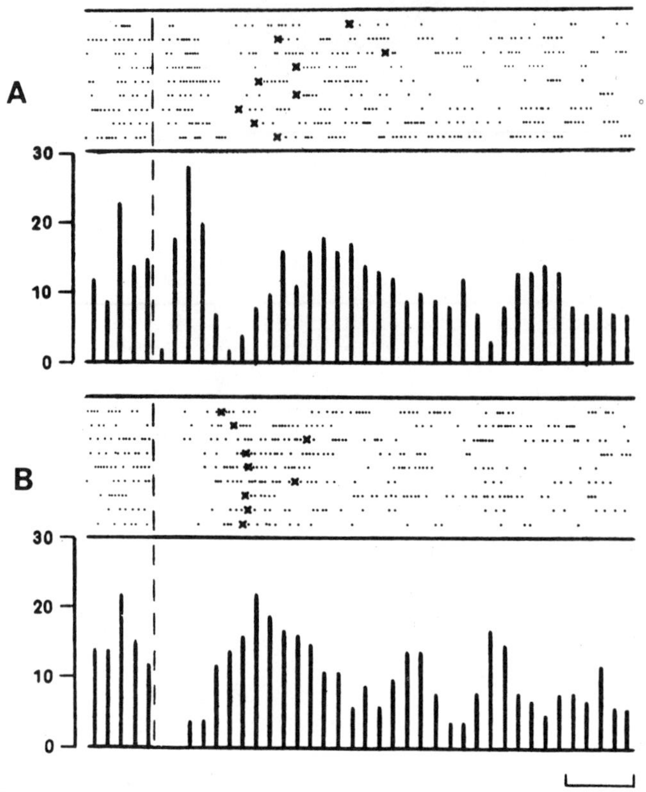

Fig. 6. Histograms and rasters of unit activity of n. ventralis anterior after trigger signal with the preceding cue display on the right (A) or on the left (B). Vertical dash line, the trigger signal onset. x,x, - moments of lever pressing. Histograms and rasters for 10 programme runs during the correct choice. The abscissa, time in s., the ordinate, number of unit discharges for $\Delta t = 200$ ms. Calibration 1 s.

No spatial selectivity was observed during the preceding instructive period. Figure 6 shows the existence of distinct differences in the activity of the same unit after the trigger signal and up to the motor act corresponding to the forthcoming lever pressing right (A) or left (B).

On the basis of the literature (Evarts et al., 1984) and the present results the following hypothetic scheme of interstructure relationships can be suggested. The striatum units are excited predominantly during the executive period of the sequence, and they in turn can suppress the activity of pallidal units. As a result, the inhibitory control of thalamic nuclei is abolished, the motor cortex can receive the required information and at this moment the motor programme is triggered.

One may suggest that the decision-making as to the side of a forthcoming response and its motor programming may have occurred at any time after the cue up to the end of the task interval since the animal was unable to reach the lever immediately after the cue. However, due to changes in the activity of neurons recorded during onset of the cue and the subsequent interval, decision-making and motor programming may occur during the cue. If this is so, within the interval, information about the cue spatial location and the corresponding programme of the forthcoming motor act is stored in short-term memory.

Our data showed the existence of spatio-selective cells in the monkey prefrontal and posterior parietal cortical areas, neostriatum and VA which reflect processing and storage of information concerning the cue location, programming and performance. Besides, there are units which reflect temporal sequences of the behavioural task in the cortical association areas. Their main properties are either fixed time of discharges or labile discharge changes, depending on the experimental sequence.

The unit activity of association cortical areas, neostriatum and VA appears to reflect two simultaneous processes connected with spatial-temporal memories. During the cue both its location and a forthcoming motor act are programmed; the temporal and spatial parameters of the latter are fully described. The high firing rate neurons in motor and anterior parietal cortex (Batuev, 1987) and globus pallidum are probably involved in motor act performance.

Thus, in most cases prefrontal and posterior parietal areas, neostriatum and VA change their neuronal activity in a similar way during the preparation of a goal-directed motor act. We suggest that different divisions of primate association systems (thalamo-cortical and subcortical) are involved in the combined integrative system which is responsible for both programming and performance on the basis of a continuous monitoring of the spatio-temporal parameters of the environment (Batuev et al., 1985).

References

Alexander, G.E., De Long, M.R., and Strick, P.L., 1986, Parallel organization of functionally segregated circuits linking basal ganglia and cortex, Ann. Rev. Neurosci., 9: 357-381.

Batuev, A.S., 1982, Programming of goal-directed behaviour and association systems of the brain, The Pavlovian J. of Biol. Sci., 17: 1-9.

Batuev, A.S., 1987, Higher integrative systems of the brain, Gordon and Breach, London.

Batuev, A.S., Shaefer, V.I., Orlov, A.A., 1985, Comparative characteristics of unit activity in the prefrontal and parietal areas during delayed performance in monkeys, Behav. Brain Res., 16: 57-70.

Batuev, A.S., Pirogov, A.A., Orlov, A.A., Shaefer, V.I., 1980, Cortical mechanisms of goal-directed motor acts in the rhesus monkey, Acta neurobiol. Experim., 40: 27-49.

Butters, N., Pandya, D., Sanders, K., Dye, P., 1971, Behavioral deficits in monkey after selective lesions within middle third of s. principalis, J. Comp. Physiol., 76: 8-14.

De Long, M.R., Georgopoulos, A.P., Crutcher, M.D., Mitchell, S.J., Richardson, R.T., Alexander, G.E., 1981, Functional organization of the basal ganglia: contributions of the single-cell recording studies, Functions of the basal ganglia, Pitman, London: 64-82.

Evarts, E.V., Shinoda, Y., Wise, S.P., 1984, Neurophysiological approaches to higher brain functions, J. Wiley and Sons, N.Y.

Fuster, J.M., 1980, The prefrontal cortex, Anatomy, physiology and neuropsychology of the frontal cortex, Raven Press, N.Y.

Hyvärinen, J., 1982, The parietal cortex of monkey and man, Springer, Berlin.

Kojima, S., Goldman Rakic, P.S., 1984, Functional analysis of spatially discriminative neurons in prefrontal cortex of rhesus monkey, Brain Res., 291: 229-240.

Kubota, K., 1975, Prefrontal unit activity during delayed response and delayed alternation performance, Jap. J. Physiol., 25: 481-493.

Kuo, J.-S., Carpenter, M.B., 1973, Organization of pallido-thalamic projections in the rhesus monkey, J. Comp. Neurol., 151: 201-236.

Mountcastle, V.B., 1976, The world around us: neuronal command functions for selective attention, Neurosci. Res. Progr. Bull., Suppl., 14: 1-43.

Niki, H., Watanabe, M., 1976, Prefrontal unit activity and delayed response: relation to cue location versus direction of response, Brain Res., 105: 79-88.

Orlov, A.A., Batuev, A.S., Shaefer, V.I., Kursina, N.P., 1984, Neuronal activity in striopallidum of monkey during the delayed spatial choice, In: The striate system and behavior in norm and pathology: 125-132.

Pirogov, A.A., Batuev, A.S., 1981, Cortical unit activity and motor act programming, 28th Int. Congr. Physiol. Sci., Adv. Physiol. Sci., 17: 239-245.

Strick, P.L., 1976, Activity of ventro-lateral thalamic nucleus neurons during arm movements, J. Neurophysiol., 39: 1032-1044.

THE STANCE PHASE AND ITS CONTROL BY SENSORY INFORMATION DURING WALKING IN CRUSTACEA

F. Clarac and A. Chrachri

Laboratoire de Neurobiologie et Physiologie comparées
Place Peyneau
33120 Arcachon, France

Summary

By definition, locomotion is a succession of steps with alternating swing and stance phases. In Crustacea, the stance phase is a slow ramp propulsive movement for which sensory control is particularly important. Spatio-temporal as well as free components can define the stance phase as opposed to the swing phase. Contact and force mechanoreceptors situated at the tip of the legs in several Decapods seemed to provide important cues to stance phase parameters. Their selective stimulation elicits a reflex which seems to be able to reset the stepping cycle. This reflex was also studied in Crayfish isolated ganglion preparations, where its role is discussed.

Introduction

When terrestrial or aquatic animals walk on a substrate, each locomotor step is composed of two parts, namely the swing phase (when the leg is moving) and the stance phase (when it stays in contact with the ground and pushes or pulls the body). By definition, locomotion is a succession of steps with alternating swing and stance phases.

These two movements are very different. Using vertebrate terminology, as did Dean (1985) when investigating the swing phase of stick insect walking, the swing phase can be said to be a "goal directed movement with ballistic component". The stance phase on the other hand is a slow ramp movement in which the forces exerted to propel the body are concentrated and for which sensory control is particularly important (Herman et al., 1976; Roberts and Roberts, 1983; Bush and Clarac, 1985).

Lower vertebrate and invertebrate preparations have been extensively studied in order to present simple models in which it is possible to describe in detail some general rules underlying motor behaviour (Grillner, 1985; Barnes and Gladen, 1985; Selverston, 1985). This is the case with the Decapod Crustacea, where in parallel to intensive behavioural studies on locomotor pattern, dissected preparations of thoracic ganglia have been used to elucidate cellular mechanisms of both central circuitry and sensory integration. Comparisons between behavioural data and central ganglion pathways will provide a very broad knowledge of locomotor processes.

After having pointed out the particular characteristics of a crustacean locomotor behavioural pattern, we will try to explain how during the stance phase, sensory receptors can control this particular motor action. To give an overview of this problem, data are presented on several different species but in particular on the Rock lobster Jasus lalandii, the crab Carcinus maenas, and the crayfish Procambarus clarkii.

Characteristics of a walking motor pattern in Decapod Crustacea

At first sight, due to their morphological diversity, their great number of legs, and their ability to walk in all directions on a substrate, it seems to be very difficult to describe the crustacean walking motor pattern. Despite their apparent complexity however, recent studies have demonstrated that Crustacea are very attractive preparations for neurobiologists for the following reasons:

i) The neuronal command is simple; about 50 motoneurons (MNs), at least in the crab, govern an entire leg musculature. Their cell bodies, ranging from 20 to 80μ in diameter are located in the hemiganglion of the thoracic segment where the leg is inserted. Using anterograde cobalt staining technique, Bevengut et al. (1983) have described the arrangement in clusters of the 35 MNs corresponding to the four proximal groups of muscles (approximately 7 to 10 MNs each for the promotor, remotor, depressor and levator muscles). A spatial segregation between functionally related MN somata can be observed within the ganglion neuropile. The innervation of the muscles of the three distal joints has been studied using neurophysiological methods and consists of a pool of 14 excitatory and inhibitory MNs. Two muscles, the stretcher and the opener are innervated by the same single excitatory MN (recordings Fig. 3 C and D).

ii) The biomechanical arrangement of the exoskeleton in each walking leg is particularly simple; at each joint the rotational axis, limited by a double condyle, moves the distal segment in only one plane (Fig. 2 A). Each leg is composed of 6 joints (7 joints in the crayfish), operating in perpendicular planes from one to the next; three controlling the antero-posterior plane of the animal and three controlling the sagittal or yaw plane (see detailed joint nomenclature in the legends of the Fig. 2). Movements are caused by pairs of antagonistic muscles, a promotor and a remotor for the thoraco-coxal joint (T-C), a levator and a depressor for the coxo-basal joint (C-B) and flexors and extensor for the mero-carpodite joint (M-C) (Fig. 1 and 2).

Although Decapod Crustacea have very different body shapes (some are elongated animals composed of a cephalothorax prolonged by an abdomen like the crayfish or the Homarus, others are enlarged animals with a flattened cephalothorax and a very reduced abdomen like the crabs; between these two extremes, several types of intermediate forms exists) the thoracic appendages are built in a similar manner in all species, with homologous muscles and similar motor and sensory innervations (Evoy and Ayers, 1982). By definition, Decapod Crustacea possess 10 legs but most of the time, whatever the biological group, only 3 pairs are efficient during walking.

iii) Although the possibility of walking in every direction implies a specific motor pattern in each given leg, stance phase and swing phase can easily be defined with EMG recordings.

A systematic study was performed with the Rock lobster, Jasus lalandii, using two different methods: the one involved the walking that occurs when the animal moves freely in an aquarium, and the other, that elicited in an animal fixed above a driven belt, the direction of which determines the direction of walking. The stance phase was measured differently in these two situations (see Fig. 3 A). During the stance phase In free walking, the leg is at a fixed point (B0) and the forces resulting from the contraction of the stance phase muscles cause the movement of the body (B0 → B1). On a treadmill, the substrate moves instead of the body and the stance phase was defined by the excursion of the tip of the leg. The distance covered corresponds to the extreme anterior point (AEP) when the leg touches down on the belt and the extreme posterior point (PEP) when the leg lifts again (L0-L1). This is in theory identical to B0-B1, the distance covered under free conditions, and corresponds to the stride length (SL).

EMG recordings showing the walking motor patterns observed in an animal under driven and free conditions are shown in Figure 1. Under driven conditions, the six recordings measure the activities of the 6 muscles governing the 3 main joints during forward walking sequences. Simultaneous recordings of the leg movement indicated that during the stance phase (leg in contact with the belt) the depressor, remotor and flexor muscles were active simultaneously. During the swing phase, the promotor, levator and extensor muscles were activated alternately. Under free conditions (fig. 1 B), the forward walking motor pattern was somewhat similar, with identical patterns of synergism and antagonism. Above the treadmill the pattern was very regular, and under free conditions abrupt changes of

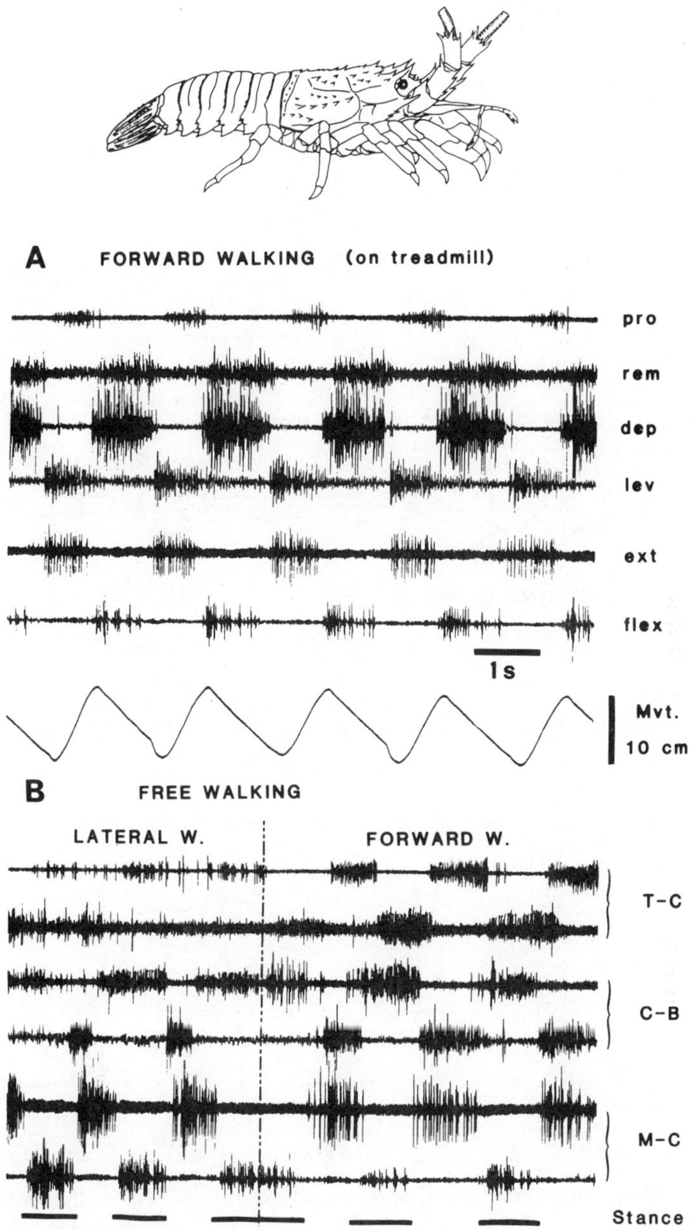

Fig. 1. Comparison of the leg motor patterns during driven walking (A) and during free walking (B) in the rock lobster Jasus lalandii. Simultaneous EMG recording of the promotor (pro) and remotor (rem) muscles of the thoracico-coxal joint (T-C), the depressor (dep) and the levator (lev) muscles of the coxo-basal (C-B) joint and the extensor (ext) and the flexor (fle), muscles of the mero-carpodite (M-C) joint. On the treadmill the leg movement was recorded at the tip (Mvt.); under free conditions, the stance phase duration is marked with a bar. A horizontal time bar (1 s.) is given in both recordings. During free walking changes from one direction to another can be abrupt as that presented here between lateral and forward walking.

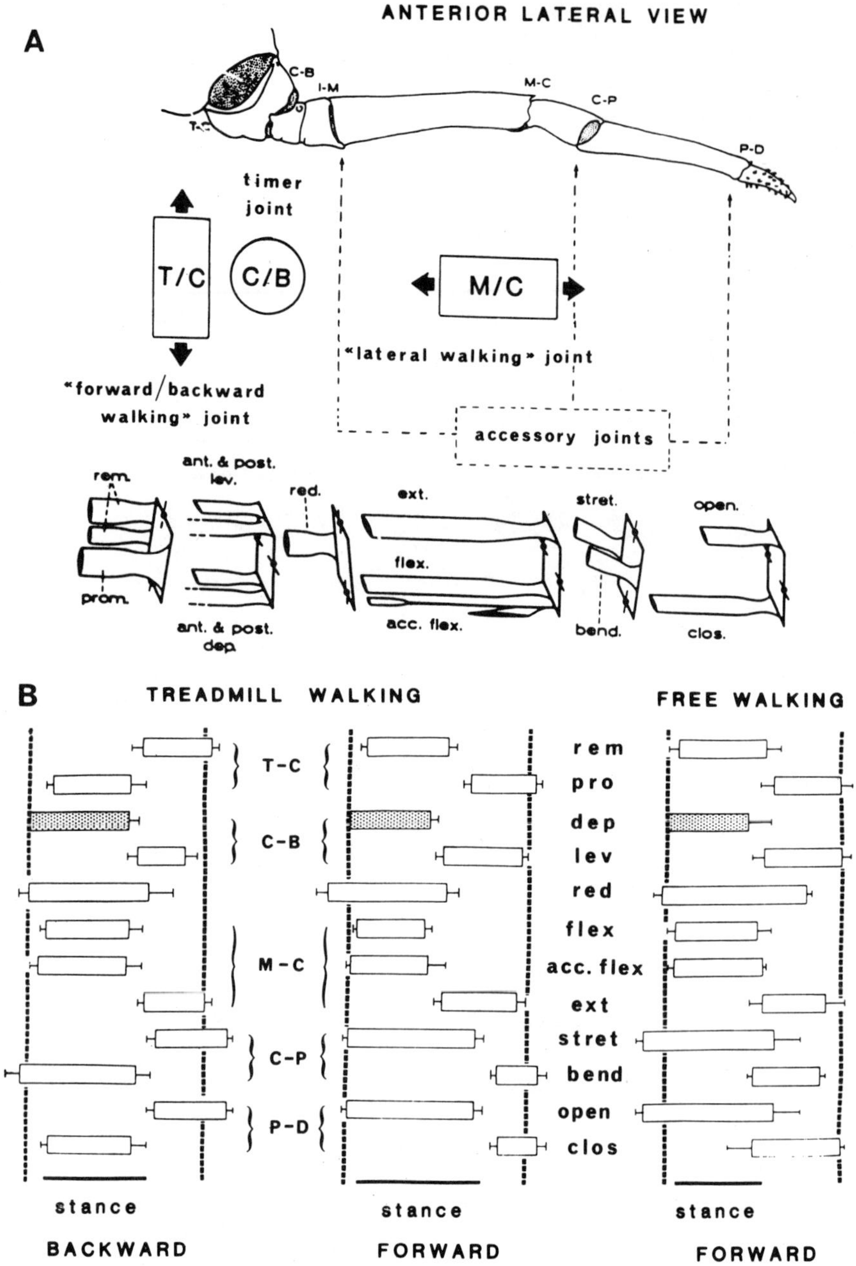

Fig. 2. Anatomical and physiological organization of a walking leg in a Decapod Crustacea. A: Scheme of the joints and muscles of the leg. Each leg is composed of six joints; three operate in the antero-posterior plane, the thoracic-coxal joint (T-C), the ischio-meropodite joint (I-M), and the carpo-propodite joint (C-P); the other three, the coxo-basipodite joint (C-B), the mero-carpopodite (M-C) and the prodactylopodite (P-D), act in a perpendicular plane, in the dorso-ventral direction. Three joints seem to be essential in organizing the direction of walking; the C-B joint as the timer joint, the T-C for antero-posterior displacements, and the M-C for lateral displacements. Underneath are schematized the joints with their condyles and the muscles operating at each joint; from proximal to distal in the leg the promotor (prom), the remotor (rem), the anterior and posterior levators (ant. and post. lev.), the anterior and posterior depressors (ant. and post. dep.) the reductors (red.), the extensor (ext.), the flexor (flex.), the accessory flexor (acc. flex.), the stretcher (stret.), the bender (bend.), the opener (open.) and the closer (clos.). B: Diagram of the leg muscle activities within the step period in forward free walking and in forward and backward driven walking. The onset of the depressor activation was taken as the beginning of each step and is represented as a shaded rectangle. Error bars are SD.

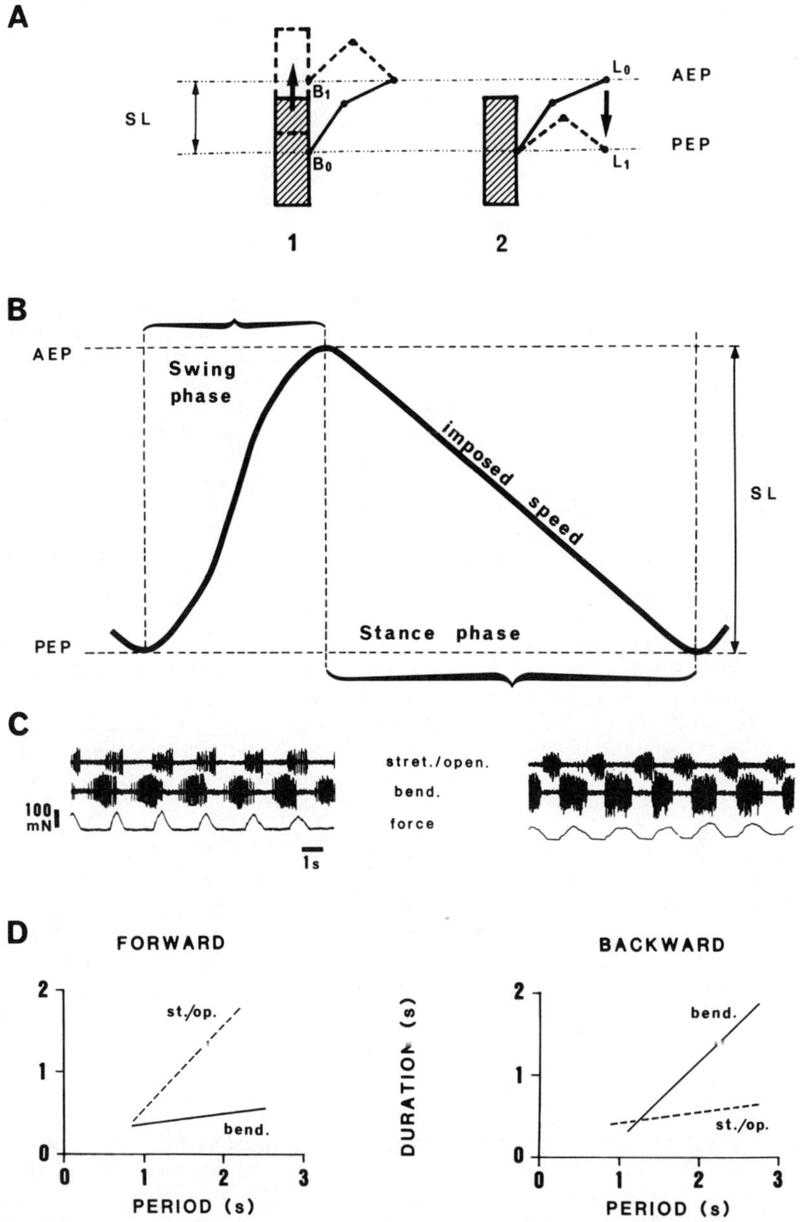

Fig. 3. Characteristics of the stance phase during walking in the rock lobster. A: Sketch illustrating that the stride length (SL) is somewhat similar during free walking (1) when the body moves from B0 to B1 and on the treadmill (2) when SL is defined as the distance between the anterior extreme position (AEP) and the posterior extreme position (PEP). B: Scheme of the leg movement on a treadmill during a given step (compare with Fig. 1 A), showing the main characteristics of the swing and stance phases. C: EMGs of the stretcher-opener and of the bender during forward and backward walking, recorded simultaneously with the forces exerted at the tip of the leg. D: Comparison between the EMG duration vs. the period in two muscles involved in both walking phases. The correlation was maximum when the muscle functioned as a stance phase muscle and minimum when it functioned as a swing phase muscle.

direction could occur as shown in fig. 1 B. First the animal walked laterally with a main activation in the C-B and M-C joint muscles, the remotor and promotor muscles being only tonically active. Then when the animal moved forward, both T-C muscles alternated and the M-C muscle bursts decreased in amplitude and in intensity. This general rule has been demonstrated in other animals (Ayers and Davis, 1977). In the rock lobster we simultaneously recorded the leg muscles in order to have a general picture of their activities during the stance and swing phases. A diagram of EMG duration is presented in Fig. 2 B where forward and backward treadmill walking and forward free walking are compared; the onset of the depressor burst being taken as reference. With both free and driven walking in the same direction, the pattern is similar with some small differences. In different and opposite directions, the pattern of synergism and antagonism between muscle activities change (Fig. 1 B, 2B). To summarize as a result of these animals being able to walk in all directions, only one muscle is always and only operating during the stance phase, the depressor (and only one is always a swing phase muscle, the levator). The other muscles, depending upon the walking direction, discharge during one or the other phase. Their position within the leg and the relative importance of the joint they govern, determine their relative importance in organizing locomotion in a given direction. For example, the remotor and the promotor are propulsive muscles generating the directional forces during, forward and backward walking respectively, the extensors and the flexors in both directions of lateral walking (for a trailing leg when it pushes, for a leading leg when it pulls). The more distal muscles are less important, since in many species, the distal part of the leg is like a strut supporting the leg. EMG activities in these muscles can be continuous in some sequences, antagonistic muscles being co-activated (Ayers and Clarac, 1978).

In Fig. 2 A, a scheme of the various joints and of their possible functional role is presented. The coxo-basal (C-B) joint is crucial in organizing both the swing and the stance phases and can be said to be the main or "timer" joint. Due to its efficiency in propulsing the legs in their respective directions, the T-C joint is essential for forward and backward walking and the M-C joint for lateral walking. The three other joints (T-M, C-P and P-D) are secondary and largely play a supporting role. (We must note, however, that this general scheme needs some corrections in the case of the crab, Carcinus maenas, which is primarily adapted for lateral walking in which the P-D joint is particularly potent in exerting forces). Previous crustacean behavioural studies give a quite complete picture of these diverse locomotor motor components, particularly the stance phase.

Behavioural properties of the stance phase during walking

Spatio-temporal as well as force components can be taken to define the stance phase as opposed to the swing phase. Systematic measurements of stride length have demonstrated that it remains invariant when different speeds are imposed by the belt. Changes in walking speed can be obtained by changing the duration of the stance phase, since the swing phase duration is more or less constant except at very slow speeds; for a Rock lobster walking above a treadmill, the period P can therefore be summarized as:

$$P = SL / V + K$$

where SL is the stride length, V the belt speed and K time constant corresponding to the swing phase and to pauses that may occur at the switch between swing and stance and vice-versa (Clarac and Chasserat, 1986). According to this formula, since SL is a constant, the stance obeys a hyperbolic law corresponding to the ratio of the stride length to the belt speed. These simple elements explain in particular why the duration of each EMG is closely related to the period when it is used as a stance phase muscle and only very slightly or not related to the period when it is a swing phase muscle. This is true both of the main muscles controlling the directional T-C and joints and of those governing accessory joints, as shown in Fig. 3 D. Fig. 3 C gives recordings of the bursting activity of the stretcher-opener and of the bender with the forces exerted during the stance phase. During forward walking, when the stretcher was a stance phase muscle, its burst duration was found to be linearly correlated with the period (R=0.906), whereas in backward walking, when it was a swing phase muscle, the coefficient of correlation was much lower (R=0.359). Similarly the bender muscle varied linearly with speed (R=0.962) when it was a stance phase muscle but much less so (R=0.389) when it was a swing phase muscle. Forces, measured with a transducer fixed at the leg tip (Fig. 3 C), were maximum in the first part of the stance phase and then decreased. Above the treadmill, the forces recorded were maximum in leg 4, less in legs 3 and 5 and legs 1 and 2 just made contact (Clarac and Cruse, 1982). A comparison between free and driven walking demonstrates that, at similar speed, similar forces can be measured, which suggests that the data obtained in studies conducted on a treadmill have a physiological significance.

To determine how this stance phase can be controlled by sensory inputs, various experimental perturbances to walking were tested (Clarac, 1985). Two examples illustrate the types of results obtained. The data obtained using the extreme slow and fast speeds of the belt, suggests the presence of a position servo-mechanism. At very slow speeds, forces exerted above the belt increased their normal value two or three times, while at very high speeds, the forces exerted became minimal. This can be explained by postulating that when the reference input representing the "desired" leg position takes on a value posterior to the actual leg position, the resulting error signal increases the posteriorly directed force (Cruse et al., 1983); on the contrary if the treadmill is going too fast, the resulting error signal is in advance of the "desired" position and results in a decrease in the same force. Barnes (1977) observed similar reactions in crayfish walking on a treadwheel.

The stride during the stance phase is also under various sensory controls; this is particularly clearcut when the ongoing movement of the neighbouring legs are modified. For example a change in the PEP of the front or in the AEP of the back leg modifies the considered stride length. If the back leg is maintained fixed in a back position, the given leg goes much further back than its normal PEP; the stride is then much longer than normal (Clarac, 1985). There exists a descending and an ascending effect in the Rock lobster; this has been used to build a model of simulation with which it is possible to reproduce various states of coordination (Chasserat and Clarac, 1986). This fits well with the systematic study by Cruse and Muller (1986) on crayfish walking on a treadmill where it was demonstrated that two coupling mechanisms occur: an anteriorly directed one is present when a leg performs a power stroke and a posteriorly directed one operates at the end of the stance phase (the onset of the swing phase) when the leg is near its PEP.

In conclusion, the stance phase, induced by the depressor discharge and depending on the direction associated with other muscle activation, is very precisely controlled by peripheral sensory loops.

Sensory information and the stance phase

Several reviews have described the different types of sensory receptors that have been found to exist in Decapod Crustacea (Wales et al., 1970; Mill, 1976; Bush and Laverack, 1982). It is apparent that the many components of leg movements are specifically coded by different sensory elements. Two types of receptors have to be considered. First, those located at a particular joint, detecting the position, the movement, or the tension exerted in this local region; it appears that those located in the proximal joints are more important than those at the distal joints, which detect the general walking activity. Secondly, those inserted along the exoskeleton that record the contact and/or the force exerted by the entire leg; especially those at the tip of the leg that respond to the nature of the substrate and in particular to vibration during contact.

Joint receptors. Each joint is equipped with one or two chordotonal organs; those like articular receptors composed of tens of cells, some of which are position sensitive and others movements detectors (some sensitive only to velocity, others to acceleration). Most of them are unidirectional and all are more active at extreme angles of the joint than in mid-position and thus at the extreme leg positions during the step cycle (during AEP and PEP) rather than in mid-position.

Each leg also contains two neuromuscular structures. In the meropodite, the myochordotonal organ resembles a chordotonal organ but is associated directly with a muscle. Its function is unclear. The TCMRO (the thoracic-coxal muscle receptor organ) , which has been studied in great detail by Bush (1981), is composed of two large afferent fibres, the large diameter of which has lent itself to an extensive intracellular analysis. Stimulated by movement of the T-C joint and stretched during remotion, one fibre is tension-sensitive (fibre T), while the other is length-sensitive (fibre S). It is controlled through two specific motor axons in a way somewhat similar to the mammalian muscle spindle.

Pure tension receptors, the apodeme tension receptors, have been found to exist in some muscles particularly in the crab. Recently Hartman (1985), studying one chordotonal organ, suggested that each of them are associated with a group of specifically tension-sensitive cells.

All these sensory receptors operate during walking and could encode various characteristics of the stance as well as of the swing phases. Unfortunately, the activities of none of these receptors have been recorded chronically due to technical difficulties; thus it

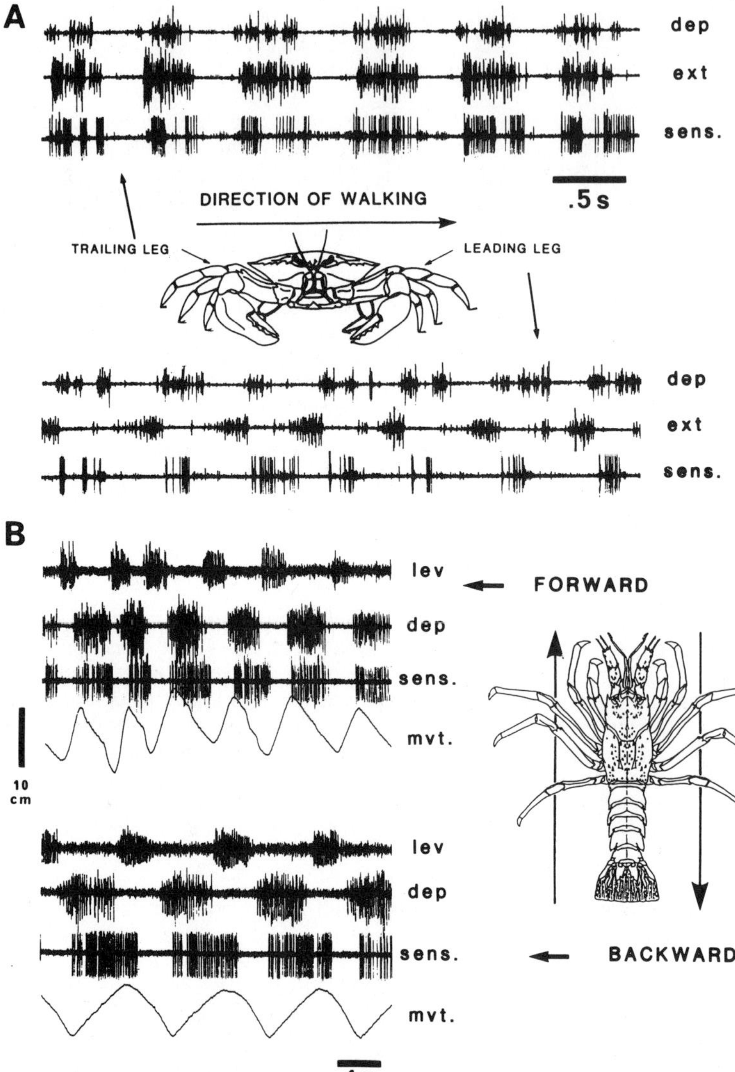

Fig. 4. Activities of a force sensitive mechanoreceptor from the dactyl, the funnel canal organ (FCO), recorded during free lateral walking in the crab Carcinus maenas and in the rock lobster walking on a treadmill. EMGs of the depressor and the extensor (in A), and of the depressor and the levator (in B), define the walking pattern. The mechanoreceptors fired with the stance phase. The movements of the leg of the rock lobster above the treadmill were also recorded during forward and backward walking.

has not been possible to test the hypotheses based on dissected preparations where sensory receptors have been activated electrophysiologically.

Cuticular receptors. This is not the case with these sensory receptors, which have been studied under both isolated and free walking conditions. Two types of structure have been described. In two cuticular depressions, in the proximal part of the leg, two groups of cells sensitive to strain of the cuticule have been called CSD1 and CSD2 (Cuticular stress detectors). Recordings during crayfish walking (Klarner and Barnes, 1986) reveal that they are specifically active during the stance phase; they detect the contact duration as well as the force exerted during that contact. They are also sensitive to the contractions of the surrounding muscles, in particular the depressor.

Similar functions can be ascribed to the funnel canal organs (FCO). Described anatomically in the crab by Gnatzy et al. (1984), they are inserted into very small pores of the cuticle. Composed of more than 20 cellular elements, of which only one or two are mechanosensitive, they exist along the leg cuticle but are concentrated in the dactyl. Those located at the extreme tip of the leg composed solely of epicuticule,are mainly contact sensitive; the other proximal receptors respond to a different component of the force exerted by this distal segment (Libersat, Zill and Clarac, 1987). Recordings from them during walking in different species show that they are specifically active during the stance phase. In Fig. 4 A a single unit recording of a distal receptor is shown during trailing when the leg pushes the body or during leading when it pulls (in each case the depressor activity reveals the stance phase). The sensory discharge is irregular and displays bursts of very high frequency (between 100 to 200 Hz in instantaneous frequency). Fig. 4 B shows similar activity recorded in the Rock lobster walking above a treadmill. The discharge had a similar frequency but seemed to be more regular; this may be attributable to the regular stepping on the substrate that occurs during driven walking and the fact that this animal walks very slowly.

Control of the stance phase by specific sensory stimulation

It has been possible not only to specifically record the FCO afferents in chronic experiments but also to stimulate them, which has made it possible to define their action on ongoing locomotory patterns.

In fixed intact animals at rest, a squeeze of the leg tip where most of the FCO are located, or an electrical stimulation of the FCO, increased the levator EMG and decreased the depressor if it was previously active; responses also occurred in the other muscle of the leg but this effect seems to be variable. Simultaneous recordings of the EMG activities of the muscles of the adjacent legs show the existence of an opposite interleg connection; in this case the depressor activity was increased and if the levator was spontaneously discharging, it stopped firing.

In a crab free to move in a large swimming pool, depressor and extensor EMGs were recorded to define the main temporal characteristic of the step cycle (in particular the onset of the depressor was taken as the reference of the onset of the period). A similar electrical stimulus which was in fixed animals decreased efficiently the depressor burst and increased the levator discharge, was able during walking to reset the rhythm (Libersat, Clarac and Zill, 1987). Moreover, whether the stimulus lengthened or shortened the step depended on the position of the stimulus within the cycle (Fig. 5). Afferent stimulation can reduce the duration of the step cycle if it occurs early in the stance phase or can increase it if the stimulus is delivered later in the stance phase. Both graphs in Fig. 5 were obtained on the same animal to compare the responses of a trailing and a leading leg. Stimuli applied during the stance phase of a leading leg resulted in a reduction in the step cycle, but this was less noticeable with the trailing leg. With both leading and trailing legs, stimuli applied during the swing phase turned on an increase in the duration of the step cycle. In all cases electrical stimulus inhibited the depressor but the response of the extensor varied. With trailing legs, its burst was always shortened, but could be increased by stimuli applied to leading legs.

In the unperturbed walking crab, cyclic input from mechanoreceptors operate during the stance phase. Under these conditions, the electrical stimulation of the FCO sensory nerve

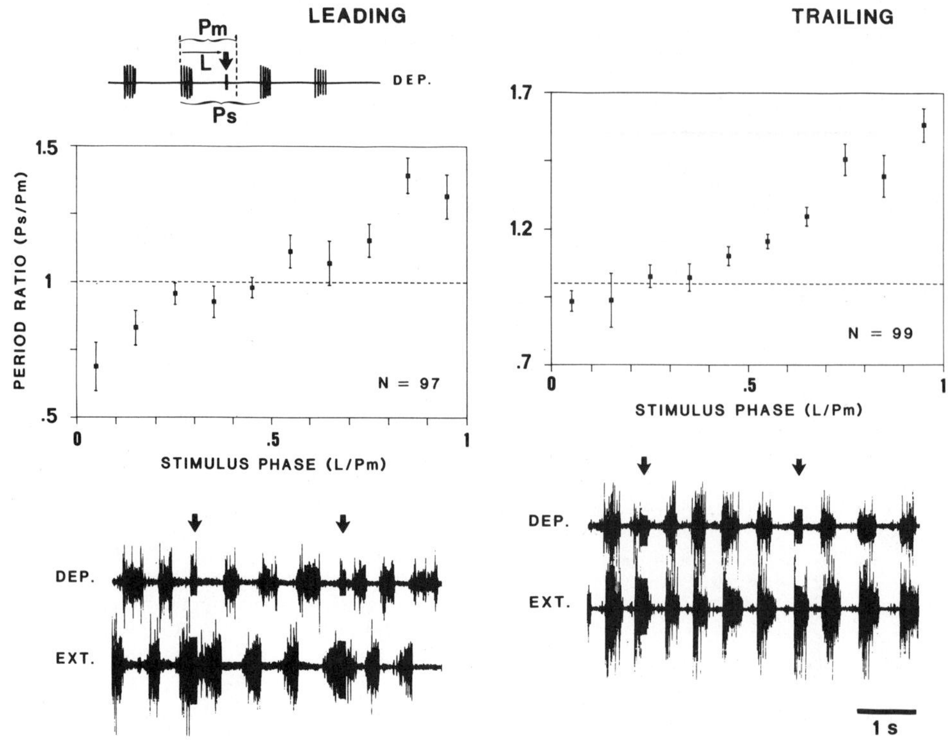

Fig. 5. Effects of electrical stimulation of the dactyl sensory nerve (which contains the FCO axons) on leading and trailing walking in the crab, Carcinus maenas. Electrical stimuli were delivered at random while myographic activity of the depressor and the extensor were recorded to define the temporal parameters of the locomotion. The onset of the depressor activation was taken as the beginning of each stepping period. The ratio of the duration of the step (PS) in which the electrical stimulus was applied to the sensory nerve of the dactyl vs. the normal step duration (Pm is the mean of 200 steps) is plotted on the ordinate; the time of occurrence of the stimulus (L) within the step cycle is plotted on the abscissa.

will be interpreted by the central nervous system in the same way as the normal sensory input. If the electrical stimulus is applied in the stance phase with the normal input, no disturbance will occur in the motor pattern; whereas if the stimulus is applied before or later, at an unusual time in the step cycle, it will induce significant modifications in the step period. In Fig. 5, the absence of any duration change in the step period, during and at the end of the stance phase (between 0.3 and 0.5 stimulus phase, Fig. 5), corresponds to the very moment where this sensory input is normally activated by the walking movements.

The FCO afferents simultaneously regulate both adjacent legs by increasing their depressor and decreasing their levator bursts. This suggests that FCO afferents are important for a given leg at the PEP during the change between stance to swing and for the two neighbouring legs at the AEP, during the switch between swing to stance. These opposite actions on its own leg and the adjacent one make FCO afferents good candidates for the regulation of the interleg pattern (Libersat, Clarac and Zill, 1987).

In the crayfish CSD2 chronic preparation, it was difficult for Klarner and Barnes (1986) to specifically stimulate these receptors. They nevertheless demonstrated that the CSD2 might contribute to regulating the interleg connections, in particular the AEP timing of a front leg connection with the AEP of a back leg.

Since it has recently become possible to approach locomotor problems at the level of the underlying cellular connectivity, we attempted on crayfish thoracic preparations to determine the afferent pathways from these FCO afferents.

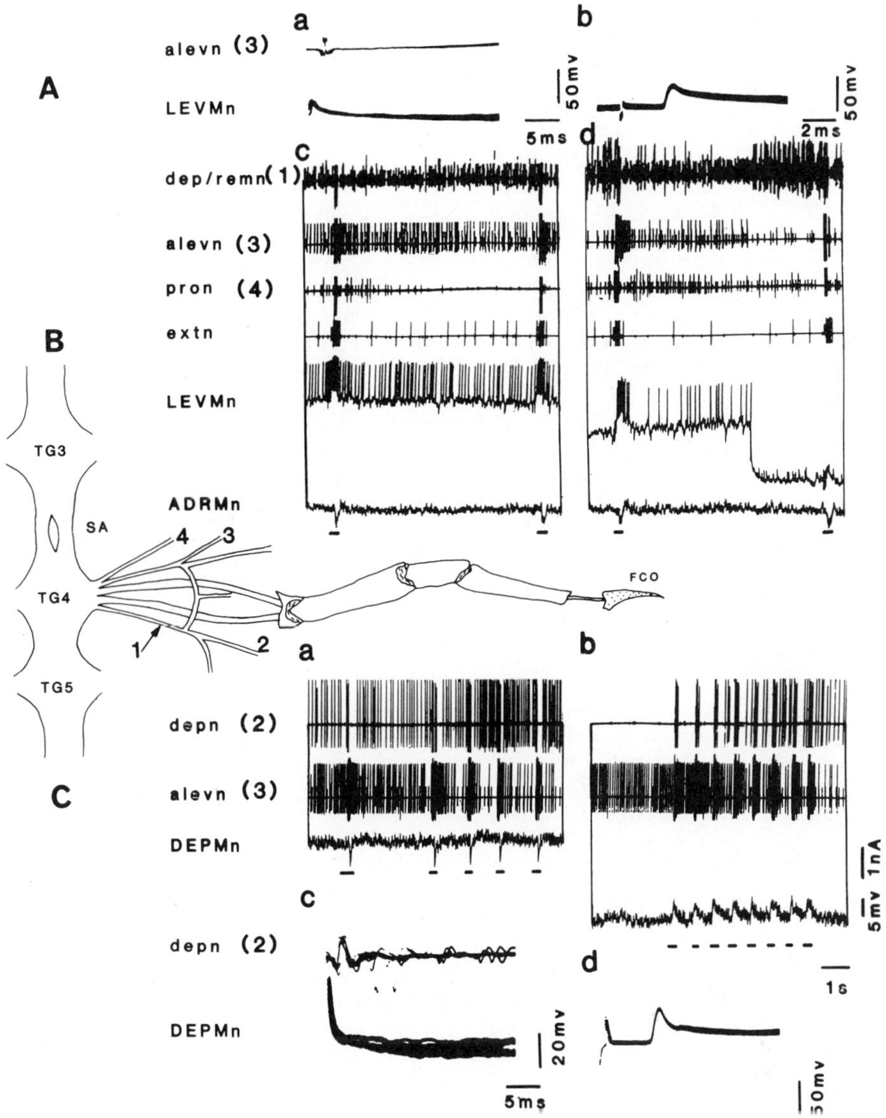

Fig. 6. Responses of some identified thoracic walking leg MNs in crayfish, to electrical stimulation of the dactyl sensory nerve which contains the FCO axons. Isolated preparation where the 4th thoracic ganglion was desheathed and the main leg nerves dissected. The distal part of the leg was preserved with the afferent of the FCO nerve. A: Intracellular activity of a levator MN (LEVMn) and a distal root MNs (ADRMs), recorded simultaneous by with the extracellular activity of the depressor/remotor (dep/remn (1)), the levator (alevn (3)), the promotor (pron (4)) and the extensor (extn) nerves. A levator MN can be identified by a spike being recorded simultaneously in the neuropile of the cell and in the motor nerve (a) and by the induction of an antidromic spike on stimulation of the motor nerve (b). In c the levator MN was depolarized (3Na); in d it was slightly depolarized (beginning of trace) and at rest (end of trace). The electrical stimulation of the FCO nerve (indicated under the different traces) induced epsps in the LEVMn and even a burst of action potentials when it was depolarized. Similar sensory stimulation induced ipsp in the ADRMn. B: Scheme of the preparation with the main nerve roots and the identification of the FCO afferents. The numbers correspond to the nerves named in A and C. C: Intracellular activity of a depressor MN recorded simultaneously with the extracellular activities of the depressor (depn (2)) and the levator nerves (alevn(3)). Similar methods to those described for the identification of the levator MN were used to identify the depressor MN (c, d). In a, the depressor MN was inhibited; this inhibition was of a chemical nature as it could be reversed by hyperpolarization (in b).

Fig. 6 gives results obtained with a preparation of this kind; ganglion 4 (TG4) was desheathed for intracellular recording and most of the motor nerves were dissected in order to record the main extracellular outputs. The distal part of the leg, and particularly the FCO afferents, was maintained intact. Several MNs were identified on comparing simultaneous extracellular nerve recordings with neuropil intracellular recordings, and applying antidromic stimulation of the appropriate motor nerves (these data are shown in the case of the levator and depressor MNs in Fig. 6 Aa, b and Fig. 6 Cc, d).

Dissection results in a preparation that is not rhythmic; some MNs fire tonically, the others are silent. Electrical stimulation of the FCO has opposite effects on levator and depressor MNs. A short stimulation train elicits a depolarization of a levator MN or a burst of spikes if it was previously depolarized: whereas a depressor MN is inhibited, and this inhibition appears to be chemical, since it can be reversed by hyperpolarization. The FCO has a broad range of effects, as distal MNs are also excited or inhibited when it is specificaly stimulated (ADRMn in 6 Ac, d).

Until now, we have not been able to induce a rhythmical response by stimulating these afferents. However, addition to the bath of some cholinergic agents such as oxotremorine or pilocarpine induces a rhythmic pattern that is very slow, but similar to a walking pattern (Chrachri and Clarac, 1987). Skorupski and Sillar (1986), working on a different crayfish preparation, obtained rhythmic activity routinely when they kept the proximal neuromuscular structure, the TCMRO, intact. This receptor may operate as the cholinergic agent inducing the regenerative properties in the motoneurons of the locomotor CPG (Sillar and Elson, 1986).

Conclusion

It is difficult to draw conclusions, since most of these studies are still in progress; nevertheless even if it is very preliminary, we would like to suggest a hypothetical scheme which combines data obtained from several different levels of analysis. We will restrict our speculation to the FCO's, CSD's and TCMRO, considering them to be among the crucial sensory elements able to control the stance phase.

According to Sillar et al. (1986), within a leg during ongoing walking the stance phase is regulated spatially and temporally through the TCMRO; the S fibre could be one of the elements of the position servo-mechanism, and the T fibre involved in the positive feedback might be necessary to reinforce the force exerted on the substrate. The FCO's, and also the CSD's, regulate this force, in particular at the AEP and at the PEP; they are crucial during the switch between the stance and the swing, and the FCO seems to also participate at the initial stage of the swing phase. Similar positive as well as negative reflex loops have been presented in both insects and cats (Pearson and Duysens, 1976; Anderson et al., 1981; Zill et al., 1981). Bassler recently (1986) pointed out that stretching a leg chordotonal proprioceptor which is extended during the stance phase can both facilitate the stance phase MNs and initiate a swing phase activity.

Sensory control of the stance phase is even more important for interleg coordination. FCO afferents and to some extent CSD information might be crucial for maintaining interleg alternation. In thoracic isolated ganglia of the crayfish, Clarac and Sillar (1985) have demonstrated that even with the TCMRO intact, the homologous muscles of different legs are active in the same phase. This does not correspond to a free walking pattern, but does occur in animals with autotomized legs (see Clarac, 1985). Thus peripheral information, in particular from the FCO, seems to be of primary importance in organizing interleg coordination.

Our knowledge of the role of sensory integration in the central generation of the locomotor pattern is still very incomplete, but nevertheless we know in what direction we have to search; a systematic comparison of data obtained at different levels of motor complexity will surely be fruitful. In particular, the fact that the legs of Decapod Crustacea are innervated by only very small number of MNs in comparison with those of the leg in Vertebrates makes it possible to compare their discharge patterns and properties in many different types of experiments.

Comparisons between isolated ganglion preparations and behavioural data suggest that, in contrast to those motor behaviours which are produced by central pattern generators producing rhythmic outputs even in the absence of sensory inputs (Selverston, 1985), the thoracic locomotor pattern depends upon the strong interaction of both central and sensory elements to elaborate a complete motor behaviour.

References

Andersson, O., Forssberg, H., Grillner, S., and Wallen, P., 1981, Peripheral feedback mechanisms acting on the central pattern generators for locomotion in fish and cat, Can. J. Physiol. & Pharmacol., 59: 713-726.

Ayers, J. L., and Clarac, F., 1978, Neuromuscular strategies underlying different behavioural acts in a multifunctional crustacean leg joint, J. comp. Physiol., 128: 81-94.

Ayers, J. L., and Davis W. J., 1977. Neuronal control of locomotion in the lobster, Homarus americanus. I. Motor programs for forward and backward walking, J. comp. Physiol., 115: 1-27.

Barnes, W. J. P., 1977, Proprioceptive influences on motor output during walking in the crayfish, J. Physiol. Paris, 73: 543-564.

Barnes, W. J. P., and Gladden, M. H., 1985, Feedback and motor control in Invertebrates and Vertebrates, Crom Helm Ltd. London, 496 pp.

Bassler, V., 1986, Afferent control of walking movements in the stick insect Cuniculina impigra, II. Reflex reversal and the release of the swing phase in the restrained foreleg, J. comp. Physiol., 158: 351-362.

Bevengut, M., Simmers, A. J., and Clarac, F., 1983, Central neuronal projections and Neuromuscular organization of the basal region of the shore crab leg, J. Comp. Neurol., 221: 185-198.

Bush, B. M. H., 1981, Non-impulse stretch receptors in crustaceans, in: "Neurones without impulses", Roberts A. and Bush B. M. H., ed. , Cambridge University Press, London, 147-176.

Bush, B. M. H., and Clarac, F., 1985, Co-ordination of motor behaviour, SEB Seminar Series 24, Cambridge University Press, Cambridge, 325 pp.

Bush, B. M. H., and Laverack, M. S.,1982, Mechanoreception, in: "The Biology of Crustacea", Atwood H. & Sandeman D. L., eds. , Academic Press, New York, 3, pp. 397-467.

Chasserat, C., and Clarac, F., 1986, Basic processes of locomotor coordination in the Rock lobster, II. Simulation of leg coupling, Biol. Cybern., 55: 171-185.

Chrachri, A., and Clarac, F., 1987, Induction of rhythmic activity in motoneurons of crayfish thoracic ganglia by cholinergic agonists, Neurosc. lett., 77: 49-54.

Clarac, F., 1985, Stepping reflexes and the sensory control of walking in Crustacea, in: "Feedback and motor control in invertrbrates and vertebrates", Barnes W. J. P., and Gladden M. H., eds. , Croom Helm Ltd., London, 379-400.

Clarac, F., and Chasserat, C., 1986, Basic processes of locomotor coordination in the Rock lobster, I. Statical analysis of walking parameters, Biol. Cybern., 55: 159-170.

Clarac, F., and Cruse, H., 1982, Comparison of forces developed by leg of the Rock lobster when walking free on a treadmill, Biol. cybern., 43: 109-114.

Clarac, F., and Sillar, K. T., 1985, Coordination between thoracic oscillators for rhythmic limb movements in the crayfish Pacifastacus leniusculus, Soc. for Neurosc., abst. 11: 510.

Cruse, H., and Muller, U., 1986, Two coupling mechainsms which determine the coordination of ipsilateral legs in the walking crayfish, J. exp. Biol., 121: 349-369.

Cruse, H., and Clarac, F., Chasserat, C., 1983, The control of walking movements in the leg of the Rock lobster, Biol. cybern., 47: 87-94.

Dean, J., 1985, Control of leg protraction in the stick insect: a targeted movement showing compensation for externally applied forces, J. comp. Physiol. A, 155: 771-781.

Evoy, W. H., and Ayers, J. L., 1982, Locomotion and control of limb movements, in: "The Biology of Crustacea", Vol.4, Neural Integration and Behavior, Sanderman D. C. and Atwood H. L., eds., Academic Press, New York, 61-105.

Gnatzy, W., Schmidt, M., and Rombke, J., 1984, Are the funnel-canal organs the "campaniform sensilla" of the shore crab Carcinus maenas (Decapoda, Crustacea) ? I. Topography, external structure and main organization, Zoomorphology, 104: 11-20.

Grillner, S., 1985, Neurobiological bases of rhythmic motor acts in vertebrates, Science, 228: 143-149.

Hartman, B. H., 1985, Tension receptors on the closed muscle apodeme in the walking legs of the blue crab, Callinectes sapidus, J. comp. Physiol. A., 157: 355-362.

Herman, R. M., Grillner, S., Stein, P. S. G., and Stuart, D. G., 1976, Neural control of locomotion in Advances in Behavioral Biology, Plenum Press, New York and London, 822 pp.

Klarner, D., and Barnes, W. J. P., 1986, The cuticular stress detector (CSD2) of the crayfish, II. Activity during walking and influences on leg coordination, J. exp. Biol., 122: 161-175.

Libersat, F., Zill, S., and Clarac, F., 1987, Single unit responses and reflex effects of force sensitive mechanoreceptors of the dactyl of the crab, J. Neurophysiol., 57 (5), 1601-1617.

Libersat, F., Clarac, F., and Zill, S., 1987, Force sensitive mechanoreceptors of the dactyl of the crab: Single unit responses during walking and evaluation of function, J. Neurophysiol., 57 (5), 1618-1637.

Mill, P. J., 1976, Structure and function of proprioceptors in the invertebrates, Chapman & Hall, London, 686 pp.

Pearson, K. G., and Duysens, J., 1976, Function of segmental reflexes in the control of stepping in cockroaches and cats in neural control of locomotion, in: Herman R. M., Grillner S., Stein P. S. G., and Stuart D. G., eds., Plenum Press, New York and London, pp. 519-537.

Roberts, A., and Roberts, B. L., 1983, Neurol origin of rhythmic movements, SEB Symposium, Cambridge University Press, 37: 503 pp.

Sandeman, D. C., 1969, The site of synaptic activity and impulse initiation in an identified motoneuron in the crab brain, J. exp. Biol., 50: 771-784.

Selverston, A. I., 1985, Model neural networks and behavior, Plenum Press, New York and London, 548 pp.

Sillar, K. T., Skorupski, P., Elson, R. C., and Bush B. M. H., 1986, Two identified afferent neurones entrain a central locomotor rhythm generator, Nature, 323: 440-443.

Sillar, K. T., and Elson, R. C., 1986, Slow active potentials in walking leg motor neurones triggered by non-spiking proprioceptive afferents in the crayfish, J. exp. Biol., in press.

Skorupski, P., and Sillar, K. T., 1986, Phase dependent reversal of reflexes mediated by the thoracicocoxal muscle receptor organ in the crayfish Pacifastacus leniusculus, J. Neurophysiol., 55 (4), 689-695.

Wales, W., Clarac, F., Dando, M. R., and Laverack, M. S., 1970, Innervation of the receptors present at various joints of the pereipods and third maxilliped of Homarus gammarus (L.) and other macruran decapods (Crustacea), Z. vergh. Physiol., 68: 345-384.

Zill, S. N., Moran, D. T., and Varela, F. G., 1981, The exoskeleton and insect proprioception, II. Reflex effects of tibial campaniform sensilla in the american cockroach, periplaneta americana, J. exp. Biol., 94, 43-55.

NEURONAL MECHANISMS CONTROLLING RHYTHMIC MOVEMENTS IN GASTROPOD MOLLUSCS

Yu. I. Arshavsky, T. G. Deliagina, I. M. Gelfand, G. N. Orlovsky,
Yu. V. Panchin, G. A. Pavlova

Institute of Problems of Information Transmission
Academy of Sciences, Moscow
Moscow State University, Moscow, USSR

Summary

We investigated the neuronal mechanisms underlying generation of the locomotor rhythm in the pedal ganglia of the marine pteropod mollusc, **Clione limacina**, and that of the feeding rhythm in the buccal ganglia of the freshwater mollusc, **Planorbis corneus**. Using a new method for isolating the identified neurons, we demonstrate that in both generators the rhythm arises due to the endogenous periodic activity of two interneuron groups (groups 7 and 8 in the **Clione** pedal ganglia and groups 1 and 2 in the **Planorbis** buccal ganglia). The appropriate phasing of the generator interneurons in the cycle is established through interactions between the two groups. In **Clione**, group 7 and 8 interneurons inhibit each other and this determines their alternating activity in different phases of the locomotor cycle. In **Planorbis**, group 1 excites group 2 and group 2 inhibits group 1. This interaction ensures the transition from the protractor phase of the feeding cycle to the retractor one; once the excitation of the group 2 neurons has been completed, a new cycle is triggered by endogenous properties of the group 1 neurons. The mechanisms of rhythm generation in the mollusc nervous system are compared with those in the nervous system of some other animals.

Introduction

"Automatic" rhythmic movements (breathing, locomotion, mastication, scratch reflex and others) play an important role in animal and human life. The neuronal mechanisms controlling these movements have been extensively investigated in vertebrates (see Orlovsky, Shik, 1976, Shik, Orlovsky, 1976; Stein, 1978; Grillner, 1981; Feldman, Grillner, 1983; Roberts, Roberts, 1983; Arshavsky et al., 1986d). Nervertheless the question as to how they are organized at the cellular level is far from having been answered. The main difficulty arises from the fact that in vertebrates an enormous number of neurons is usually involved in the control of any motor act, so that the role of any single cell is negligible.

Studies on the cellular basis of motor activity have been more successful when dealing with invertebrates (in particular, the gastropod molluscs; see Kandel, 1976; Selverston, 1985), the nervous systems of which contain considerably fewer neurons. On the other hand, a number of the principles on which nervous control is based seem to be common to both vertebrate and invertebrate animals.

In the present study, we investigated the cellular mechanisms generating the locomotor rhythm in the marine pteropod mollusc **Clione limacina** and the feeding rhythm in the freshwater mollusc **Planorbis corneus**.

Results

Control of locomotion in **Clione**

Clione is widely distributed over the northern seas. It is a plankton animal, which maintains itself at a particular depth by making continuous rhythmic (1-5 Hz) wing movements (Fig. 1). These wing movements are controlled by the pedal ganglia (PED in Fig. 2) (Vagner, 1885; Sakharov, 1960; Satterlie et al., 1985). During swimming, two bursts per cycle can be observed in the wing nerve electroneurogram (ENG), corresponding to the wing movements in dorsal and ventral directions (D and V phases of the cycle) (Arshavsky et al., 1985a).

The locomotor rhythm in **Clione** (as in other animals, see Delcomyn, 1980; Grillner, 1981) is of central origin since it can be generated by the isolated pedal ganglia ("fictitious swimming"; see also Satterlie, 1985; Satterlie, Spencer, 1985). Each pedal ganglion contains about 400 neurons, about 50 of which exhibit periodical activity related to the swim rhythm generated in the isolated pedal ganglia (Arshavsky et al., 1985b). About 30 of these "rhythmic" neurons are motoneurons, that is, they send axons into the wing nerves (Fig. 3). Roughly half of the motoneurons are active in the D-phase of the cycle. These motoneurons were divided into group 1 (including the large 1A motoneuron which could be identified visually, Fig. 3B) and group 3. Other motoneurons are active in the V-phase. They were divided into groups 2, 4, 6 and 10. The neurons of group 2 include the large 2A neuron. Motoneurons of the same group (except for groups 6 and 10) are electrically coupled to each other (see Fig. 11). When motoneurons of a given phase are firing, the antagonistic ones are inhibited (Fig. 4).

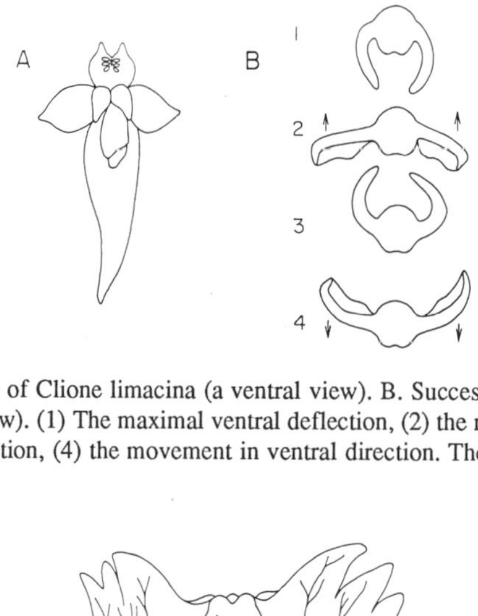

Fig. 1. A. Schematic drawing of Clione limacina (a ventral view). B. Successive wing positions during the locomotor cycle (a frontal view). (1) The maximal ventral deflection, (2) the movement in dorsal deflection; (3) the maximal dorsal deflection, (4) the movement in ventral direction. The intervals between frames are about 100 ms.

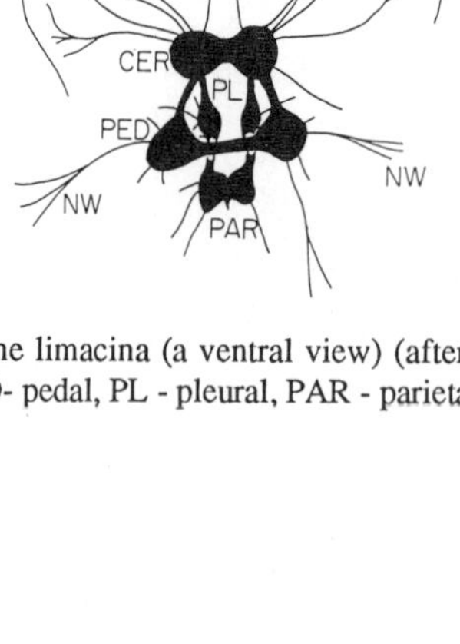

Fig. 2. The nervous system of Clione limacina (a ventral view) (afterVagner, 1885). The ganglia shown: BUC - buccal, CER - cerebral; PED- pedal, PL - pleural, PAR - parietal. The nerves supplying wings (NW) are also shown.

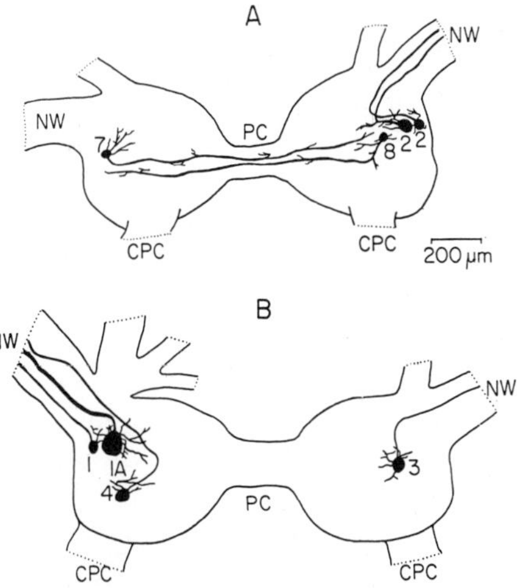

Fig. 3. Morphology of pedal neurons of different groups as revealed by intracellular staining with Lucifer yellow. A: The ventral view of pedal ganglia. B: The dorsal view. 1A - the 1A motoneuron; 1, 2, 3, 4-motoneurons of corresponding groups; 7, 8 - interneurons of groups 7 and 8 (see test). NW - nerve of the wing, PC - pedal commissure, CPC - cerebropedal connective.

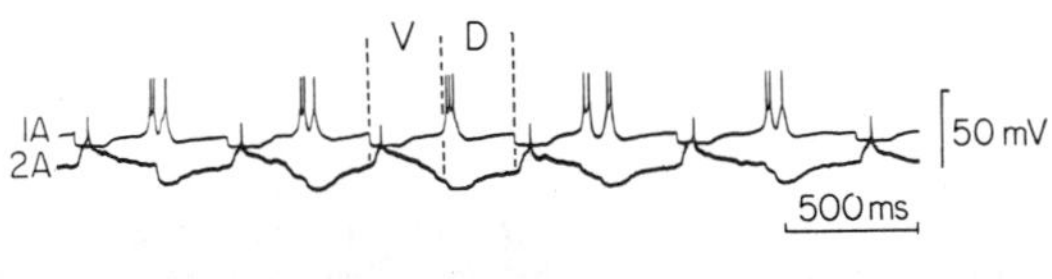

Fig. 4. Activity of the 1A and 2A motoneurons in the isolated pedal ganglia preparation during "fictitious swimming". D- and V-phase of the swim cycle are indicated.

Besides the motoneurons, the locomotor rhythm generator contains about 20 interneurons, that is, cells whose axons do not leave the pedal ganglia (Fig. 3). The majority of the interneurons generate one prolonged (50-150 ms) action potential per cycle (Fig. 5). These interneurons can be divided into two antagonistic groups, 7 and 8, firing in the D and V phases, respectively. With the excitation of one of the groups, an IPSP appears in the cells of the antagonistic groups (Fig. 5A). The interneurons of a given group (both from one ganglion and from different pedal ganglia) are electrically coupled to each other (Fig. 5C) as well as with some motoneurons which are active in the same phase (see fig. 11). All the group 7 interneurons generate the action potential simultaneously, as was demonstrated by recording pairs of these cells (Fig. 5B). On the other hand, group 8 interneurons can be divided into two subgroups ("early" neurons, 8e and "delayed" neurons, 8d); the 8d cells generate the action potential which a latency of 100-150 ms (Fig. 5D), which may be due to the fact that the 8d neurons have a higher threshold. The 8e interneurons are always active during "fictitious swimming". On the contrary, the 8d interneurons are active only during more intense "swimming" (when the swim rhythm is faster and the motoneuron discharges stronger).

Which neurons are responsible for locomotor rhythm generation? First of all, we wanted to find out whether these are motoneurons or interneurons. By using several approaches we found that the rhythm is generated by the interneurons. In experiments involving photoinactivation of the motoneurons, we demonstrated directly that the interneurons alone can generate the locomotor rhythm (Arshavsky et al., 1985e). In these experiments, the motoneurons were filled with Lucifer yellow through the transected wing nerves (for methods see Stewart, 1978, 1981). The nerve contains axons of all the motoneurons of the wing muscles, and the dye filled all (or at least the majority) of these cells (Fig. 6A). The ganglia were then illuminated with intense blue light which resulted in the inactivation of the stained cells (Selverston, Miller, 1980). Fig. 6B illustrates the interneuronal activity recorded after the inactivation of the motoneurons. One can see that the pattern of activity did not differ from that which is to be observed in a normal ganglion: the neuron periodically generated the prolonged action potential alternating with the IPSP (cf. Fig. 6B and Fig. 5).

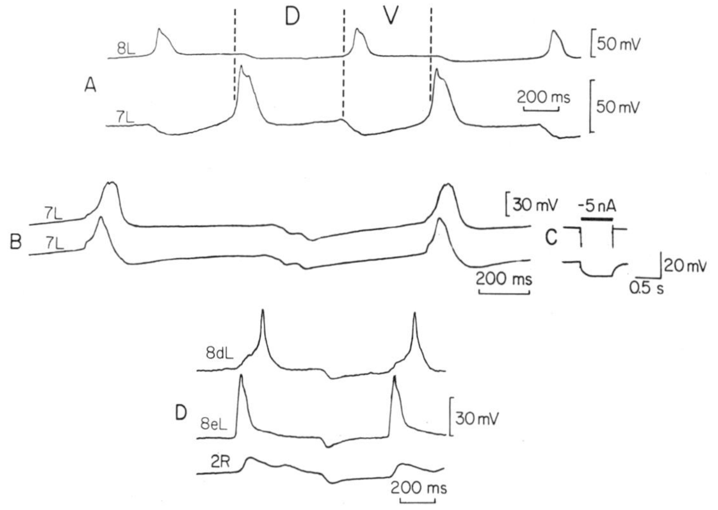

Fig. 5. Characterization of group 7 and 8 interneurons. A: Simultaneous recording of group 8 and 7 neurons. B: Synchronous activity of two group 7 neurons. Electrical connection between these neurons is demonstrated in C, where the negative pulse of current injected into one of the neurons evoked a negative deflection of the membrane potential in the other neuron. D: "Delayed" and "early" neurons of group 8 recorded with the group 2 motoneuron. In this and subsequent figures, the number near each trace indicates the group of a neurons as well as the ganglion (L, left; R, right) from which the neuron was recorded. The period of current injection is marked by a solid line, near which the strength and polarity of the current are indicated.

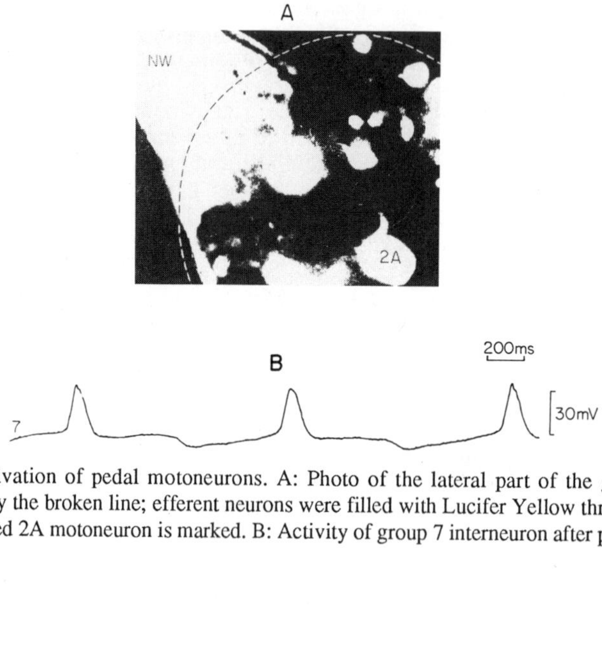

Fig. 6. Photoinactivation of pedal motoneurons. A: Photo of the lateral part of the ganglion. Ganglion contour is shown by the broken line; efferent neurons were filled with Lucifer Yellow through the wingnerve (NW). The identified 2A motoneuron is marked. B: Activity of group 7 interneuron after photo inactivation of motoneurons.

The interneurons of groups 7 and 8 are the main source of the rhythmic postsynaptic potentials in all the rhythmic neurons of the pedal ganglia (Arshavsky et al., 1985c). In Fig. 7, two antagonistic motoneurons and a group 7 interneuron were recorded simultaneously. Injection of a depolarizing current into one of the motoneurons (3L in Fig. 7A), in the absence of "swimming", evoked a series of spikes, but these spikes were not accompanied by any postsynaptic potentials in the antagonistic motoneuron (2R). Due to the electrical coupling between the 3L motoneuron and the group 7 interneuron, the latter generated an action potential. Only after this did a full-value IPSP occur in the 2R motoneuron (Fig. 7A). Direct depolarization of the group 7 interneuron resulted in its excitation, and both the EPSP in the 3L motoneuron and IPSP in the 2R motoneuron occurred immediately (Fig. 7B). Besides controlling the motoneurons, the group 7 and 8 interneurons inhibited each other, which determined their reciprocal, alternating activity.

Is the rhythmic activity a property of individual interneurons of groups 7 and 8, or do the rhythmic oscillation arise due to interaction within these groups of cells? To answer this question, we used a new method for isolating individual neurons by means of the same microelectrode as that used for intracellular recording (Dyakonova, 1985: Arshavsky et al., 1986c). With this method, it is possible to isolate both visually identifiable neurons and those which can be identified only after their activity has been recorded. After ganglia have been treated with proteolytic enzymes (pronase and trypsin), mechanical connections between neurons become weaker, and any neuron, together with its processes, can be extracted from a ganglion by moving the microelectrode in the horizontal plane (Fig. 8). The condition of isolated neurons can be proved by checking that their membrane and action potentials have remained unchanged for many hours. This neuron isolation method is a powerful means of studying the neuronal mechanisms of rhythm generation. During extraction of a cell, its synaptic connections with other cells are disrupted, and therefore, one can find out to what extent the behaviour of a cell (i.e., its periodic discharges) is determined by its own properties, and to what extent by interaction with other cells. This was precisely the method used to directly demonstrate the capacity of group 7 and 8 interneurons to endogenously generate periodic discharges (Arshavsky et al., 1986c). Fig. 9A shows the initial activity of two group 7 interneurons from the right and left pedal ganglia. One can see periodic discharges of neurons (one per cycle) and mid-cycle IPSPs. In the course of the neuron extraction (fig. 9B, the upper trace) the IPSPs diminished in amplitude and then entirely disappeared. The discharge frequency of the neuron increased in the course of the extraction due to cessation of the inhibitory input, and, probably, to some deformation of the neuron membrane. Within 20-40 s of extraction, the neuron discharge frequency decreased somewhat and then remained at a constant level for many hours (Fig. 9C). Each discharge of the neuron was preceded by a smoothly increasing depolarization (pacemaker potential). The same result was obtained with group 8 interneurons. The discharge frequency of isolated interneurons depends on the level of the membrane potential. By de- and hyperpolarizing a neuron, we could cover the whole range of the locomotor frequencies (0.5-5 Hz). These findings demonstrate that the locomotor rhythm arises from the endogenous periodic activity of individual interneurons.

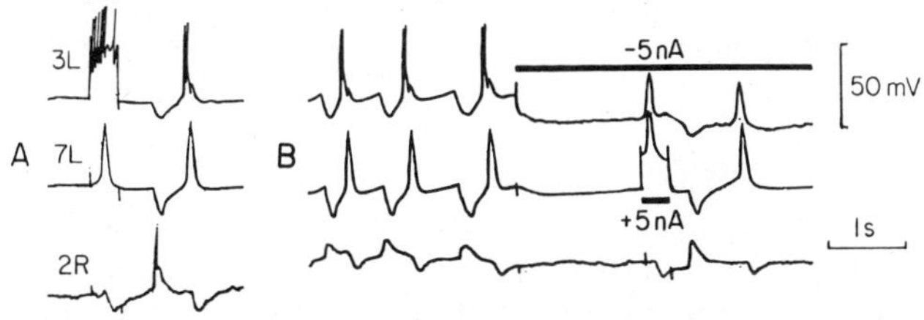

Fig. 7. Relationship between postsynaptic potentials in group 2 and 3 neurons and excitation of group 7 neurons. A: Excitation of the group 7 neurons was elicited indirectly, by depolarizing the type 3 motoneuron. B: Excitation of this neuron was elicited by its depolarization. In A, swimming was absent; in B it was suppressed by continuous hyperpolarization of the group 3 neuron.

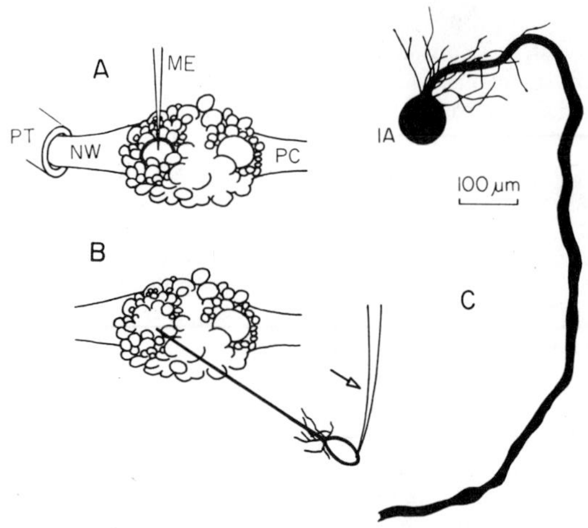

Fig. 8. Method of neuron isolation. A: Schematic drawing of a pedal ganglion (the second ganglion is not shown), NW - nerve of the wing sucked into a polyethylene tube (PT), PC - pedal commissure, ME - microelectrode. B: Extracting of a neuron from the ganglion by means of a microelectrode. C: The isolated 1A efferent neuron (drawing was made under a microscope after withdrawing the electrode from the neuron).

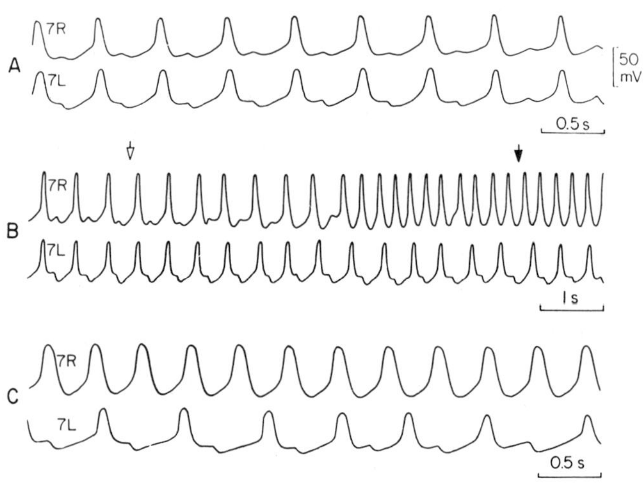

Fig. 9. Isolation of a group 7 interneuron. A: Initial activity of two group 7 neurons from the right and left pedal ganglia. B: Extraction of the neuron from the right ganglion. The white arrow marks the beginning of the extraction; the black arrow marks the moment of the total extraction of the neuron with its processes from the ganglion. C: Activity of the group 7 neuron within 10 minutes of its isolation. The lower trace is the activity of the group 7 neuron which remained in the left ganglion.

Besides groups 7 and 8, a further group of interneurons (12) was detected in the pedal ganglia (Arshavsky et al., 1985d). Group 12 interneurons are non-spiking cells generating long-lasting "plateau" potentials. They are involved in the rhythmic process only during intense "swimming" together with subgroup 8d interneurons (see above). Under the influence of subgroup 8d interneurons, group 12 neurons shift to the depolarized state; they remain in this state until the beginning of the D-phase, when they shift to the initial state under influence of group 7 interneurons (Fig. 10A). In the depolarized state, group 12

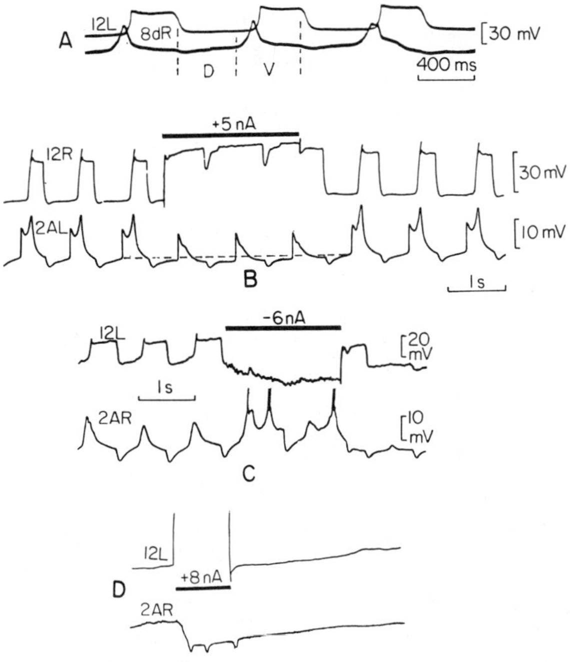

Fig. 10. Characterization of group 12 interneurons. A: Simultaneous recording of a group 12 neuron and a "delayed" group 8 neuron. B-D: Simultaneous recordings of a group 12 interneuron and a 2A motoneuron active in the V-phase of the cycle. B: Depolarization of a group 12 interneuron resulted in hyperpolarization and a decrease in the EPSP in the 2A motoneuron. C: Hyperpolarization of a group 12 interneuron resulted in an excitation of the 2A motoneuron. D: Depolarization of a group 12 interneuron, in the absence of swimming, resulted in both hyperpolarization of the 2A motoneuron and the appearance of a series of the IPSPs correlating with the activation of group 7 interneurons. All recordings are from different experiments.

neurons inhibit V-phase neurons (Fig. 10B, C) and excite D-phase neurons (Fig. 10D). Thus, they promote the transition from the V-phase of the cycle to the D-phase.

A scheme of the locomotor generator of **Clione** is shown in Fig. 11. The locomotor rhythm is determined by the endogenous periodic discharges of the group 7 and 8 interneurons. Neurons of these two groups inhibit each other, which determines their alternating activity. During intense "swimming", an additional group of interneurons (12), promoting the transition from one phase of the cycle to another, is involved. Group 7 and 8 interneurons control motoneurons: they excite motoneurons with the same phase and inhibit antagonistic ones.

Control of feeding movements in **Planorbis**

The freshwater snail **Planorbis corneus**, like many other gastropod molluscs, performs rhythmic movements with the radula and other parts of the buccal apparatus during feeding. The movements are controlled by buccal ganglia. Fig. 12A shows a preparation of the buccal mass with the buccal ganglia, capable of co-ordinated rhythmic movements (Fig. 12B). A cycle of radula movement consists of three phases: quiescence (Q), protraction (P) and retraction (R). In intact animals, the radula is brought into contact with a food substrate during the P-phase. Then, in the R-phase, the radula scratches food fragments and transports them to the oesophagus. Activity in the radular nerve (n1) corresponds mainly to the P-phase, and that in the dorsobuccal nerve (n2), to the R-phase. Similar activity is observed in these nerves in completely isolated buccal ganglia (Fig. 13). According to an approximate estimation, each buccal ganglion contains 500 neurons, about 10% of them being "rhythmic" ones, i.e. they exhibit periodic activity related to the feeding rhythm (like the neurons shown in Fig. 13). Depending on the pattern of electrical activity and the phase of this activity in the feeding cycle, all rhythmic neurons were divided into 7 groups. Five of these groups (3, 4, 5, 7, 9) comprised rather large neurons (30-60 μm in diameter). It was demonstrated with Lucifer yellow staining that they sent axons into peripheral nerves, i.e., these cells were efferent neurons (motoneurons). Neurons of two other groups (1 and 2) were small cells (15-20 μm in diameter). They were interneurons since, after intracellular staining, their axons were not found in peripheral nerves.

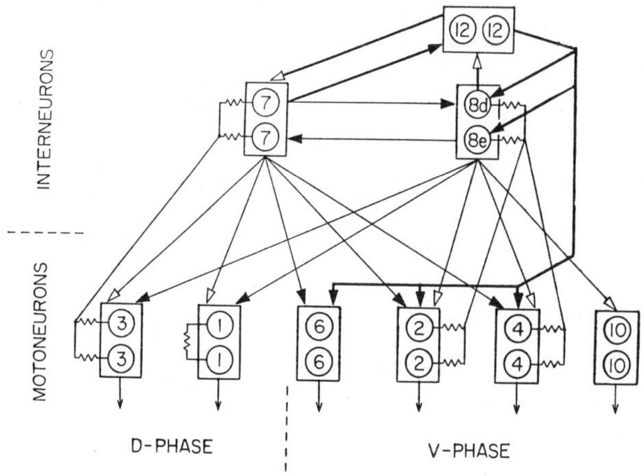

Fig. 11. Locomotor generator of Clione. Electrical connections between neurons are shown by resistor symbols, excitatory chemical connections by white arrows, inhibitory connections by black arrows. Figures in the circles are the neuron groups. Thin lines show those connections between neurons which are functioning during both "weak" and "intense" swimming; thick lines show the connection functioning during "intense" swimming only.

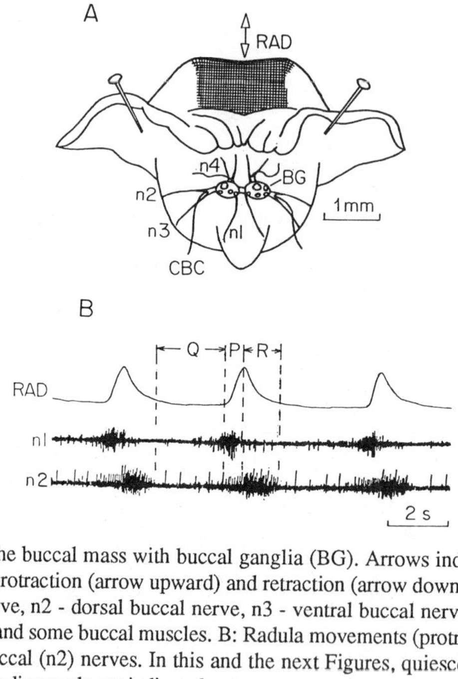

Fig. 12. A: Preparation of the buccal mass with buccal ganglia (BG). Arrows indicate the direction to radula (RAD) movements during protraction (arrow upward) and retraction (arrow downward). CBC - cerebrobuccal connective, n1 - radular nerve, n2 - dorsal buccal nerve, n3 - ventral buccal nerve, n4 - nerves supplying the oesophagus, salivary gland and some buccal muscles. B: Radula movements (protraction - upwards) and ENGs in radular (n1) and dorsobuccal (n2) nerves. In this and the next Figures, quiescence (Q), protractor (P) and retractor (R) phases of the feeding cycle are indicated.

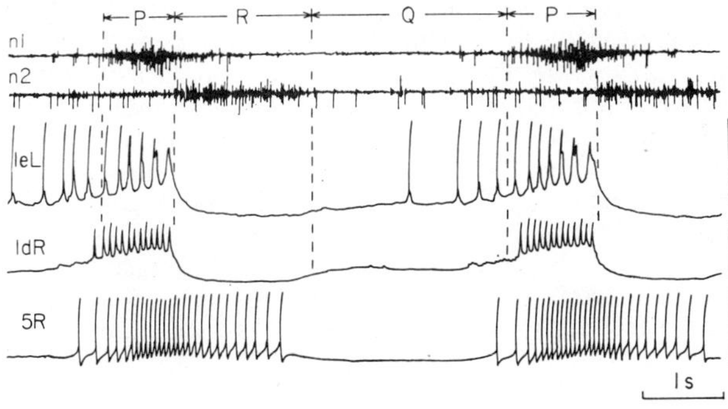

Fig. 13. Generation of the feeding rhythm in the isolated buccal ganglia. Simultaneous recording of subgroup 1e and 1d interneurons and a group 5 efferent neurons as well as ENGs in radular (n1) and dorsobuccal (n2) nerves.

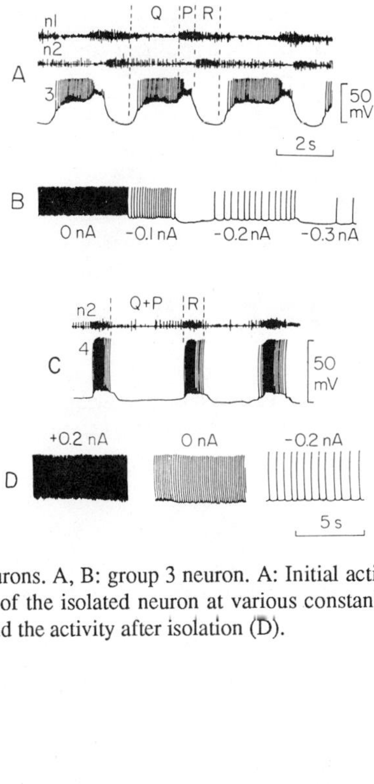

Fig. 14. Isolation of efferent neurons. A, B: group 3 neuron. A: Initial activity of the neuron during feeding rhythm generation, B: Activity of the isolated neuron at various constant current injected. C, D: Group 4 neuron; the initial activity (C) and the activity after isolation (D).

To ascertain which of the neuron groups were responsible for the rhythm generation, we used the same method as for studying the locomotor generator of **Clione**, i.e. we isolated individual cells (Arshavsky et al., 1986a). After being extracted from the buccal ganglion, the overwhelming majority of motoneurons did not generate any slow rhythm comparable with that of feeding. Fig. 14 shows the result of isolation of two motoneurons: one of group 3 active in the Q- and P-phase (A, B) and one of group 4 active in the R-phase (C, D). When isolated, both neurons generated spikes at a fairly constant frequency, the latter being determined by a polarizing current. But whatever the level of the membrane potential, they did not generate rhythmic bursts like those they generated before isolation. One can conclude that these motoneurons do not belong to the feeding rhythm generator; their periodical activity is not determined by their own properties but rather by the signals they receive from other cells.

A quite different result was obtained when group 1 and 2 neurons were isolated. Fig. 15 illustrates isolation of the group 1 neuron. Before isolation, this neuron was gradually depolarized throughout the Q- and P-phases and strongly hyperpolarized in the R-phase. In the course of depolarization, it generated spikes with an increasing frequency (see also Fig. 13). On being extracted from the ganglion, the neuron exhibited a similar rhythmic activity, i.e., a slow depolarization accompanied by spike discharge acceleration, followed by a rapid hyperpolarization (Fig. 15, B-D). The rhythm depended on the value of a polarizing current injected into the neurons; at a certain current we could obtain a similar pattern of rhythmic activity to the initial one. One can conclude that the rhythmic oscillations observed in the group 1 neurons are of endogenous origin. Therefore, these neurons can be said to be a part of the feeding rhythm generator rather than simply executors of driving commands.

Group 2 neurons were also found to possess the generator property. An example of the isolation of such a neuron is shown in Fig. 16. Before its extraction from the ganglion, the neuron was active in the R-phase of the feeding cycle. It generated M-shaped depolarization waves with spike-like oscillations on the wave top. A similar pattern of activity was observed in the isolated neuron (Fig. 16, B-D).

Thus, in the buccal ganglia of **Planorbis**, two groups of interneurons were found to exist which, on being isolated, were capable of generating a similar rhythmic activity to that they exhibited prior to isolation. The hypothesis that the endogenous cyclic activity of

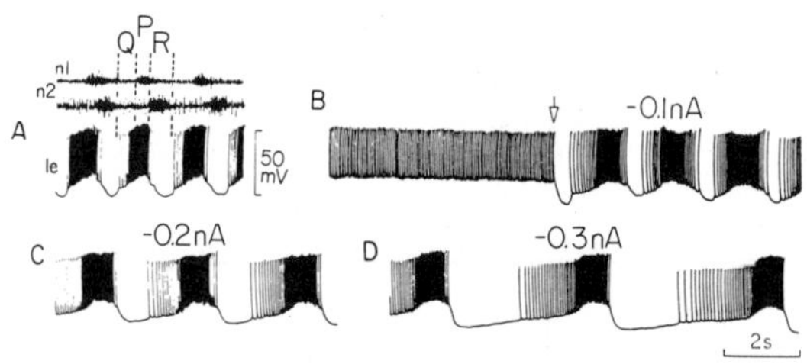

Fig. 15. Isolation of a group 1 neuron. A: Initial activity of the neuron during feeding rhythm generation. B-D: Activity of the isolated neuron at various constant polarizing current injected. In B, the arrow marks the switching-on of the current.

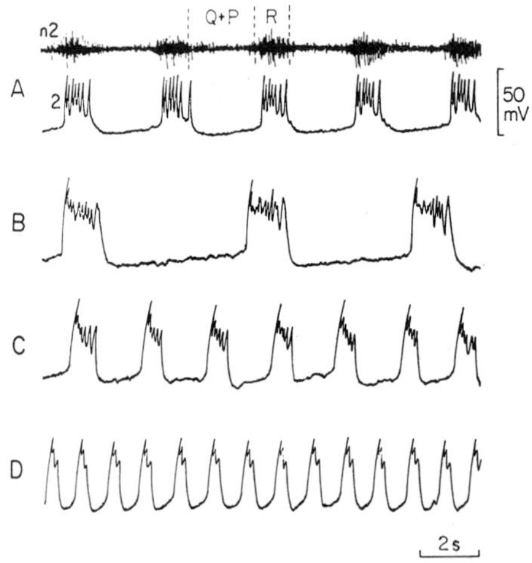

Fig. 16. Isolation of a group 2 neuron. A: Initial activity. B-D: Activity of the isolated neuron hyperpolarized by the constant current of 0.5 nA (B) and of 0.3 nA (C) and in the absence of polarization (D).

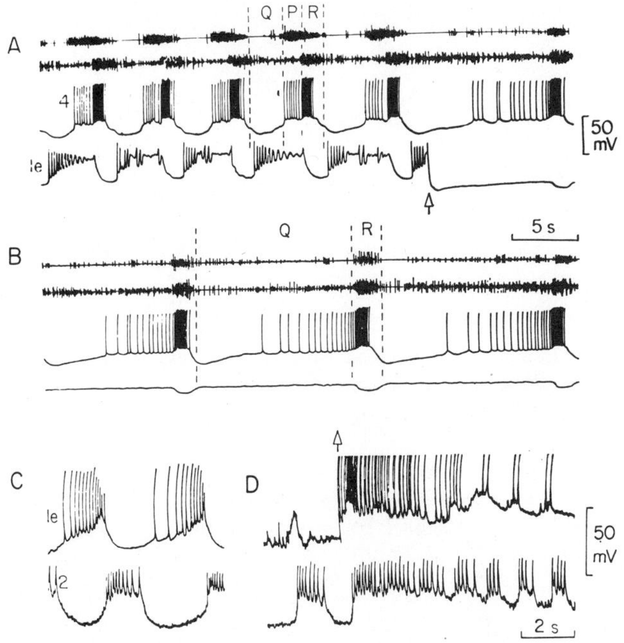

Fig. 17. Interaction of the generator neurons in buccal ganglia. A, B: Simultaneous recording of group 4 and subgroup 1e neurons (B is the continuation of A). The arrow marks the switching-on of the hyperpolarization current (0.5 nA). C, D: Simultaneous recording of subgroup 1e and group 2 neurons. C: Initial activity. Then the 1e neuron was hyperpolarized by a constant current (0.5 nA). D: Switching-off of the current evoked an intense excitation of the 1e neuron which resulted in an excitation of the group 2 neuron.

some buccal neurons may constitute the basis of the feeding rhythm generation in gastropods was first advanced by Merickel and Geay (1980) and by Rose and Benjamin (1981). Using the new method for isolating the identified neurons, we have now obtained the direct proof of this hypothesis.

Since the feeding rhythm generator contains two groups of neurons with endogenous periodic activity, a question arises concerning the co-ordination of these groups. We found that group 1 neurons excite the group 2 neurons . Fig. 17A shows the activity of group 1 and 4 neurons during generation of the feeding rhythm (the group 4 neuron monitors the activity of group 2 neurons with which it is electrically coupled, see Arshavsky et al., 1986a). There is an EPSP in the group 4 neuron (and therefore, in group 2 neurons) in the P-phase of the cycle. This EPSP is evoked by group 1 neurons. Injection of a strong hyperpolarizing current into a group 1 neuron led to inactivation of all the neurons in this group and to elimination of the P-phase (Fig. 17B). As a result, the P-phase EPSP in the group 4 neuron, which preceded the R-phase EPSP, disappeared, and the rhythm became slower. Under these conditions, the feeding rhythm is entirely determined by the endogenous activity of group 2 neurons. From the experiment illustrated in Fig 17A, B one can conclude that the excitatory action of group 1 neurons upon group 2 and 4 neurons results in their being excited earlier and in the rhythm being accelerated. By simultaneously recording the group 1 and 2 neurons we directly demonstrated this excitatory action. As shown in Fig. 17C, D, excitation of a group 1 neuron (postinhibitory rebound) evoked intense firing of the group 2 neuron.

As shown in Fig. 15, the isolated group 1 neuron, after reaching a high level of depolarization, spontaneously shifts to a hyperpolarized state. However, in the rhythm generating system of the buccal ganglia , group 1 neurons perform this transition under the influence of group 2 neurons, producing the IPSP in them. The IPSP can be easily observed

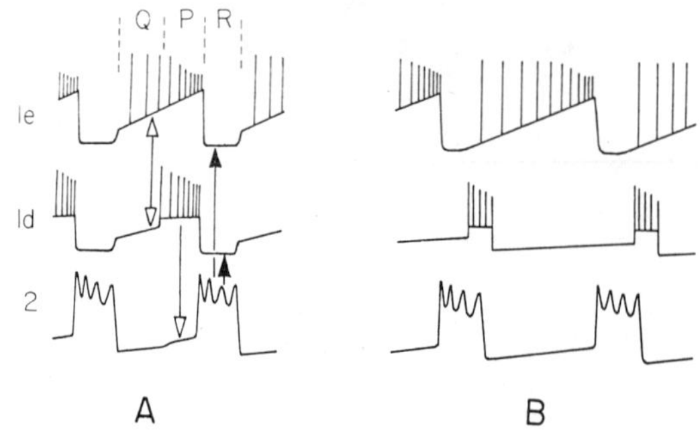

Fig. 18. A: A scheme illustrating activity in the main interneuron groups (1e, 1d and 2) during the feeding cycle. White arrows indicate excitatory influences and black arrows, inhibitory ones. B: A scheme illustrating activity of isolated interneurons of the same groups.

when group 1 is inactivated by hyperpolarizing one of the cells of this group (Fig. 17B). A similar pattern of phases, responsible for generation of the feeding rhythm in the buccal ganglia, has been found to exist in another freshwater snail, **Lymnaea** (Elliott, Benjamin, 1985).

The experimental findings described above led us to the following model of the feeding rhythm generator in **Planorbis** (Fig. 18).Group 1 cells possess the property of generating a slowly rising depolarization accompanied by an increasing spike activity. Depending on the moment of the spike discharge onset, the group 1 neurons can be divided into two subgroups: an early group (1e), those in which activity begins in the Q-phase and a late group (1d), those in which activity coincides completely with the P-phase (Fig. 13) (for details sees Arshavsky et al., 1986 a, b). The group 1 neurons (mainly 1d) produce an excitatory drive for the group 2 neurons. The latter, on being excited, generate an M-shaped depolarizing potential which determines the duration of the R-phase. The group 2 neurons inhibit the group 1 neurons and force their membrane to shift to the hyperpolarized state, which gives rise to a new cycle. Besides acting upon one another, the neurons of subgroup 1d and group 2 exert action upon corresponding motoneurons, and thus coordinate the activity of the buccal muscles.

Conclusion

The organization of the locomotor generator of **Clione** and that of the feeding rhythm generator of **Planorbis** have two features in common: (i) The rhythm of the whole system is generated by the endogenous rhythmic activity of a part of its neurons, and (ii) The appropriate phasing of different neuron groups within the cycle is established by interaction. The endogenous rhythmic activity of some of the neurons provides the basis for the activity of most of the rhythm generators studied up to date in invertebrate animals: in the pyloric generator of the lobster stomatogastric ganglion (Maynard, Selverston, 1975; Selverston, Miller, 1980; Miller, Selverston, 1982a, b); in the generators of the cardiac rhythm in the leech (Calabrese, 1979; Stent et al., 1979) and in crustaceans (Tasaki, Cooke, 1979a, b; Yamagishi, Ebara, 1985). We expect this list to be lengthened before long, thanks to the new cell isolation method.

The interactions between two endogenously oscillating neuron groups in the pedal ganglia of **Clione** and those in the buccal ganglia of **Planorbis** are different. In **Clione** (at least during weak swimming) the interaction is symmetrical: groups 7 and 8 inhibit each other. A similar organization was observed in the spinal mechanism controlling swimming in the **Xenopus** embryo (Khan, Roberts, 1982; Roberts, Kahn, 1982; Soffe et al., 1984; Dale, 1985): the left and right generators in each segment inhibit one another, which results in their reciprocal, alternating activity. On the contrary, in **Planorbis** the interaction between the groups 1 and 2, constituting the feeding rhythm generator, is asymmetrical: group 1 excites group 2 while group 2 inhibits group 1. Models of this kind, where the activity of the first neuronal population increases and then ceases due to the excitation of the second population of inhibitiory neurons, have been previously suggested to explain the rhythmic activity in various systems of vertebrate animals: in the respiratory medullar centre (Bradley et al., 1975 Cohen, Feldman, 1977; Euler, 1977, 1983; Wyman, 1977; Cohen, 1979); in the spinal centre controlling scratching and stepping movements (Berkinblit et al., 1978a, b), etc. In such models, it was suggested that the reason for the slowly rising activity in the first neuronal population might be the regenerative mutual excitation process. However, in the present study we have demonstrated that the slowly rising activity may be rhythmically generated in individual cells due to a special property of the membrane. It is possible that a similar mechanism may also be responsible for rhythm generation in vertebrates. This suggestion is supported by the discovery in brain slices of endogenous rhythmic activity in some cells of the medullar respiratory centre, (Dekin et al., 1985).

References

Arshavsky, Yu. I., Beloozerova, I. N., Orlovsky, G. N., Panchin, Yu. V., and Pavlova, G. A., 1985a, Control of locomotion in marine mollusc Clione limacina. I, Efferent activity during actual and fictitious swimming, Exp. Brain Res., 58: 255-262.

Arshavsky, Yu. I., Beloozerova, I. N., Orlovsky, G. N., Panchin, Yu. V., and Pavlova, G. A., 1985b, Control of locomotion in marine mollusc Clione limacina. II, Rhythmic neurons of pedal ganglia, Exp. Brain Res., 58: 263-272.

Arshavsky, Yu. I., Beloozerova, I. N., Orlovsky, G. N., Panchin, Yu. V., and Pavlova, G. A., 1985c, Control of locomotion in marine mollusc Clione limacina. III, On the origin of locomotory rhythm, Exp. Brain Res., 58: 273-284.

Arshavsky, Yu.I ., Beloozerova, I. N., Orlovsky, G. N., Panchin, Yu. V.,and Pavlova, G. A., 1985d, Control of locomotion in marine mollusc Clione limacina, Role of type 12 interneurons, Exp. Brain Res., 58: 285-293.

Arshavsky, Yu. I., Orlovsky, G. N., and Panchin, Yu. V., 1985e, Control of locomotion in marine mollusc Clione limacina. V, Photoinactivation of efferent neurons, Exp. Brain Res., 59: 203-205.

Arshavsky, Yu. I., Deliagina, T. G., Orlovsky, G. N., and Panchin, Yu. V., 1986a, Endogenous periodic activity of neurons of the feeding rhythm generator in the freshwater snail, Planorbis corneus. Proc. Acad. Sci. of USSR, 291: 473-475, in Russian.

Arshavsky, Yu. I., Deliagina, T. G., Orlovsky, G. N., and Panchin, Yu. V., 1986b, Organization of the feeding rhythm generator in the freshwater snail, Planorbis corneus, Proc. Acad. Sci. of USSR, 291: 728-731, in Russian.

Arshavsky, Yu. I., Deliagina, T. G., Orlovsky, G. N., Panchin, Yu. V., Pavlova, G. A., and Popova L. B., 1986c, Control of locomotion in marine mollusc Clione limacina. VI, Activity of isolated neurons of pedal ganglia, Exp. Brain Res., 63: 106-112.

Arshavsky, Yu. I., Gelfand, I. M., and Orlovsky, G. N.,1986d, Cerebellum and Rhythmical Movements, Springer-Verlag, Berlin

Berkinblit, M. B., Deliagina, T. G., Feldman, A. G., Gelfand,I. M., and Orlovsky, G. N., 1978a, Generation of scratching. I., Activity of spinal interneurons during scratching, J. Neurophysiol., 41: 1040-1057.

Berkinblit, M. B., Deliagina, T. G., Feldman, A. G., Gelfand, I. M., and Orlovsky, G. N., 1978b, Generation of scratching. II. Nonregular regimes of generation, J. Neurophysiol., 41: 1058-1069.

Bradley, G. W., Euler, C. von., Marttila, I., and Roos, B., 1975, A model of the central and reflex inhibition of inspiration in the cat, Biol. Cybern., 19: 105-116.

Calabrese, R. L., 1979, The roles of endogenous membrane properties and synaptic interaction in generating the heartbeat rhythm of the leech, Hirudo medicinalis, J. Exp. Biol., 82: 163-176.

Cohen, M. I., 1979, Neurogenesis of respiratory rhythm in the mammal, Physiol. Rev., 59: 1105-1173.

Cohen, M. I., and Feldman, J. L., 1977, Models of respiratory phase-switching Fed. Proc., 36: 236-2374.

Dale, N., 1985, Reciprocal inhibitory interneurons in the Xenopus embryo spinal cord, J. Physiol., 363: 61-70.

Dekin, M. S., Richerson, G. B., and Getting, P. A., 1985, Thyrotropin-releasing hormone induces rhythmic bursting in neurons of the nucleus tractus solitarius, Science, 229: 67-69.
Delcomyn, F., 1980, Neural basis of rhythmic behavior in animals, Science, 210: 492-498.
Dyakonova, T. L., 1985, Two types of neurons differing by plastic properties: study of ionic mechanisms, Pavlov J. Higher Nerv. Activity, 35: 552-560, in Russian.
Elliott, C. J. H., and Benjamin, P. R., 1985, Interactions of pattern-generating interneurons controlling feeding in Lymnaea stagnalis, J. Neurophysiol., 54: 1396-1411.
Euler, C. von, 1977, The functional organization of the respiratory phase switching mechanisms, Fed. Proc., 36: 2375-2380.
Euler, C. Von, 1983, On the origin and pattern control of breathing rhythmicity in mammals, in: Roberts A., Roberts B., eds. , "Neural origin of rhythmic movements", Symp. Soc. Exp. Biol., pp. 469-485.
Feldman, J. L., and Grillner, S., 1983, Control of vertebrate respiration and locomotion: a brief account, Physiologist, 26: 310-316.
Grillner, S., 1981, Control of locomotion in bipeds, tetrapods, and fish, in: "Handbook of Physiology", The Nervous System, v.2, Motor Control, Ed. by V. Brooks, Baltimore: Warverly Press, p. 1179-1236.
Kahn, J. A., and Roberts A., 1982, Experiments on the central pattern generator for swimming in amphibian embryos, Phil.Trans. R. Soc., 296B: 229-243.
Kandel, E. R., 1976, The Cellular Basis of Behaviour, Freeman a. Co., San Francisco.
Maynard, D. M., and Selverston, A. I., 1975, Organization of the stomatogastric ganglion of the spiny lobster, IV. The pyloric system, J. Comp. Physiol., 100: 161-182.
Merickel, M., and Gray, R., 1980, Investigation of burst generation by the electrically coupled cyberchron network in the snail Helisoma using a single-electrode voltage clamp, J. Neurobiol., 11: 73-102.
Miller, J. P., and Selverston, A. I., 1982a, Mechanisms underlying pattern generation in lobster stomatogastric ganglion as determined by selective inactivation of identified neurons, II Oscillatory properties of pyloric neurons, J. Neurophysiol., 48: 1378-1391.
Miller, J. P., and Selverston, A. I., 1982b, Mechanisms underlying pattern generation in lobster stomatogastric ganglion as determined by selective inactivation of identified neurons, IV. Network properties of pyloric system, J.Neurophysiol., 48: 1416-1432.
Orlovsky, G. N., and Shik, M. L., 1976, Control of locomotion: a neurophysiological analysis of the cat locomotor system. Internat, Rev. Physiol., Neurophysiol, 110: 281-317.
Roberts, A., and Kahn, J. A., 1982, Intracellular recordings from spinal neurons during "swimming" in paralysed amphibian embryos, Phil. Trans. Roy. Soc., 296B: 213-228.
Roberts, A., Roberts, B. L., eds., 1983, Neural origin of rhythmic movements, Cambridge Univ. Press.
Rose, R. M., and Benjamin, P. R., 1981, Interneuronal control of feeding in the pond snail Lymnaea stagnalis, II. The interneuronal mechanism generating feeding cycles, J. Exp. Biol., 92: 203-228.
Sakharov, D. A., 1960, On rhythmic activity of pedal ganglia in the pteropodial mollusc Clione limacina, Biol. Sci., 3: 60-62 , in Russian.
Satterlie, R.A., 1985, Reciprocal inhibition and postinhibitory rebound produce reverberation in a locomotor pattern generator, Science, 229: 402-404.
Satterlie, R. A., LaBarbera, M., and Spencer, A. N., 1985, Swimming in the pteropod mollusc, Clione limacina. I. Behaviour and morphology, J. Exp. Biol., 116: 189-204.
Satterlie, R. A., and Spencer, A. N., 1985, Swimming in the pteropod mollusc, Clione limacina. II, Physiology. J. Exp. Biol., 116: 205-222.
Selverston, A. I., ed., 1985, Model neural networks and behavior, Plenum Press, New York.
Selverston, A. I., and Miller, J. P., 1980, Mechnisms underlying pattern generation in lobster stomatogastric ganglion as determined by selective inactivation of identified neurons, I. Pyloric system, J. Neurophysiol., 44: 1102-1121.

Shik, M. L., and Orlovsky, G. N., 1976, Neurophysiology of locomotor automation Physiol. Revs., 56: 465-501.
Soffe, S. R., Clarke, J. D. W., and Roberts, A., 1984, Activity of commissural interneurons in spinal cord of Xenopus embryos, J. Neurophysiol., 51: 1257-1267.
Stein, P. G., 1978, Motor systems, with specific reference to the control of locomotion, Ann. Rev. Neurosci., 1: 61-81.
Stent, G. S., Thompson, W. J., and Calabrese, R. L., 1979, Neural control of heartbeat in the leech and in some other invertbrates, Physiol. Rev., 59: 101-136.
Stewart, W. W., 1978, Functional connection between cells as revealed by dye-coupling with a highly fluorescent naphthalimide tracer, Cell, 14: 741-759.
Stewart, W. W., 1981, Lucifer dyes-highly fluorescent dyes for biological tracing, Nature, 292: 17-21.
Tazaki, K., and Cooke, I. M., 1979a, Spontaneous electrical activity and interaction of large and small cells in cardiac ganglion of the crab, Portunos sanguinolentus, J. Neurophysiol., 42: 975-999.
Tazaki, K., and Cooke, I. M., 1979b, Isolation and characterization of slow depolarizing responses of cardiac ganglion neurons in the crab, Portunos sanguinolentus, J. Neurophysiol., 42: 1000-1021.
Vagner, N. P.,1885, Invertebrates in the White Sea, in Russian.
Wyman, R. J., 1977, Neural generation of the breathing rhythm, Rev. Physiol., 39: 417-448.
Yamagishi, H., and Ebara, A.,1985, Spontaneous activity and pacemaker property of neurons in the cardiac ganglion of an isopod, Ligia exotica, Comp. Biochem. Physiol., 81A: 55-62.

ON THE BULBOSPINAL LOCOMOTOR COLUMN IN THE CAT

O.V. Kazennikov, V.A. Selionov, and M.L. Shik

Institute of Information Transmission Problems
Academy of Sciences of the USSR, Moscow

Summary

Repetitive (30 to 60 pps) stimutation of the appropriate fibers at the medullary or spinal level elicits stepping. At both levels, there exist neurons which give synaptic responses even to single stimuli applied to the locomotor or stepping point. Most responses in the proximity of this stimulation site occur within 5-10 ms. Less than half of these responses are monosynaptic. This suggests the existence of strong excitatory interactions between neurons. Nevertheless the polysynaptic propagation of the activity evoked by a single stimulus was found to show a steep decrement and a significant temporal dispersion (up to 10 and even 30 ms). Further propagation of this activity is due to conduction along the fiber bundle.

Fibers of the stepping strip in the dorsolateral funiculus excite the stepping generator through the propriospinal neurons sending their axons into the ventrolateral funiculus.

Stepping can be elicited both by the direct bulbospinal pathway and via propriospinal neurons.

*
* *

The cat is not capable of walking spontaneously when the forebrain and diencephalon have been removed. However this mesencephalic preparation can walk or run on the moving belt of a treadmill when certain points of the brain stem are stimulated. One of these points was termed the mesencephalic locomotor region. It is located in the lateral tegmentum of the caudal midbrain. The other low threshold point was located in the lateral tegmentum of the caudal medulla (Fig. 1A) (Shik and Yagodnitsyn, 1977).

Repetitive (30 to 60 pps) electrical stimulation of either the medullary or the mesencephalic locomotor point can elicit pure, well co-ordinated locomotion. Stimulation along the narrow strip between these two points (pontobulbar locomotor strip) can also induce locomotion, although often with various concomitant side effects (Mori et al., 1977). Stimulation of points in the locomotor region of the brain stem on one side (left or right) is followed by stepping of both the ipsi- and contralateral limbs (Shik and Yagodnitsyn 1977).

Stimulation of the appropriate fasciculus in the dorsolateral funiculus of the spinal cord (Fig. 2) also elicits stepping, however in this case only the ipsilateral hind limb is involved (Lennard and Stein, 1977; Stein, 1978; Yamaguchi, 1981). This fasciculus was termed the stepping strip. Left and right strips must be stimulated simultaneously to elicit coordinated (reciprocal) stepping in both hind limbs (the threshold usually decreases 1.5-2.0 times when stimulation is bilateral).

Stimulation of the locomotor region of the brain stem may activate the spinal stepping generators through both the direct bulbospinal pathway and the propriospinal system. The role of the direct pathway originating in the medial reticular formation and located in the ventrolateral funiculus is well known (Eidelberg, 1981; Orlovsky, 1970;

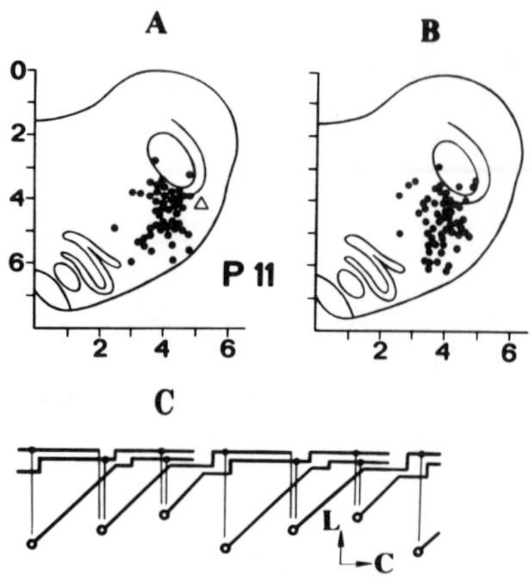

Fig.1. Location of neurons situated 1 to 4 mm rostral to the obex which responded synaptically to stimulation of the medullary locomotor point (triangle) and were invaded antidromically by stimuli delivered to this point (A) or to the locomotor point caudal (4 mm on average) to the obex (B). In this set of experiments currents of up to 30 μA were used. Projection onto frontal section of the medulla at the level 3 mm rostral to the obex. Twelve neurons were invaded antidromically from both rostral and caudal locomotor points. Intracellular recordings. Numbers near the coordinates are in mm. C - A scheme of arrangement of the medullary locomotor strip and column.

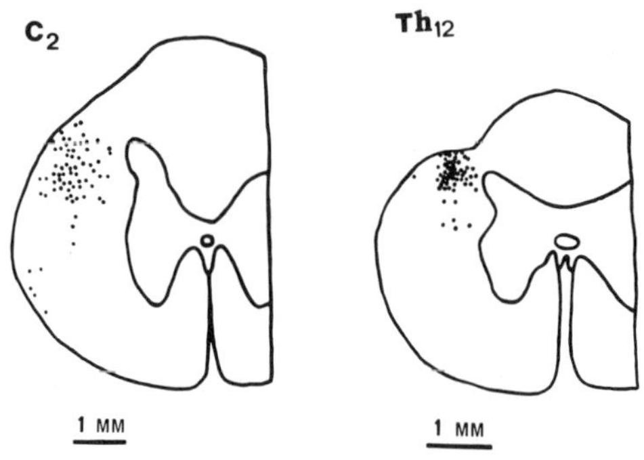

Fig. 2. Location of stepping points at levels C2 and Th 12.

Shimamura et al., 1982; Steeves and Jordan, 1980; Feldman et al. 1985). Therefore we propose to consider here only the experimental data relating to the role of the propriospinal system.

In lower vertebrates, swimming can be elicited after local lesion of the gray matter of the spinal cord (Williams et al., 1984) or when transmission in chemical synapses is interrupted by an extensive division of the spinal cord (McClellan, 1984; Rovainen, 1985). This means that synaptic retransformation of the signal is not necessary and that direct transmission is sufficient at least in animals such as the stingray or lamprey. Nevertheless, propriospinal neurons maintain longitudinal (Grillner, 1974; Rovainen, 1985) and bilateral (Clarke and Roberts, 1984; Soffe and Roberts, 1982) co-ordination during swimming in lower vertebrates. The role of these neurons in the control of mammalian locomotion is not yet understood. One can assume that they form the longitudinal column, and activation of this column is essential for locomotor effect. This might explain why single stimuli delivered to the locomotor point elicit no motor effects in the cat, and why repetitive stimulation is necessary to evoke stepping. Jerks can be produced by stimulation of the direct pathway whereas stepping is dependent on the intervention of the propriospinal neurons.

Neurons which give synaptic responses to microstimulation of the locomotor region of the brain stem or the spinal stepping strip in the cat, are to be found in the medulla and are distributed along the spinal cord as well. In particular, such neurons exist at the upper cervical and low thoracic levels while the generators of stepping movements in the limbs are situated in other divisions of the spinal cord, namely in the cervical and lumbar enlargements, in the case the fore- and hindlimbs, respectively. In this paper we suggest that at least some of these neurons form the bulbospinal column and participate in locomotion.

1. Microstimulation of the medullary locomotor strip evokes synaptic responses in neurons of both the lateral and medial (2 mm from the sagittal plane was accepted as the limit) reticular formation (Fig. 3, Fig. 4). In both cases, not only mono-, but also di-, oligo- and polysynaptic responses occur (Mori et al., 1986; Selionov and Shik, 1981; Selionov and Shik, 1984). Less than half of all these responses are monosynaptic ones: the medians of the unidimensional distribution of latencies of responses with the firing index ≥0.5 were 2.2 ms

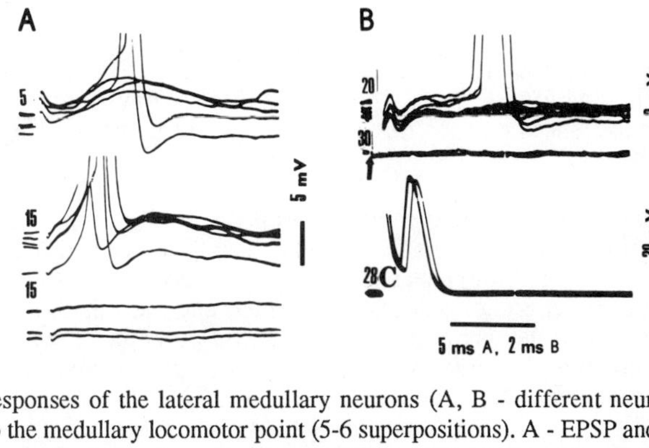

Fig. 3. Synaptic responses of the lateral medullary neurons (A, B - different neurons) evoked by single stimuli delivered to the medullary locomotor point (5-6 superpositions). A - EPSP and AP of synaptic origin, B - the orthodromic response (the top curve) to stimulation of the locomotor point rostral to the obex and antidromic response to stimulation of the locomotor point caudal to the obex. Numbers on the left designate current, μA.

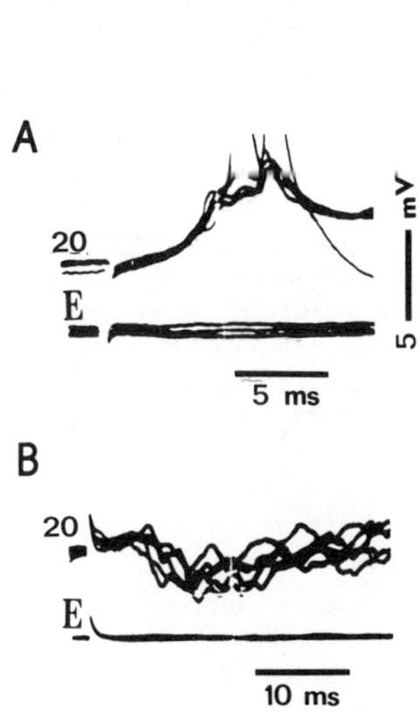

Fig. 4. Synaptic responses (A and B - different neurons) to stimulation of the medullary locomotor point. A - EPSP and AP of synaptic origin in the medial medullary neuron, B - IPSP in the lateral medullary neuron. Numbers on the left designate current, μA.

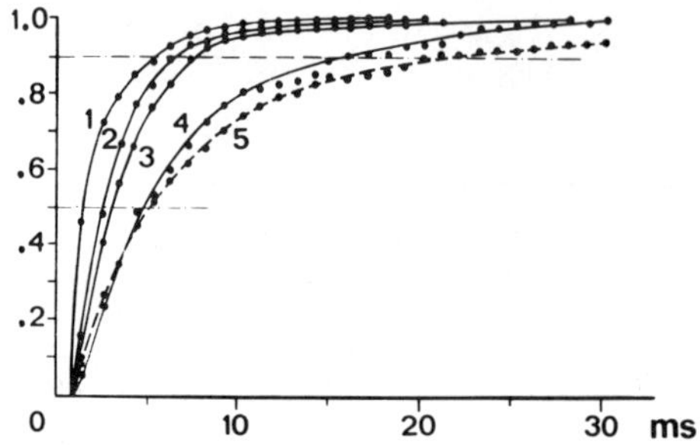

Fig. 5. Normalized unidimensional distributions of the latencies of synaptic responses with firing index ≥ 0.5 evoked by single or paired stimuli (interstimulus interval 2-3 ms), current ≤ 20 μA. Abscissa - time in ms, ordinate - fraction of responses. Curves 1-4 - extracellular recordings, curve 5 - intracellular recordings. Curves 1 and 5: stimuli were applied to the medullary locomotor point, recordings were made from the medullary neurons, N_1= 111, N_5=127, mean distance between stimulation and recording points x_1=x_5= 2 mm. Curves 2 and 3 - recordings from neurons at level Th 13, stimulation of the stepping point in the dorsolateral funiculus rostral to the recording point, x_2= 8 mm, x_3= 24 mm, N_2= 284, N_3= 172. Curve 4 - stimulation of the medullary locomotor point, recording from neurons at the level C2, x_4= 27 mm, N_4= 106.

with neurons studied by extracellular recording (Fig. 5, curve 1) and even reached 5.9 ms with the population investigated by intracellular recording (Fig. 5 curve 5) .The long latency of the latter responses may be due to the sample of neurons being modified by the intracellular recording.

Even shortlasting reverberating activity occurring after a single stimulus means that excitatory interaction between neurons is relatively strong. Later activity disappears in the vicinity of the site of stimulation : perhaps longlasting reverberation is impossible since the number of interconnected excitatory neurons is limited and their thresholds have increased during some milliseconds after generation of the action potential. Polysynaptic excitatory postsynaptic potentials may become subthreshold due to progressive temporal dispersion after each successive synaptic transmission.

Finally, the involvement of inhibitory neurons may lead to cessation of the reverberation (Fig. 4B), (Selionov and Shik, 1984.) In any case, although the activity near the stimulation site (the medullary locomotor point) ceases, it propagates (with decrement) polysynaptically both to the medial reticular formation and to the neurons of the upper cervical segments.

2. Most of the lateral reticular neurons (contrary to the medial ones) are propriomedullary (Brodal, 1957). Only axons of those which are located caudally to the obex can reach the first three upper cervical segments when they terminate in Rexed laminae VII to IX (Feldman, et al., 1985). (More rostral reticulospinal neurons project only to lamina IX caudally to C3). At level C2-C3 neurons are to be found which can be driven synaptically by stimulating the medullary locomotor strip (Kazennikov et al., 1979). (It is possible that some of these neurons belong to the group of propriospinal neurons located at the C3-C4 level which control the forelimb (Lundberg, 1979). However, their firing index is usually lower, and the mean latency and its fluctuation larger (often more than 2 ms) than near the site of stimulation.

In the vicinity of the stimulation site, 90% of responses occur within 5 ms of the stimulus whereas at the C2-C3 level they have a latency of 17 ms (Fig. 5, curves 1 and 4), although the mode of the histogram of latencies shifts only 2 ms. It is noteworthy that many of the C2 neurons can be excited by stimulation of both the ipsi-and contralateral medullary locomotor strips, as well as from the midbrain locomotor region (Kazennikov and Shik, 1983).

We assume that the activity propagates mainly polysynaptically from the medullary locomotor point to the C2-C3 level. In fact, the distribution of latencies (4) in Fig. 5 shifts significantly, to the right in relation to distribution (3) , although the distance between the stimulation and recording sites is nearly the same.

3. We have used the term medullary "locomotor column" to designate those lateral medullary neurons which gave synaptic responses to single (or paired) stimuli applied to the medullary locomotor strip. About a third of these neurons send their axons into the locomotor strip, mainly in the caudal direction (Fig. 1A, B), (Selionov and Shik, 1984). The length of these axons averages 4 to 9 mm, and they join the strip 1-3 mm caudally to the parent cell body. Thus, the medullary locomotor strip includes axons mediating interactions between columnar neurons (Fig 1 C). (However, according to Garcia-Rill , (Garcia-Rill, 1986; Garcia-Rill et al., 1983), the locomotor strip corresponds to Probst's tract, i.e. to fibers originating in primary afferent neurons of the mesencephalic nucleus of the trigeminal nerve). These neurons are intermingled with many other reticular cells which are not responsive to stimulation of the locomotor strip. On the other hand, the strip may contain longer axons originating from cells which do not belong to the locomotor column.

4. When the frequency of stimulation of the medullary locomotor point was increased from 1-4 pps to 20-40 pps, those C2-C3 neurons which responded with a firing index of 0.1-0.3 increased their response to 0.4-0.6 (Fig. 6), (Kazennikov and Shik, 1983). This facilitation can be essential to trigger locomotion. Propagation of the activity seems to be more successful when repetitive stimulation is delivered. The decrease in activity along the polysynaptic way from the medulla to the upper cervical segments may underlie the locomotion threshold phenomenon.

Frequency potentiation is not restricted to the locomotor column alone. A similar potentiation can be observed in the synaptic responses (including both excitatory and inhibitory postsynaptic potentials) of medial medullary neurons to stimulation of the mid-sagittal pontine site which suppresses evoked locomotion (Mori et al., 1986).

Even single stimuli applied to the medullary locomotor strip are thus polysynaptically propagated along the medulla, and reach the medial reticular formation and the upper cervical segments. This process becomes more effective when repetitive stimulation is employed.

5. In the next set of experiments, three stepping points 7-10 mm apart were found to exist in the same dorsolateral funiculus at low thoracic level (Fig. 7). The middle point was then destroyed (Kazennikov et al., 1983a). Thereafter the caudal point usually became ineffective whereas stimulation of the rostral one elicited stepping of the hindlimb as before (Fig. 7B). Hence, the fibers of this strip cannot activate the stepping generator directly. What then is the relay?

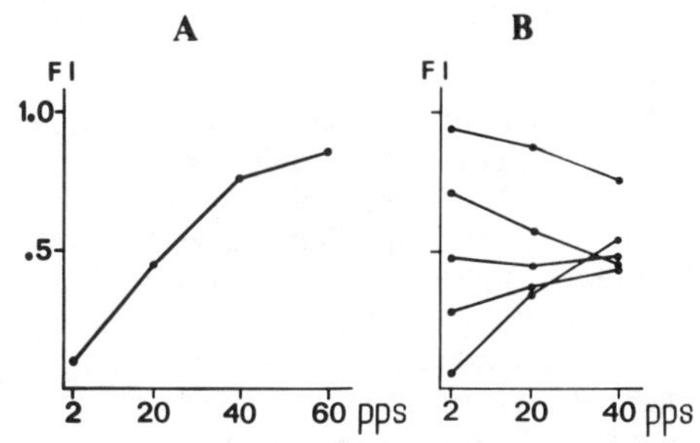

Fig. 6. Relationship between the firing index (ordinate) of synaptic responses of neurons at level C2 and frequency of stimulation (abscissa) of the medullary locomotor point. A - Responses of a neurons, current 20 μA. B - Responses of 95 neurons, distributed into 5 groups on the basis of firing indexes when frequency of stimulation was 2 pps.

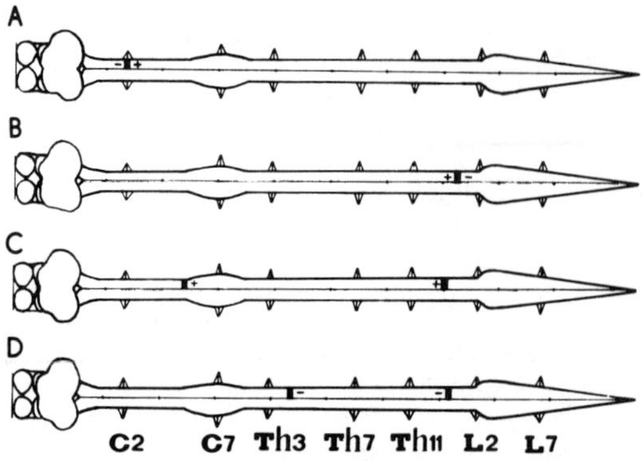

Fig. 7. Scheme of experiments after of the stepping strip in the dorsolateral funiculus. A black square indicates the site of coagulation, + a stepping point which remained effective after the coagulation, - a point which became ineffective.

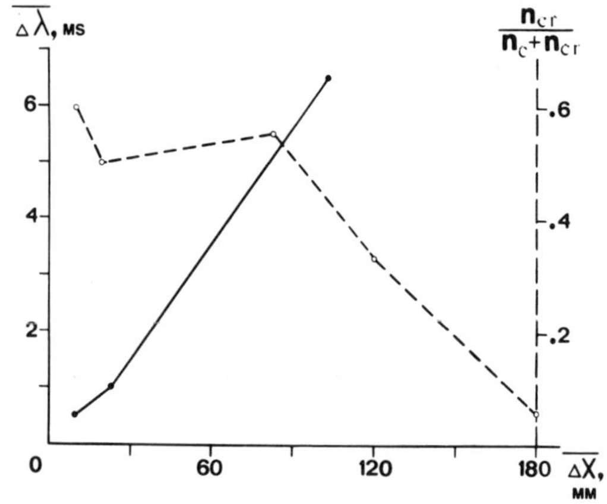

Fig. 8. Dependence of the occurrence of synaptic responses with firing indexes ≥ 0.5 (broken line, N=104) and of the difference between latencies of responses (unbroken line, n=21, the ordinate on the left) to stimulation of the far and near stepping points, both being rostral to the recording site at the level Th 13, on the distance between these points (abscissa, mm). Ordinate on the right - ratio of number of neurons responsive to stimulation of both points and sum of neurons which responded to stimulation of the closer point and neurons which responded to stimulation of both points. Single or double stimuli up to 20 μA were employed. The mean distance between the near stepping point and the recording site was 8 mm, the average latency of responses to stimulation of this point was 3 ms. Extracellular recordings.

Lesion of the middle point in the similar triplet at the upper cervical level abolished stepping when the rostral point was stimulated while stimulation of the caudal point elicited stepping as previously (Fig. 7A). This means that the brain stem is not used as a relay. Propriospinal neurons were the last candidate. Indeed, we found in the same stepping strip a point at the upper cervical level, another point at the low thoracic level and a third one between them. When both the rostral and caudal points were destroyed, the middle one was effective as before (Fig. 7 C). However, when in order to decrease the distance between the extreme points, we took the rostral one at the upper thoracic level and then, after lesion of both boundary points, stimulated the middle one, stepping was no longer elicited (Fig 7 D). When the length of the isolated part of the stepping strip was too short (as for example in the last experiment) stepping did not occur, possibly because the number of excited propriospinal neurons was too small. The regenerative process operating in the column did not propagate across the borders of the intact interval of the strip. At least, experiments B and

C suggest that axons of postulated propriospinal neurons were not situated in the dorsolateral funiculus.

It has not yet been established where the fibers of the stepping strip in the dorsolateral funiculus arise. It is even possible that in different parts of the strip, these fibers may originate from different types of cells.

6. We recorded synaptic responses of neurons in the low thoracic division to single or double stimuli applied 5 to 10 mm apart to the stepping strip. Half of these neurons were excited from a distance of 30 mm and a third responded even when the stimulation point was 120 mm from the recording site (Fig. 8), (Kazennikov et al., 1985). The responses were mainly oligo- and polysynaptic, although monosynaptic ones also sometimes occurred. The mean velocity of propagation of a signal along the strip was about 20 m/sec.

More than a third of the neurons which give synaptic responses to stimulation of the stepping strip send their axons into the ventrolateral funiculus (but not into the dorsolateral one), mainly in the caudal direction (Kazennikov et al., 1985). Those axons which are more than 6-8 mm long are not less than 30-40 mm long. The characteristic length of the fibers of the stepping strip in the dorsolateral funiculus is probably even greater. What role is played by the propriospinal neurons which give synaptic responses to stimulation of the stepping strip and send their axons into the ventrolateral funiculus? Are they similar to the medial medullary reticulospinal neurons, or rather to the lateral propriomedullary neurons?

7. What is the role of the spinal stepping strip and column when locomotion is evoked by stimulation of the brain stem ?. The strip is not necessary, since brain stem stimulation elicits locomotion even after bilateral lesion of the strip (Kazennikov et al., 1983a; Yamaguchi, 1986). On the other hand, the neurons distributed along the spinal axis which give synaptic responses to single stimuli delivered to the stepping strip (these neurons amount to only a small fraction of all the spinal neurons in the cat) may perhaps be excited when the locomotor region of the brain stem is stimulated. In particular, half of the neurons of the low thoracic division excited by stimulation of the stepping strip were also responsive to a short train of stimuli applied to the locomotor region of the brain stem (Kazennilov et al., 1985).

Excitation of columnar neurons may contribute to the local activation of generators of cyclic activity of both trunk and limb muscles or participate in propagating activity along the column. It is also possible that responsive propriospinal neurons may be distributed along the neural axis which, through their long axons (running in the ventral quadrant), directly activate the stepping generator. Lastly, they may mediate interactions between adjacent segments.

Repetitive stimulation of appropriate fasciculi connected with medullary and spinal divisions of the locomotor column elicits stepping. Those axons of the medullary strip which originate from lateral medullary neurons are relatively short, whereas the fibers in the spinal strip are long. The axons of the spinal strip originate from (unknown) noncolumnary neurons, whereas the axons of the spinal column neurons are located in the ventrolateral funiculus and do not join the stepping strip (stimulation of the strip does not evoke antidromic responses in these neurons). It is not known how activity is transmitted from the medullary division of the locomotor column to the spinal one. It is noteworthy that neurons of the upper cervical segments which give synaptic responses to stimulation of the stepping strip are usually not excitable from the locomotor region of the brain stem.

8. In both divisions of the locomotor column, activity does not cease abruptly in the proximity of the site of stimulation. The duration of evoked activity and the dimensions of the region where synaptic responses occur depend on the stimulation frequency. Both the duration and area of sustained activity increase, while the current necessary to elicit stepping diminishes when the stimulation frequency is increased. Nevertheless, at least in the spinal cord, synaptic retransmission ceases at a distance of about 8-10 mm from the site of stimulation. Any further propagation is due to conduction along fibers. Indeed, distribution (2) and (3) in Fig. 5 almost coincide, although the mean distances between stimulation and recording sites are 8 and 24 mm, respectively.

One can assume that the relative weight of long connections and of local excitatory interaction between columnar neurons are adapted to two different types of movements. In fact, during voluntary movements only certain muscles are active whereas during synergies (in particular, locomotor ones), many different muscles (and their motoneurons and premotoneurons) are appropriately linked.

The steep temporal and spatial decrease in the propagation of activity evoked by a single stimulus may contribute to the control of separate muscles (or even muscle portions) or suitable muscle groups using the neural axis as a continuous sheet.

References

Brodal, A., 1957, The Reticular Formation of the Brain Stem, Oliver and Boyd, Edinburg.

Clarke, J. D. W., and Roberts, A., 1984, Interneurons in the Xenopus embryo spinal cord: sensory excitation and activity during swimming, J. Physiol., 354: 345-362.

Eidelberg, E.,1981, Consequences of spinal cord lesions upon motor function, with special reference to locomotor activity, Progr. Neurobiol., 17: 185-202.

Feldman, J. L., Loewy, A. D., and Speck, D. F.,1985, Projections from the ventral respiratory group to phrenic and intercostal motoneurons in cat: an autoradiographic study, J. Neurosci., 5: 1993-2000.

Garcia-Rill, E., 1986, The basal ganglia and the locomotor regions, Brain Res. Rev., 11: 47-63.

Garcia-Rill, E., Skinner, R. D., and Gilmore, S. A., 1983, Owings R; Connections of the mesencephalic locomotor region (MLR). II. Afferents and efferents, Brain Res. Bull., 10: 63-71.

Grillner S. 1974. On the generation of locomotion in the spinal dogfish. Exp. Brain Res., 20: 159-170.

Kazennikov, O. V., Selionov, V. A., Shik, M. L., and Yakovleva, G. V., 1979, Neurons of upper cervical segments responding to stimulation of the bulbar "locomotor strip", Neurophysiology (Kiev), 11: 245-253.

Kazennikov, O. V., Shik, M. L., and Yakovleva, G. V., 1983a, Stepping movements elicited by stimulation of the dorsolateral funiculus in the cat spinal cord, Bull. Exper. Biol. Med. (Moscow), 96: 8-110.

Kazennikov, O. V., Shik, M. L., and Yakovleva, G. V., 1983b, Responses of neurons of upper cervical segments in cat to stimulation of brain stem locomotor region with different frequences, Neurophysiology (Kiev), 15: 355-361.

Kazennikov, O. V., Shik, M. L., and Yakovleva, G. V. 1985, Synaptic responses of propriospinal neurons to stimulation of the stepping strip in the cat dorsolateral funiculus, Neurophysiology (Kiev), 17: 270-278.

Lennard, P. R., and Stein, P. S. G., 1977, Swimming movements elicited by electrical stimulation of turtle spinal cord. I . Low-spinal and intact preparation, J. Neurophysiol., 40: 768-778.

Lundberg, A., 1979, Integration in a propriospinal motor center controlling the forelimb in the cat. in: Integration in the Nervous System, A Symposium in Honor of D.P.C. Lloyd and R. Lorente de No, H. Asanuma, V.J. Wilson, eds. , 47-65, Igaku-Shoin, Tokyo.

McClellan, A. D., 1984, Descending control and sensory gating of "fictive" swimming and turning responses elicited in an in vitro preparation of the lamprey brain stem/spinal cord, Brain Res., 302: 151-162.

Mori, S., Selionov, V. A., and Shik, M. L., 1986, Responses of medullary neurons to stimulation of the locomotor and inhibitory points of the lower brain stem, Neurophysiology (Kiev), 18: 525-533.

Mori, S., Shik, M. L., and Yagodnitsyn, A. S., 1977, Role of pontine tegmentum for locomotor control in mesencephalic cat, J. Neurophysiol., 40: 284-295.

Orlovsky, G. N., 1970, Work of the reticulspinal neurons during locomotion, Biophysics, 15: 728-734.

Rovainen, C. M., 1985, Effects of groups of propriospinal interneurons on fictive swimming in the isolated spinal cord of the lamprey, J. Neurophysiol., 54: 959-977.

Selionov, V. A., and Shik, M. L., 1981, Responses of medullary neurons to microstimulation of the "locomotor strip" in cat, Neurophysiology (Kiev), 13: 275-282.

Selionov, V. A., and Shik, M. L., 1984, Medullary locomotor strip and column in the cat, Neurosci., 13: 1267-1278.

Shik, M. L., and Yagodnitsyn, A. S., 1977, The pontobulbar "locomotor strip", Neurophysiology (Kiev), 9: 95-97.

Shimamura, M., Kogure, I., and Wada, S. -I., 1982, Reticular neuron activities associated with locomotion in thalamic cats, Brain Res., 231: 51-62.

Soffe, S. R., and Roberts, A., 1982, Tonic and phasic synaptic input to spinal cord motoneurons during fictive locomotion in frog embryos, J. Neurophysiol., 48: 1279-1288.

Steeves, J. D., and Jordan, L. M., 1980, Localization of a descending pathway in the spinal cord which is necessary for controlled treadmill locomotion, Neurosci. Lett., 20: 283-288.

Stein, P. S. G., 1978, Swimming movements elicited by electrical stimulation of the turtle spinal cord: the high spinal preparation. J. Comp. Physiol., 124: 203-210.

Williams, B. J., and Livingston, C. A., and Leonard, R. B., 1984, Spinal cord pathways involved in initiation of swimming in the stingray, Dasyatis sabina: spinal cord stimulation and lesions, J. Neurophysiol., 51: 578-591.

Yamaguchi, T., 1981, Fictive stepping evoked by electrical stimulation of the white matter of the cervical cord in decerebrate cats, J. Physiol. Soc., Japan, 43: p. 303, Abstr. N 108.

Yamaguchi, T., 1986, Descending pathway eliciting forelimb stepping in the lateral funiculus: experimental studies with stimulation and lesion of the cervical cord in decerebrate cats, Brain Res., 379: 125-136.

ANALYSIS OF THE PATTERN OF ACTIVITY IN "KNEE FLEXOR" MOTONEURONS DURING LOCOMOTION IN THE CAT

C. Perret, J.-M. Cabelguen and D. Orsal

Département de Neurophysiologie comparée
Institut des Neurosciences
Université Paris 6, 9 quai Saint-Bernard
75230 Paris, Cedex 05, France

Summary

During fictive locomotion of the thalamic cat, the pattern of membrane potential variations in motoneurons of the posterior biceps-semitendinosus muscle is complex. Periods of chloride dependent inhibition, alternating with excitatory periods, were identified during the second part of the flexor phase and in the mid-extensor phase. These excitatory and inhibitory influences were shown to be selectively controlled by tonic afferent inputs from the ipsilateral hindlimb. The results are discussed in relation with the organization of the central pattern generators for hindlimb and forelimb locomotor movements.

Introduction

The study of the pattern of efferent activity in a limb during locomotion is an approach to the organization of the central pattern generator for locomotion (CPG), to the interactions between this CPG and peripheral influences, and to the relationship between the activations of the various muscle groups and the mechanical characteristics of the movement. It is also a means of testing the operation of neuronal networks shown to exist in earlier anatomical and electrophysiological investigations.

An electromyographic analysis in the intact cat led Engberg and Lundberg (1969; see also Lundberg, 1981) to hypothesize that the complex patterns of activity observed in some muscles, e.g. in ankle and knee flexors, during locomotion may result from modifications in the simple output of a network with a half-center organization by phasic reflexes. In further EMG studies performed during locomotor activity in the intact animal (Rasmussen et al., 1978; Forssberg et al., 1980; see also Wentink, 1976 in the dog), in the thalamic cat (Perret and Cabelguen, 1976, 1980; Perret, 1983), in the mesencephalic cat (Gambarian et at., 1969; Grillner and Zangger, 1975, 1984) and in the spinal cat (Edgerton et al., 1976; Forssberg et al., 1980), a complex activity pattern was usually found in knee flexor muscles, especially in the posterior biceps-semitendinosus (PBSt). During the step cycle, this muscle displayed one burst at the beginning of the flexor phase and another one, which was larger during fast locomotion, at the beginning of the extensor phase. In some cases, a single burst was observed but it occurred during either the flexor or the extensor phase. A flexor-like activation could occur in the intact cat (Engberg and Lundberg, 1969), in the thalamic cat (Perret and Cabelguen, 1976) and in the spinal cat (Forssberg and Grillner, 1973; Edgerton et al., 1976; Grillner and Zangger, 1979). Conversely, an extensor-like activity was observed either during fast locomotion (Gambarian et al., 1969), or after deafferentation of the limb in the thalamic (Perret and Cabelguen, 1976), mesencephalic (Grillner and Zangger, 1975,1984) or spinal cat (Perret and Cabelguen, 1976; see also Grillner and Zangger, 1979). Similar complex or variable patterns were observed in the PBSt nerve during fictive locomotion in the thalamic cat (Perret and Cabelguen, 1976, 1980; Perret 1983, 1986; Orsal et al., 1986) and in the acute spinal cat (Grillner and Zangger, 1979), demonstrating their central origin.

The aim of this paper is to briefly present our recent results concerning the respective participation of excitatory and inhibitory synaptic inputs in the central locomotor drive of PBSt motoneurons. These will be discussed in connection with data obtained at the forelimb level in the case of muscles (shoulder flexors) which display an efferent pattern similar to the PBST one. Hypotheses about the organization of the CPG network for locomotion will be proposed.

Results

Our experiments were performed on thalamic paralyzed cats (for methods, see Orsal et al., 1986). Locomotor-like activities were recorded from various hindlimb muscle nerves ("fictive locomotion", see Perret, 1983). They occurred either spontaneously or during electrical stimulation of the subthalamic region. Intracellular d.c. recordings were carried out using glass micropipettes, filled with 3 mol/l potassium chloride, connected to an electronic device allowing simultaneous injection of chloride ions, which was done by passing a 10-30 nA hyperpolarizing d.c. current through the microelectrode for 1 to 5 minutes. The recorded membrane potential variations will be described in relation to the phases of the fictive locomotor cycle defined by the alternating nerve bursts in hip flexor (sartorius) and knee extensor (vasti) muscle nerves. The efferent activity was also recorded from the nerve of the rectus femoris, a hip flexor-knee extensor muscle, antagonist of the PBSt (knee flexor-hip extensor). One should note that the nerve branches supplying a given muscle (such as sartorius) or a given muscle group (such as posterior biceps and semitendinosus) exhibit the same efferent pattern during fictive locomotion in the thalamic cat.

During fictive locomotion, all impaled PBSt alpha-motoneurons, without exception, displayed membrane potential oscillations (range 5 to 15 mv peak to peak) with a locomotor periodicity. An example is illustrated in Fig. 1. Two depolarizing waves could be seen within each fictive locomotor cycle (Fig. 1 A) : the first one peaked at the beginning of the flexor phase and was related to the flexor-like burst of the PBSt nerve; the other one peaked at the

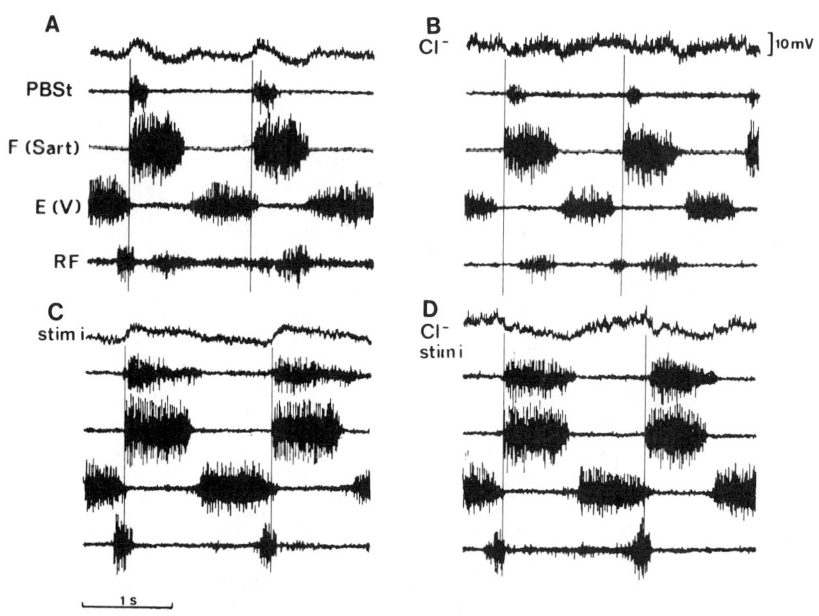

Fig. 1. Intracellular d.c. recording on a posterior biceps-semitendinosus motoneuron during fictive locomotion in the thalamic cat before (A, C) and after (B, D) intracellular chloride injection (2 mn, 30 nA), before (A, B) or during (B, D) slight tonic pinching of the ipsilateral toes (stim i). Lower traces are electroneurograms of posterior biceps-semitendinosus (PBSt), sartorius (Sart), vasti (V), rectus femoris (RF). F: flexor; E: extensor. Locomotor activity was induced by stimulation of the subthalamic locomotor region.

beginning of the vasti nerve burst and was related to the extensor-like burst of the posterior biceps-semitendinosus nerve when present (see Fig. 3 A). The amplitude of these depolarizations increased with the amplitude of the related PBSt nerve burst. Some of the recorded motoneurons fired action potentials during each of these depolarizing waves. The membrane potential reached maximum polarities during the last part of the flexor and extensor phases, i.e. with the same timing as the bursts of activity in the rectus femoris nerve when present (see Fig. 3 A). All the impaled PBSt motoneurons had similar complex membrane potential oscillations corresponding to components occurring at a precise time during the cycle, and the changes consisted only of changes in the amplitude of these components but not in their timing.

It was impossible to relate the membrane potential variations to excitatory or inhibitory synaptic events, mainly because the level of the resting potential was unknown. However, in an attempt to assess the relative contribution of excitatory and inhibitory synaptic inputs to the periodic changes in motoneuron membrane potential, since postsynaptic inhibition in motoneurons involves chloride ions (Coombs et al., 1955), we observed the effects of chloride ion injection in PBSt motoneurons during fictive locomotion. It induced changes in the membrane potential variations (Fig. 1 B): the repolarizing waves first flattened and were then progressively replaced by depolarizing waves with an opposite time course. If the injection lasted long enough, the new depolarizations could be so large that the original ones were no longer visible. In that case, the variations in the membrane potential were opposite to the original ones. Thus, the PBSt motoneurons could show two new depolarizing waves within each fictive step cycle : one during the extensor phase with a maximum at its end, the other during the second part of the flexor phase, instead of the two maximum polarities previously observed at this point in the fictive locomotor cycle. Changes in the amplitude of these additional depolarizing waves were associated with changes in the amplitude of the simultaneous antagonistic nerve burst and more generally, to spontaneous or induced (see below) changes in the fictive locomotor pattern.

Several observations strongly suggest that intracellular injections of hyperpolarizing current reduce and reverse the hyperpolarizing phases of the membrane potential by increasing the intracellular activity of chloride ions and that chloride-dependent inhibitory potentials take place during the interburst interval. (1) These effects were observed only with KCl microelectrodes. (2) Currents up to 30 nA had no immediate effect, but within 2-5 minutes led to a progressive flattening and reversal of the repolarizing waves. (3) These waves were associated with intense synaptic noise. (4) The observed effects persisted for several minutes after current injection was turned off. The original pattern of membrane potential variations was then progressively re-established. (5) The additional depolarizing waves had a shape which closely mirrored the original repolarizing waves.

The existence of a phasic inhibitory drive from the CPG to motoneurons during the fictive step cycle has also been demonstrated using intravenous administration of strychnine, which can remove the interburst hyperpolarization of flexor and extensor motoneurons without abolishing their rhythmic discharge (see Jordan, 1983). This result also indicates that the depolarizations responsible for bursting activity correspond to true excitatory postsynaptic influences. During intracellular chloride injection, repolarizing waves were reversed but, when they were not too large, the original depolarizing events could remain visible. This difference between the effects of current injection on repolarizing and depolarizing waves provide further indirect support for the existence of phasic excitatory synaptic inputs to motoneurons alternating with inhibitory ones during the fictive step cycle.

The decrease in the depolarizations observed during the chloride injection could be explained by the summation of the depolarizing waves with larger and overlapping reversed hyperpolarizing ones. This overlap between excitatory and inhibitory potentials may explain why no significant changes in the membrane input resistance were observed between the depolarized and hyperpolarized phases of the step cycle (unpublished results; see also Jordan, 1983; Shefchyk and Jordan, 1985).

The spontaneous variations in the PBSt efferent activity could be reproduced by changing the level of tonic afferent inflow. Deafferentation of the ipsilateral hindlimb as well as tonic exteroceptive stimulations applied to any part of the body except this limb usually

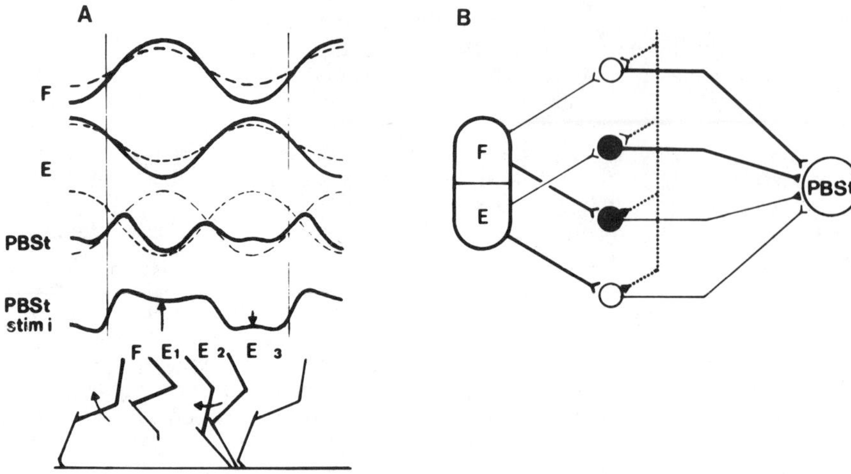

Fig. 2. Interpretation of the PBSt motoneuron activity during locomotion. A: Schematic diagram of the potential variations in flexor (F), extensor (E) and PBSt motoneurons before (PBSt) or during (PBSt stim i) ipsilateral tonic exteroceptive stimulation. The dashed lines indicate the assumed basic flexor and extensor drives. Thick lines in F, E and PBST stim i represent the membrane potential variations during the afferent stimulation. Arrows in PBST stim i underline the main changes (compare with the variations before stimulation: PBSt). Below, stick diagram of four phases of the step cycle; arrows indicate the direction of the leg movement in the phases where main PBSt activations occur. B: Diagram of the pathways from the rhythmic part of the locomotor CPG (F E) to a PBSt motoneuron (PBSt). Forked terminals are excitatory. Black triangles: inhibitory terminals. Dotted lines: ipsilateral exteroceptive inputs.

decreased the PBSt activity during the flexor phase and increased that during the extensor phase (Perret and Cabelguen, 1976, 1980; Perret, 1983). On the contrary, tonic ipsilateral afferent stimulation, e.g. slight pinching of the ipsilateral toes (Fig. 3 B), increased or brought about a flexor-like activity in the PBSt nerve and decreased or suppressed the activity during the extensor phase. Opposite changes were observed in the rectus femorus nerve. Correspondingly, in PBSt motoneurons a depolarizing wave of the membrane potential occurred during the whole flexor phase, due to a reduction of the hyperpolarization previously present in the late part of this phase (Fig. 1 C). After intracellular chloride injection (Fig. 1 D), the amplitude of the reversed hyperpolarizing wave occurring during the flexor phase was decreased by the stimulation and the amplitude of the one occurring during the extensor phase was increased.

Discussion

Several authors have shown that the periodic oscillations of the membrane potential in flexor and extensor hindlimb motoneurons during fictive locomotion in thalamic cats (Perret, 1983; Orsal et al., 1986), mesencephalic (see Jordan, 1983; Shefchyk and Jordan, 1985) and spinal cats (Edgerton et al., 1976; Chandler et al., 1984) are produced by rhythmic alternating excitatory and inhibitory synaptic inputs. Our present study extends these results to knee flexor motoneurons by demonstrating that the complex membrane potential variations in PBSt motoneurons result from excitatory and inhibitory synaptic influences during both flexor and extensor phases of the fictive locomotor cycle. Furthermore, it shows that these synaptic influences are controlled by tonic exteroceptive inputs from the limb. The average time courses of the membrane potential oscillations of flexor, extensor and PBSt alpha-motoneurons during fictive locomotion are summarized in Fig. 2 A. They closely parallel that of the activity in the related nerves (Fig. 3 A. C). It is noteworthy that similar time courses are found at the forelimb level in elbow flexor, elbow extensor and shoulder flexor (e.g. teres major) alpha-motoneurons and muscle nerves respectively, during fictive locomotion (Fig. 3 B, D; see also Cabelguen et al., 1981).

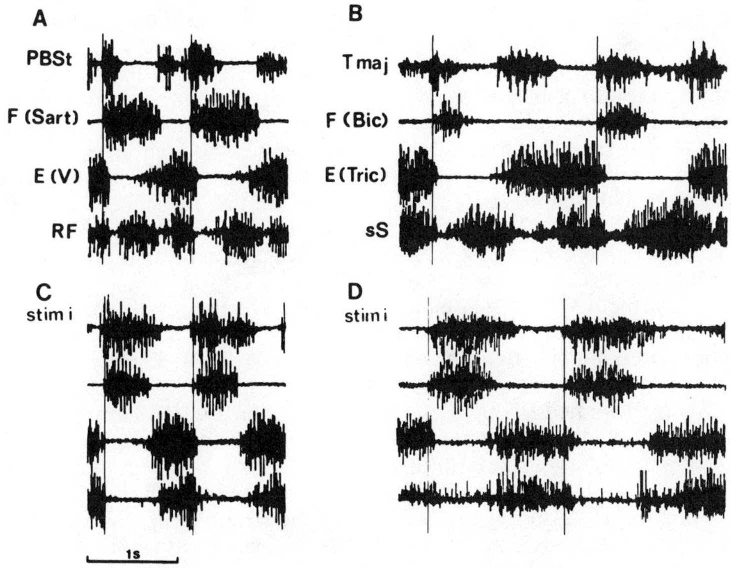

Fig. 3. Efferent activity in various hindlimb (A, B) and forelimb (C, D) nerves during fictive locomotion in the thalamic cat, before (A, C) or during (B, D) slight tonic pinching of the ipsilateral toes. Electroneurograms of posterior biceps-semitendinosus (PBSt), sartorius (Sart), vast (V), rectus femoris (RF), teres major (T maj), biceps (Bic), triceps (Tri), subscapularis (sS), flexor (F), extensor (E).

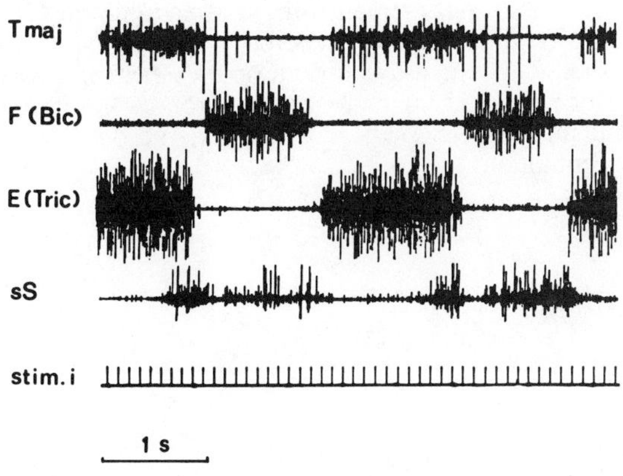

Fig. 4. Modulation of the flexor reflex in teres major nerve (T maj) during fictive locomotion. Flexor reflexes are induced by repetitive electrical stimulation (10 Hz) of the skin of the ipsilateral foot. Lower traces are electroneurograms of biceps (Bic), triceps (Tri), subscapularis (sS). F : flexor; E: extensor.

Several findings led us to propose a model neuronal network for the generation of the pattern of activity in knee flexor motoneurons during fictive locomotion.

First, when the duration of the flexor and extensor phases of the locomotor cycle remains constant, i.e. during regular activity of the part of the CPG involved in the production of the locomotor rhythm, the pattern of PBSt activity can vary either spontaneously or during afferent stimulations which can act without resetting the rhythm. This suggests the existence of a neuronal network interposed between a rhythm generating part of the locomotor CPG (Fig. 3 B, F E) and the motoneurons (Fig. 3 B, PBSt). The pattern of activity in knee flexor motoneurons during fictive locomotion can be understood as the result of interactions between the central locomotor drive and afferent or central (e.g. descending) influences (Fig. 3 B, dotted lines) on interneurons of this output network. For instance, during the flexor phase, excitation of PBSt motoneurons can be induced by the influences of the CPG on interneurons receiving exteroceptive inputs. The PBSt activation is larger during tonic external stimulation, which increases the level of excitability of these interneurons and decreases the activity in the pathway responsible for the inhibition during the flexor phase. During the extensor phase, a reciprocal scheme accounts for the depolarization and firing of PBSt motoneurons mainly at the beginning and at the end of the extensor phase, and for the variations in the activity which can occur during the latter phase.

Secondly, the rhythmic modulation of efferent responses and motoneuron PSPs in response to phasic peripheral stimulation during the locomotor cycle can be out of phase with the motoneuron locomotor activations (Forssberg et al., 1977; Duysens and Loeb, 1980; Perret and Cabelguen, 1980; Forssberg, 1981; Drew and Rossignol, 1985; Schomburg et al., 1981). This is illustrated in Fig. 4 in a shoulder flexor (Fig. 4). During phasic repetitive stimulation of the distal part of the ipsilateral forelimb, flexor reflex responses were evoked in the shoulder flexor nerve. Their amplitude was larger at the beginning of the extensor phase, decreased during the maximum of the extensor-like burst in the nerve and increased after the end of this burst, during the first half of the flexor phase. This indicates that the increases in the reflex responses do not result from the depolarization of the motoneurons but from changes in the excitability of interneurons of segmental reflex pathways like those activated by flexor reflex afferents. It can be postulated that these interneurons belong to the pattern generating part of the CPG.

Thirdly, the characteristics (time course, amplitude) of the inhibitory waves in PBSt motoneurons are highly correlated with those of the excitatory waves displayed simultaneously in rectus femoris motoneurons. This suggests that the same part of the CPG for locomotion simultaneously activates these motoneurons and inhibitory interneurons to PBSt motoneurons. The inhibitory elements might be Ia inhibitory interneurons since they are rhythmically activated during fictive locomotion (Feldman and Orlovsky, 1975; see Jordan, 1983). One should note that a contribution by other interneurons to the control of inhibitory and excitatory synaptic inputs to motoneurons is not impossible since recurrent (see Jordan, 1983) and presynaptic (Baev and Kostyuk, 1982) influences have been show to occur during fictive locomotion.

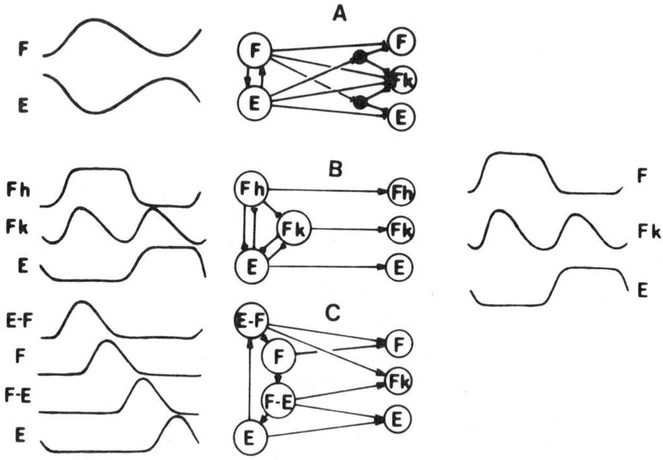

Fig. 5. Three models of the CPG for locomotion. A: half-center rhythm generating part, premotoneuronal pattern generating part, motoneurons. B: multi-unit generator. C: ring generator. On the left, shape of the output of each central rhythmic element. On the right, shape of the membrane potential variations in a flexor (F), an extensor (E) and a knee flexor (Fk) motoneuron. Fh: hip flexor. See text.

The model can be extended to explain the variable and sometimes complex efferent activities of some knee extensors (rectus femoris), shoulder extensors (sub-scapularis, spinatus) and distal (digit) muscles (see Perret and Cabelguen, 1980; Fleshman et al., 1984). Moreover, one should note that the PBSt and rectus femoris efferent patterns during scratching and their variations under tonic afferent influences display some similarities with the locomotor ones (Deliagina et al., 1981).

Our results show that inhibitory synaptic inputs in motoneurons play a role in the elaboration of the final efferent pattern which can thus be explained by the existence of a simple (half-center) oscillator and organized connections with motoneurons (Fig. 5 A). Various other kinds of CPGs for locomotion can account for the complexity of the locomotor pattern : oscillators connected in a complex way and corresponding to each muscle group (see Grillner, 1981), coupled oscillators corresponding to each phase of the cycle (ring hypothesis, see Shik and Orlovsky, 1976). They should be modified to account also for the inhibitory influences on motoneurons and for the phasic gating of reflex pathways.

Finally, it is important to determine whether if the efferent pattern of knee flexors analysed during fictive locomotion in the thalamic cat has a functional significance. It is noteworthy that a complex and variable pattern of activity was observed not only in these biarticular hindlimb muscles, which have opposite functions at hip and knee joints, but also in the forelimb, in shoulder flexors which are monoarticular muscles. This might be related to the fact that the knee and shoulder joints display two phases of flexion during the locomotor cycle (Fig. 2 A). In the E2 phase, knee (shoulder) flexion is required to produce a backward movement of the leg, i.e. propulsion, a role which is devoted to extensors of the other joints. At the end of the E3 phase, knee (shoulder) flexion produces elevation of the more distal parts of the limb, thus playing a role which is that of the flexors of the other joints. This duality is also shown in the reflex connections since the PBSt participates in the flexor reflex in spite of its Ia excitatory connections with "pure" extensors (see Perret and Cabelguen, 1975). Therefore, the classification of the muscles as flexors or extensors is inadequate in the case of locomotion, where different muscle groups are activated in each phase of the step cycle. One should note that other classifications (on the basis of the muscle fibre composition, their phylogenetic origin, the location of the motoneuron pools, the participation in standing, scratching, etc) can be established, which might help to understand how a simple basic organization of the central pattern generator for locomotion has been adapted to the differences in the anatomy of the fore- and hindlimbs.

With their complex and variable pattern of activity related to the transition phases of the step cycle, the knee and shoulder muscles emerge as the main elements in the adaptation of locomotor movements. For instance, their extensor-like burst is more clearly detectable when the velocity increases (for references, see Introduction). Unfortunately, the modifications in their pattern of activity related to changes in the conditions (slope, load, etc) of locomotion at a given speed have not yet been studied. These muscles are likely to be a main target of the descending control. Corresponding, discharges of lumbar-projecting rubrospinal neurons related to changes in the efferent pattern of PBSt and rectus femoris have been reported (Arshavsky et al., 1986).

References

Arshavsky, Y. I., Orlovsky, G. N., Perret, C., 1986, Activity of rubrospinal neurons during locomotion and scratching, Behav. Brain Res., in press.

Baev, K. V., Kostyuk P. G., 1982, Polarization of primary afferent terminals of lumbosacral cord elicited by the activity of spinal generator, Neuroscience, 7: 1401-1409.

Cabelguen, J. M., Orsal, D., Perret, C., and Zattara, M., 1981, Central pattern of forelimb and hindlimb locomotor activities in the cat, in: Szentagothai, J., Palkovits, M., and Hamori, J., eds. , "Advances in physiological sciences", vol. 1, Pergamon Press/ Akademiai Kiado, Budapest, pp. 199-211.

Chandler, S. H., Baker, L. L., and Golberg, L. J., 1984, Characterisation of synaptic potentials in hindlimb extensor motoneurons during L-DOPA-induced fictive locomotion in acute and chronic spinal cats, Brain Res., 303: 91-100.

Coombs, J. S., Eccles, J. C., and Fatt, P., 1955, The specific ionic conductances and the ionic movements across the motoneuronal membrane that produce the inhibitory postsynaptic potential, J. Physiol. (Lond), 130: 326-373.

Deliagina, T. G., Orlovsky, G. N., and Perret, C., 1981, Efferent activity during fictitious scratch reflex in the cat, J. Neurophysiol., 45: 595-604.

Drew, T., and Rossignol, S., 1985, Forelimb responses to cutaneous nerve stimulation in intact cats, Brain Res., 329: 323-328.

Duysens, J., and Loeb, G. E., 1980, Modulation of ipsi and contralateral reflex responses in unrestrained walking cats, J. Neurophysiol., 44, 1024-1037.

Edgerton, V. R., Grillner, S., Sjöström A., and Zangger, P. 1976, Central generation of locomotion in vertebrates, in: Herman R. M., Grillner S., Stein P., and Stuart D. G., eds. , "Neural Control of Locomotion", Plenum Press, New York, 439-464.

Endberg, I., and Lundberg, A., 1969, An electromyographic analysis of muscular activity in the hindlimb of the cat during unrestrained locomotion, Acta Physiol. Scand., 75: 614-630.

Feldman, A. G., and Orlovsky, G. N., 1975, Activity of interneurons mediating reciprocal Ia inhibition during locomotion, Brain Res., 84: 181-194.

Fleshman, J. W., Lev-Tov, A., and Burke, R. E., 1984, Peripheral and central control of flexor digitorum longus and flexor hallucis longus motoneurons : the synaptic basis of functional diversity, Exp. Brain Res., 54: 133-149.

Forssberg, H., 1981, Phasic gating of cutaneous reflexes during locomotion, in: Taylor A., Prochavska A., eds. , "Muscle receptors and movement", Macmillan, London, 403-412.

Forssberg, H., and Grillner, S., 1973, The locomotion of the acute spinal cat injected with clonidine i.v., Brain Res., 50: 184-186.

Forssberg, H., Grillner, S., and Halbertsma, J., 1980, The locomotion of the low spinal cat, I. Coordination within a hindlimb, Acta Physiol. Scand., 108: 269-281.

Forssberg, H., Grillner, S., and Rossignol, S., 1977, Phasic gain control of reflexes from dorsum of the paw during spinal locomotion, Brain Res., 132: 121-139.

Gambarian, P. P., Orlovsky, G. N., Protopopova, T. Y., Severin, F. V. and Shik, M. L., 1971, The activity of muscles during different gaits and adaptative changes of moving organs in family Felidae, Morphology and ecology of vertebrates, Proc. Inst. Zool. Acad. Sci. USSR, 48: 220-239.

Grillner, S., 1981, Control of locomotion in bipeds, tetrapods, and fish, in: Brooks V. B., ed. , "Handbook of physiology", Sect I, vol. 2, Waverly Press, Baltimore, 1179-1236.

Grillner, S., and Zangger, P., 1975, How detailed is the central pattern generation for locomotion, Brain Res., 88: 367-371.

Grillner, S., and Zangger, P., 1979, On the spinal generation of locomotion in the low spinal cat, Exp. Brain Res., 34: 241-261.

Grillner, S., and Zangger, P., 1984, The effect of dorsal root transection on the efferent motor pattern in the cat's hindlimb during locomotion, Acta Phisiol. Scand., 120: 393-405.

Jordan, L. M., 1983, Factors determining motoneuron rhythmicity during fictive locomotion, in: Roberts A., Roberts B., eds. , "Neural origin of rhythmic movements", Cambridge University Press, Cambridge, 423-444.

Lundberg, A., 1981, Half-centres revisited, in: Szentagothai J., Palkovits M., Hamori J., eds. , "Advances in physiological sciences," vol. 1; Pergamon Press/Akademiai Kiado, Budapest, 155-167.

Orsal, D., Perret, C., and Cabelguen, J. M., 1986, Evidence of rhythmic inhibitory synaptic influences in hindlimb motoneurons during fictive locomotion in the thalamic cat, Exp. Brain Res., 64: 217-224.

Perret, C., 1983, Centrally generated pattern of motoneuron activity during locomotion in the cat, in: Roberts A., Roberts B., eds. , "Neural origin of rhythmic movements", Cambridge University Press, Cambridge, 405-422.

Perret, C., 1986, Patterns of fore- and hindlimb motoneuron activity during fictive locomotion in the cat, in: Grillner S., Herman R., Stein P. S. G., Stuart D. G., eds.,"Neurobiology of vertebrate locomotion", Macmillan Press, Basingstone, in press.

Perret, C., and Cabelguen, J. M., 1976, Central and reflex participation in the timing of locomotor activations of a bifunctional muscle, the semi-tendinosus, in the cat, Brain Res., 106: 390-395.

Perret, C., and Cabelguen, J. M., 1980, Main characteristics of the hindlimb locomotor cycle in the decorticate cat with special reference to bifunctional muscles, Brain Res., 187: 333-352.

Rasmussen, S., Chan, A. K., and Goslow, G. E. Jr., 1978, The cat step cycle: electromyographic patterns for hindlimb muscles during posture and unrestrained locomotion, J. Morph., 155: 253-270.

Schomburg, E. D., Behrends, H. B., and Steffens, H., 1981, Changes in segmental and propriospinal reflex pathways during spinal locomotion, in: Taylor A., Prochaska A., eds. , "Muscle receptors and movements", Macmillan, Londonn, 413-425.

Shefchyk, S. J., and Jordan, L. M., 1985, Motoneuron input-resistance changes during fictive locomotion produced by stimulation of the mesencephalic locomotor region, J. Neurophysiol., 54: 1101-1108.

Shik, M. L., and Orlovsky, G. N., 1976, Neurophysiology of locomotor automatism, Physiol. Rev., 56: 465-501.

Wentink, G. H., 1976, The action of the hind limb musculature of the dog in walking, Acta anat., 96: 70-80.

MODULATION OF LOCOMOTOR ACTIVITY INDUCED BY NATURAL STIMULATION OF THE VESTIBULAR SYSTEM

V.V. Marlinsky, D.A. Vasilenko and Th. I. Tsyntsabadze

A.A. Bogomoletz Institute of Physiology
Academy of Sciences of the Ukrainian SSR
Kiev, USSR

Summary

Vestibular influences upon spontaneous locomotor muscle activity and that evoked by electrical stimulation (5-25 μA) of mesencephalic structures were studied in decerebrated guinea pigs. Natural stimulation of the vestibular system was performed by fixed and cyclic sinusoidal tilts about a longitudinal axis with an amplitude of ± 20° and a frequency range of 0.02-0.4 Hz. Electromyographic (EMG) locomotor discharges in forelimb extensors (m. triceps brachii) corresponding to the stance phase increased and discharges in forelimb flexors (m. biceps brachii) corresponding to the swing phase decreased during ipsilateral tilt, while the reverse changes in EMG activity were observed during contralateral tilt. Changes in EMG activity in hindlimb extensors (m.soleus) demonstrated the same trend as in the forelimbs. Activity in hindlimb flexors (m.tibialis anterior) had two periods of enhancement: one during ipsilateral tilt, coinciding with an increase in the homolateral extensor activity, and a second one during contralateral tilt. Changes in locomotor EMG discharges during sinusoidal tilting led to in phase displacement of the animal (in forelimb extensors 80-60°, in forelimb flexors, 60-40°). It is suggested that the vestibular system may control spinal motor output in parallel with the spinal locomotor generator. The observed changes in locomotor muscle activity are discussed in terms of vestibulospinal and vestibulo-reticulospinal effects.

Introduction

At present, information about the influences exerted by the vestibular system upon the activity of limb muscles in terrestrial vertebrates is scarce in connection with certain aspects. Detailed analysis of vestibular modulation of postural tonic muscle reactions has been reported. At the same time contradictory data have been published concerning the role of vestibular effects in the regulation of patterned cyclic movements, in particular locomotion. Data about vestibulo-spinal influences under free fall conditions applied to the interpretation of cyclic processes during jumping and running have led to the suggestion that vestibular effects are of great importance for the organization and maintenance of precise temporal interrelations between the activity of various muscles during locomotion (Melville Jones et al., 1973). Nevertheless, experimental findings concerning the depression of dynamic reactions of vestibulo-spinal neurones during locomotor rhythmicity have been taken to be evidence that the participation of the vestibular system in the control of locomotion is very limited (Arshavsky et al., 1984; Orlovsky et al., 1982). It should be emphasized that in both cases, these conclusions were based mainly on indirect data. There are also contradictions between the results and interpretations of experiments using electrical stimulation of Deiters' nucleus during locomotion (Orlovsky, 1972, Russeland Zajac, 1979). Therefore the aim of the present study was to analyse changes occurring in the locomotor activity of guinea pig limb muscles during natural stimulation of the vestibular system. Stationary and cyclic tilts of the animal about its longitudinal axis were used. Special attention was paid to evaluating the dynamic characteristics of changes in motor activities.

Methods

Decerebrated guinea pigs weighing 600-800 g were used in these experiments. Surgical operations were performed under Hexenal (100 mg/kg intraperitoneal) or Kalipsol (Ketalor, Ketamine, 50 mg/kg intramuscular) anaesthesia with atropine + promedol premedication. The animal was fixed rigidly in the frame of the device for adequate vestibular stimulation by means of a stereotaxic headholder and lumbar vertebral clamps. With this device, it was possible either to displace the animal about its longitudinal axis with fixation of stationary shift (± 20°) or to tilt it in a cyclic manner (sinusoidally) with frequencies of 0.02, 0.04, 0.08, 0.2 and 0.4 Hz and an amplitude of ± 20°. Immobility of the head with respect to the trunk was preserved during displacements to prevent the development of neck reflexes. The extremities were allowed to hang down freely without support.

In some experiments, alternating locomotor movements of limbs occurred spontaneously when anaesthesia was reduced (see below), and in other experiments locomotor activity was evoked by electrical stimulation of the mesencephalic locomotor region (MLR). The zone of the guinea pig brain stem with the lowest stimulation threshold was located in the plane A 2.0-2.5 between the ventromedial areas of the nucleus colliculus inferior and the dorsolateral border of the periaqueductal grey matter immediately dorsal to the nucleus cuneiformis (Tsintsabadze, 1972). The MLR was stimulated by a glass micropipette filled with Wood's alloy (tip diameter 15-25 μm) with current pulses of 5-25 μA, 0.5 ms in duration at a frequency of 50/s; a reference electrode was fixed in the occipital bone.

Electromyogram (EMG) were recorded bilaterally from forelimb m.triceps (Tr) and m. biceps (Bic) and from hindlimb m. soleus (Sol) and m. tibialis anterior (Tib). Bipolar bichrome wire electrodes (diameter 0.2 mm, insulation-free tip of 0.5 mm) were used. The recorded activity was passed through a low frequency filter (150 Hz), rectified and integrated with a time constant of 0.16 s.

Phase-frequency characteristics for the modulated locomotor activity were obtained by Fourier analysis. Sinusoidal displacement of the animal was regarded as an input signal and the fundamental harmonic component of locomotor EMG activity changes as an output one.

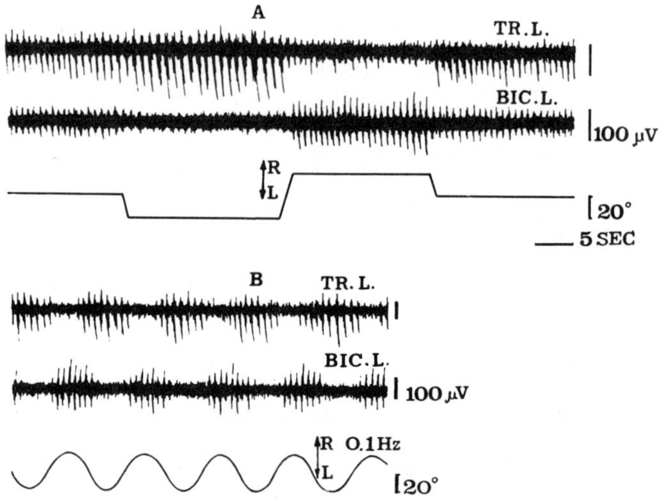

Fig. 1. Changes in spontaneous locomotor activity of the guinea pig forelimb muscles during tilts about the longitudinal axis. A - fixed tilts; B - sinusoidal tilting. Top to bottom: electromyograms (EMG) recorded from m. triceps (Tr.) and m. biceps (Bic.) of the left (L.) forelimb: position of the animal (R - tilt to the right. L - to the left). EMG is not rectified.

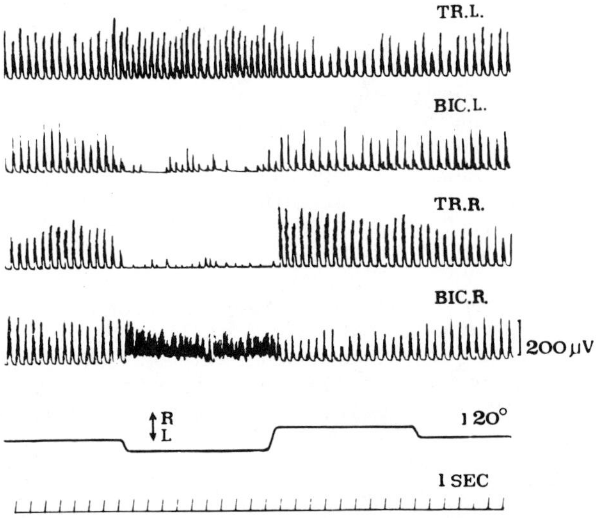

Fig. 2. Changes in locomotor activity of the guinea pig forelimb muscles evoked by mesencephalic locomotor region (MLR) stimulation during fixed tilts about the longitudinal axis. Top to bottom: EMG recorded from antagonistic muscles of the left (L.) and right (R.) forelimbs; position of the animal; time scale. Other abbreviations as in Fig. 1.

Results

1. Spontaneous locomotor activity

Rhythmic movements of limbs and the corresponding alternation of EMG discharges in limb muscles could occur in intercollicularly decerebrated animals for 20-60 min within 2.5-3 hours of Hexenal injection. The amplitude of discharges of extensor and flexor muscles which corresponded to the stance and swing phases of the locomotor cycle varied from one case to another within the range of 100-500 µV. The frequency of locomotor activities varied somewhat from one animal to another, the mean value being approximately 1 cycle per s.

When the animal was fixed in a tilted position, the amplitude of Tr locomotor discharges on the side of the shift increased noticeably while the amplitude of Bic discharges decreased. The reverse changes occurred on the contralateral (lifted) side: the amplitude of Tr discharges decreased and that of Bic discharges increased (Fig. 1 A). The relative intensity of these changes varied from one experiment to another and some trends could be observed during the same experiment. The mean increase or decrease reached 30-50 % of the initial discharge amplitude. When the orientation of the animal was changed, some fluctuations of the frequency of locomotion could occur, but they were non-significant. Alternation of the activity between antagonistic muscles was not significantly disturbed.

Cyclic displacements of the animal during spontaneous locomotor rhythmicity were accompanied by cyclic changes in the intensity of extensor and flexor muscle discharges. The period of these changes corresponded to the tilting period. The direction of discharge changes was identical to that observed during fixed shifts. Tr discharges were increased on the inclining side and depressed on the opposite side. Opposite patterns of changes in Bic activities were observed. The relative extent of these changes was noticeably higher than during fixed shifts: the amplitude of discharges usually increased or decreased approximately two-fold as compared with the initial values. Durations of flexor and extensor activity (i.e. swing and stance) and their alternation were not influenced to any appreciable extent.

2. Evoked locomotor activity

Stimulation of the MLR in pre-collicularly decerebrated guinea pigs was applied to evoke long periods (lasting at least a few minutes) of locomotor activity. The EMG pattern of this activity was similar to that observed during spontaneous locomotion, but the frequency of movements was higher (usually 2-4 Hz). The evoked locomotion always began with forelimb movements. Hindlimb movements were usually less stable than forelimb movements. Only those cases were chosen in which the evoked locomotion and its parameters were suitably steady: mechanical displacements started only after initiation of locomotion and stabilization of its characteristics.

Stationary and cyclic displacements of an animal about its longitudinal axis caused substantial modulation of locomotor EMG discharges. The pattern of this modulation was similar to that observed in the case of spontaneous locomotion. After a fixed 20° shift, the Tr discharges corresponding to the stance phase on the inclined side were enhanced and accompanied by a parallel decrease in Tr discharges on the contralateral side. At the same time opposite changes occurred with discharges from the flexor muscles (Bic) active in the swing phase. It should be noted that usually such a "pure" pattern was observed only on the side contralateral to the MLF stimulation side. Significant modifications in Bic discharges could be observed in the limb homolateral to the side of the stimulation. When the animal was tilted in the opposite direction, Bic discharges were prolonged that resulted in tonic contraction of this muscle lasting the whole duration of the shift. This protracted Bic activity was usually accompanied by complete inhibition of locomotor rhythmic activity in the Tr of the same limb.

The direction of cyclic modulation of locomotor discharges during sinusoidal tilting corresponded to that occurring under fixed shift conditions: the intensity of extensor discharges was increased on the side of the tilt and accompanied by a parallel depression of flexor activity, and the reverse effects were observed in antagonistic muscles of the opposite lifted limb. The pattern of activity alternation was not affected in the muscles of the limb contralateral to the side of MLR stimulation (Fig. 3). At the same time, this alternation could be considerably disturbed in the limb homolateral to the side of stimulation (Fig. 3). During the half-period of the shift to the opposite direction, Bic discharges in this limb could be prolonged and their duration exceeeded the usual period of locomotor rhythmicity. Parallel to

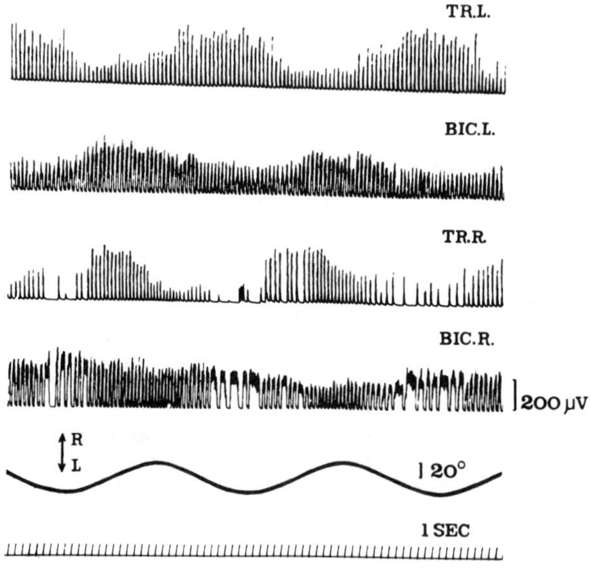

Fig. 3. Changes in locomotor activity of forelimb muscles evoked by MLR stimulation during cyclic tilting of the animal. Abbreviations as in Fig. 2.

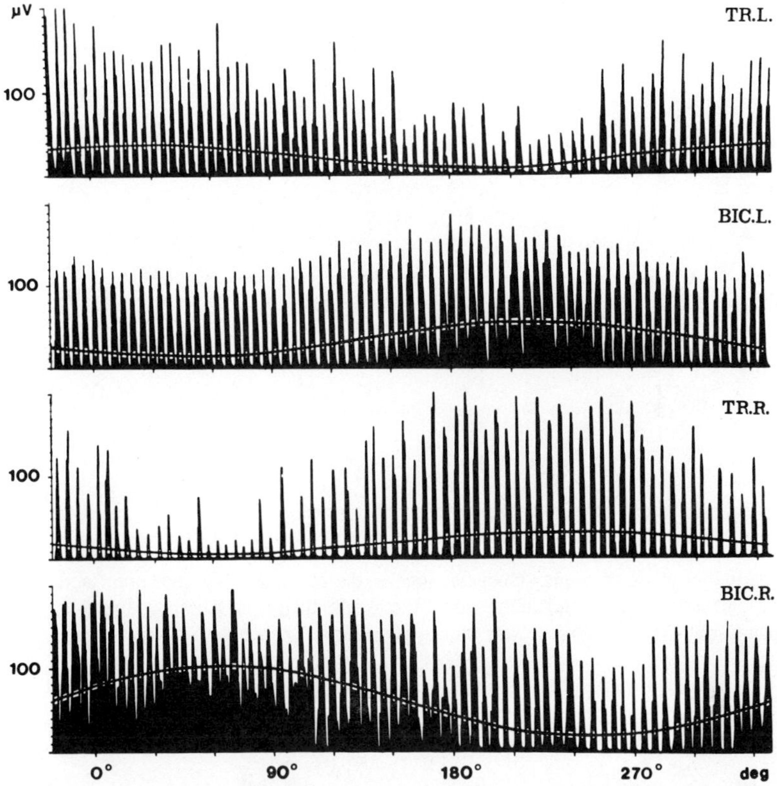

Fig. 4. Changes in locomotor activity of forelimb muscles evoked by MLR stimulation during one cycle of sinusoidal tilting (frequency 0.04 Hz) and fundamental harmonic of these changes. Graphs of fundamental harmonic are imposed on EMG activities. Abscissa: position of the animal; 90° corresponds to maximal tilt to the left. 270° - to the right; ordinate: amplitude of EMG, uV. Other abbreviations as in Figs. 1-3.

these prolonged Bic contractions, the activity of Tr in this limb was completely depressed (Fig. 3).

The modulation of locomotor activities caused by cyclic displacements took place throughout the whole frequency range of the mechanical stimulation. The fundamental harmonics of the EMG changes indicated that its period was in all cases identical to that of the input signal, i.e. the cyclic displacement period (Fig. 4). It should be mentioned that the amplitude of the fundamental harmonic varied from one experiment to another: this characteristic was found to vary depending on the muscle under analysis. Expecially large differences were found to exist in the amplitude of this harmonic when the activity was compared between the Bic muscles ipsi- and contralateral to the side of MLR stimulation. However, the phasic characteristics of the fundamental harmonic were fairly stable at all the frequencies of mechanical stimulation used in the present experiments.

Logarithmic phase-frequency characteristics of the changes in locomotor EMG activity showed the existence of a phase lead related to the displacement of the animal throughout the whole range of frequencies used. The phase lead of Tr discharge changes reached approximately 80° at 0.02 Hz and decreased slightly during the increase in frequency to 60° at 0.4 Hz. The corresponding values for Bic discharge changes were somewhat lower (approximately 60° and 40°). (Fig. 5). Changes in the activity of homologous muscles of the right and left extremities were opposite in phase.

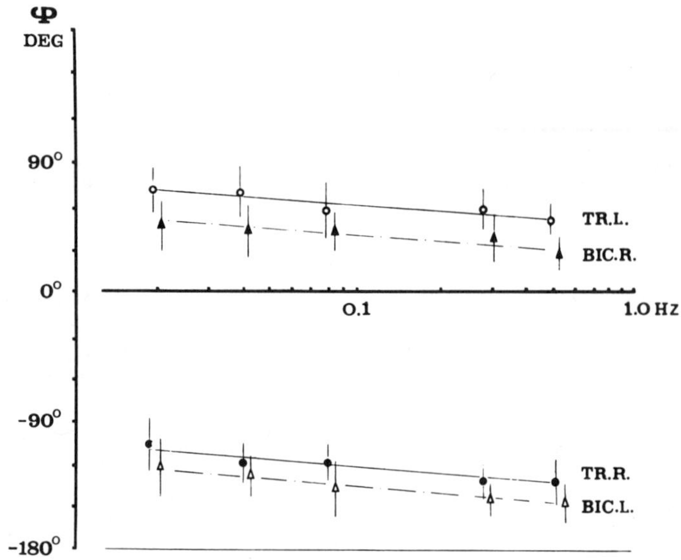

Fig. 5. Logarithmic phase-frequency characteristics of the changes in the locomotor activity of forelimb muscles during cyclic tilting of the animal about the longitudinal axis. Abscissa: frequency of tilting. Hz; ordinate: phase shift. Degrees; vertical bars: standard deviations. Other abbreviations as in Figs. 1-4.

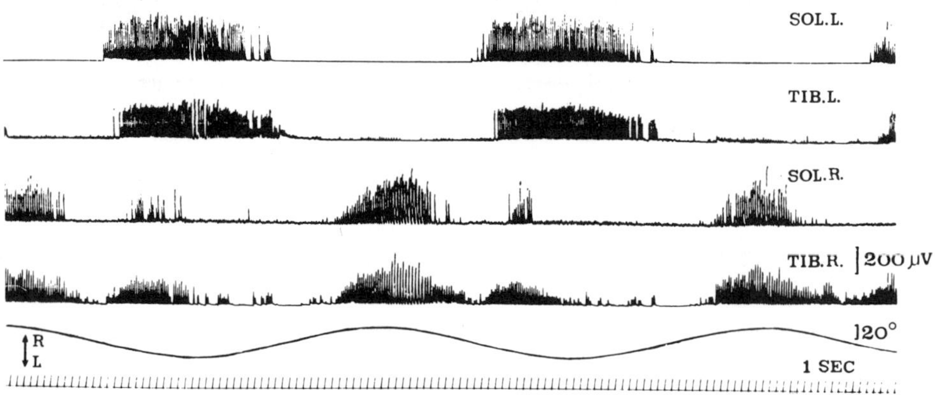

Fig. 6. Changes in MLR-evoked locomotor activity of hindlimb muscles during cyclic tilting of the animal about the longitudinal axis. EMGs were recorded from m. soleus (Sol.) and m. tibialis anterior (Tib.). Other abbreviations as in Fig. 3.

The modification of the evoked locomotor activity of hindlimb muscles differed in certain respects from that observed in the forelimbs. The direction of changes in the discharge of hindlimb extensors (Sol) during vestibular stimulation was the same as in the forelimbs (Tr), but the intensity of modulation was higher. As a rule, rhythmic discharges of Sol were recorded only when the animal was in a horizontal position or was tilted to the side of the recording. When the animal was shifted to the opposite side, Sol rhythmic activity was intensely (in most cases completely) depressed (Fig. 6). Changes in hindlimb flexor muscle (Tib) activity were not the reverse of the extensor activity discharges as was observed in the forelimbs. Rhythmic discharges in flexors occurred in parallel with homolateral extensor discharges, i.e. during the tilt to the side of the muscle recorded. They were almost absent during shifts to the opposite side (Fig. 6). Thus tilting to the side opposite to the studied hindlimb muscles could result in substantial depression of locomotor rhythmicity. Nevertheless it should be noted that in the Tib homolateral to the MLR stimulation site, two periods of activity could be observed during one cycle of displacement. One period during the decline of the recorded side that was described above might be accompanied by the second one during the beginning of the tilt to the opposite side (Fig. 6). The intensity of Tib rhythmic discharges during the second period was usually less then in the first period. During the second period, Tib discharges could be followed by a relatively short, unstable burst of rhythmic discharges in the Sol of the same limb.

Discussion

The data obtained here indicate that natural stimulation of the vestibular system consisting of tilting the animal about its longitudinal axis exerts intense modulatory effects on locomotor muscular activity. This modulation was observed during both spontaneous and MLR-evoked locomotor activities, both in extensor and flexor muscles. The most common feature of this modulation consisted of an increase in extensor activity corresponding to the stance phase on the inclined side and in a decrease in this activity on the lifted side.

Modulation of extensor discharges is associated with reciprocal modulation of flexor activities corresponding to the swing phase. Flexor discharges are depressed on the inclined side and increased on the contralateral side. The adaptive nature of such changes is obvious. Their direction coincides with the direction of changes in the activity of antagonistic muscles during static maintenance of posture (Magnus, 1924; Roberts, 1978) and agrees with the existing data concerning the pattern of vestibulofugal synaptic actions on motoneurones of these muscles (Grillner and Hingo, 1972; Shapovalov, 1975; Wilson, 1972). Amplitude modulation of locomotor rhythmic discharges can be regarded as the major aspect of vestibular influence. Usually the frequency of these discharges is not affected; if some changes in this parameter do occur, they are relatively weak. The alternation of activity between antagonistic muscles is not changed either. In cases where changes in alternation do take place, they are probably due to factors not directly connected with vestibular influences (see below). Thus, the main part of vestibulofugal actions is exerted upon the neurones which provide spinal motor output: to a considerably lesser extent, the interneuronal networks that generate locomotor rhythmicity are also affected. This functional specialization is reflected first of all in a pattern of organization of vestibulo-spinal pathways projecting mainly to the ventromedial regions of the spinal grey (Kuypers, 1964; Nyberg-Hansen and Mascitty, 1963), whereas the interneurons closely related to generation of rhythmic locomotor activity are localized predominantly more dorsally and laterally, in the intermediate zone and at the base of the dorsal horn (Boyer et al., 1979; Jankowska et al., 1967).

As was mentioned above, distortions of locomotor rhythmicity, which were absent in spontaneously moving animals, were often observed under similar tilting conditions during MLR-evoked locomotion. They developed only on the side ipsilateral to midbrain stimulation. This suggests that this phenomenon may result from hyperactivation of the flexor motoneuronal population. MLR stimulation largely increases tonic flexor muscle activity especially on the stimulation side (Tsintsabadze, 1987). An additional increase in flexor activity during tilting to the opposite side can result in an overexcitation of flexor motoneurones, in prolonged contraction of flexor muscles and in corresponding reflex processes at segmental level which disturb the regular alternation of flexor and extensor activity.

Some part of the effects originating during natural stimulation of the vestibular system seems to be transmitted to spinal neurones, in particular to flexor motoneurones, via reticulo-spinal pathways involved in the activity due to well developed connections between vestibular and reticular nuclei (Brodal et al., 1962; Wilson and Yoshida, 1969; Wilson et al., 1970). Transreticular transmission of vestibular effects might be responsible for relatively small but obvious changes in the frequency characteristics of locomotor activity, since an enhancement of flexor activity correlates with some increase in the frequency of alternating discharges. It is known that reticulo-spinal pathways, especially their lateral divisions project to the main site of locomotor generator interneuronal populations (Kuypers, 1964; Nybert-Hansen, 1965).

There is a tendency for greater changes to occur in the intensity of motor discharges observed during cyclic displacements than during fixed shifts, although precise measurements are difficult to obtain, expecially in case of MLR-evoked locomotion accompanied by an artificial increase in flexor activity on the side of stimulation. This peculiarity of amplitude modulation indicates that both tonic effects from otolith receptors and phasic effects connected with dynamic activation of the receptors of semicircular canals and otolith apparatus are decisive factors affecting the intensity of locomotor discharges. Therefore, adaptive modulation of locomotor activity in limb muscles may occur both during changes in the supporting surface relief, resulting in variations in the animal's position in relation to the vector of the gravity, and during variations in the movement trajectory that result in the occurrence of accelerations acting upon the animal.

The possibility of applying Fourier analysis to the study of dynamic properties of the vestibular system and its components has been discussed in the literature (Berthoz and Anderson, 1971; Bilotto et al., 1982). Precise analysis of amplitude changes was difficult in the present study because of the presence of factors which changed the level of muscle discharges during the experiment and could not be controlled (fluctuations in excitability of segmental interneuronal populations during long lasting periods of locomotor activity, changes in the intensity of descending influences during prolonged MLR stimulation, etc.). At the same time, the frequency characteristics of the changes in the locomotor EMG activity were fairly stable and reproducible from one experiment to another. The frequency of the fundamental harmonic of the changes in both extensor and flexor activities was equal to the frequency of the modulating signal (tilting) throughout the range of frequencies studied. The observed phase lead of forelimb extensor locomotor activity changes is similar to the corresponding value reported for postural tonic activity of these muscles in the cat tilted about its longitudinal axis in the same range of frequencies (Berthoz and Anderson, 1971). The somewhat lower values obtained for the phase lead of changes in flexor muscle activity can probably be related to the above mentioned circumstance: efforts in flexors are essentially mediated by the reticular formation. Besides, inhibitory effects in flexor motoneurons mediated by vestibulofugal activity are transmitted through additional synaptic relays at segmental level (Melda et al., 1975; Shapovalov et al., 1966). Thus the reason for differences in phase lead values may be that different neuronal networks transmit vestibular effects to the motoneurones of different muscle groups.

Certain difficulties arise while interpreting the changes in hindlimb locomotor muscle activity during tilting. The modulation of extensor discharges in hindlimb and forelimb extensors is very similar but it is greater in the former. The difference may arise from the greater development of oligosynaptic vestibulo-spinal projections to the hindlimb motoneurones as compared with the forelimb ones (Shapovalov, 1975; Wilson and Yoshida, 1969). The modification of hindlimb flexor activity was quite different: an increase in flexor discharges occurred in parallel with that of homolateral extensor discharges. The possibility cannot be excluded that in this case the greater flexor activation may be due to segmental stretch reflexes evoked by enhanced extensor activity rather than to the vestibulofugal influences themselves, although the intervention of facilitating influences of the vestibular system on lumbar flexor motoneurones is possible (Grillner and Hongo, 1972; Shapovalov et al., 1966). The second period of increased flexor activity in the hindlimb ipsilateral to MLR stimulation corresponds to analogous changes in the homolateral forelimb. This probably results from vestibulofugal action itself. This second period of flexor activity may in turn enhance rhythmic activity in antagonists of the same extremity via stretch reflexes.

Thus, the main proportion of the locomotor activity changes observed during natural vestibular stimulation as the result of tilting about a longitudinal axis can be satisfactorily described in terms of interaction between the vestibulo- and reticulo-spinal effects. The integration of influences from both supraspinal and spinal neuronal systems occurs principally at the level of the motor output according to Sherrington's "common pathway" principle (Sherrington, 1906). This mechanism provides a parallelism between the action of the locomotor generator, the segmental reflex arcs and the vestibular system; it may constitute a possible basis for the fast adaptive reorganization of muscle activity during real locomotion.

References

Arshavaki, Yu.I., Galfand, I.M., Orlovsky, G.N., 1984, Cerebellum and control of rhythmic movements (Russ.). Moskow. Neuka.

Esyev, K.V., Degtiarenko, A.M., Zavadskaya, T.V., Mostyuk, P.G., 1979, Activity of lumbar spinal cord interneurones during fictive locomotion of thalamic cats. Neirofiziologija (Russ.), 11: 329-337.

Berthoz, A., Anderson, J.H., 1971, Frequency analysis of vestibular influences on extensor motoneurons. I. Response to tilt in forelimb extensors. Brain Res., 34: 370-375.

Bilotto, G., Goldberg, J., Peterson, G.W., Wilson, V.J., 1982, Dynamic properties of vestibular reflexes in the decerebrate cat. Exp. Brain Res., 47: 343-352.

Brodal, A., Pompeiano, O., Wolberg, F., 1962, The vestibular nuclei and their connections. Anatomy and functional correlations, Springfield, Ch. C. Thomas.

Grillner, S., Hongo, T., 1972, Vestibulospinal effects on motoneurons in the lumbosacral cord. In: Progr. Brain Res., V.37, Basic aspects of central vestibular mechanisms. Ed.: A. Brodal, Elsevier, Amsterdam-London-New York.

Jankowska, E., Jukes, M.G.M., Lund, S., Lundberg, A., 1967, The effect of DOPA on the spinal cord. 6. Half-centre organization of interneurons transmitting effects from the flexor reflex afferents. Acta physiol. scand., 70: 389-402.

Kuypers, II.G.J.M., 1964, The descending pathways to the spinal cord, their anatomy and function. In: Progr. Brain Res., V. 11, Organization of the spinal cord. Ed.: J.C. Eccles, J.P. Schade, Elsevier, Amsterdam.

Maada, M., Maunz, R.A., Wilson, V.J., 1975, Labyrinthine influences on cat forelimb motoneurones. Exp. Brain Res., 22: 69-86.

Magnus, R., 1924, Körperstellung, Springer, Berlin.

Mellwill Jones, G., Watt, D.G.D., Rossignol, G., 1973, Eight nerve contributions to the synthesis of locomotor control. In: Control of posture and locomotion. Ed.: R.D. Stein, Plenum, New York, p. 579-597.

Nyberg-Hansen, R., 1965, Sites and mode of termination of reticulo-spinal fibers in the cat. An experimental study with silver impregnation methods. J. comp. Neurol., 124: 71-100.

Nyberg-Hansen, R., Mascitty, T.A., 1963, Sites and mode of termination of fibres of the vestibulospinal tract in the cat. An experimental study with silver impregnation methods. J. comp. Neurol., 122: 369-387.

Orlovsky, G.N., 1972, The effect of different descending systems on flexor and extensor activity during locomotion. Neirofiziologija (Russ.), 4: 311-316.

Roberts, T.D.M., 1978, Neurophysiology of postural mechanisms. 2-nd ed., Butterworths, London-Boston.

Russel, D.F., Zajac, F.R., 1979, Effect of stimulating Deiters' nucleus and medial longitudinal fasciculus on the timing of the fictive locomotor rhythm induced in cats by DOPA. Brain Res., 177: 588-592.

Shapovalov, A.I., 1975, Neurones and synapses of supraspinal motor systems (Russ.)., Nauka, Leningrad

Shapovalov, A.I., Kurchavy, G.G., Strogonova, M.P., 1966, Synaptic mechanisms of vestibulo-spinal actions on a-motoneurones. Fiziol. Zh. SSSR (Russ.), 52: 1401-1409.

Sherrington, C.S., 1906, The integrative action of the nervous system. Yale University Press, New Haven.

Skinner, R.D., Rommel, R.S., 1970, Monosynaptic inputs to lumbar interneurons from lateral vestibulospinal tract and the medial longitudinal fasciculus. Neurosci. Lett., 10: 259-264.

Tsintsabadze, Th.I., 1987, Mesencephalic locomotor region in guinea pig. Neirofiziologija (Russ.), 19: 270-273.

Wilson, V.J., 1972, Vestibular influences on alpha motoneurons in the cervical and thoracic cord. In: Progr. Brain Res., V.37, Basic aspects of central vestibular mechanisms. Ed.: A. Brodal, Elsevier, Amsterdam-London-New York., p. 233-242.

Wilson, V.J., Yoshida, M., 1969, Comparison of effects of stimulation of Deiters' nucleus and medial longitudinal fasciculus on neck, forelimb and hindlimb motoneurons. J. Neurophyziol., 32: 743-758.

Wilson, V.Y., Yoshida, M., Schor, R.H., 1970, Supraspinal monosynaptic excitation and inhibition of thoracic back motoneurons. Exp. Brain Res., 11: 282-295.

PRESYNAPTIC CONTROL OF AFFERENT INPUT DURING REAL LOCOMOTION IN RATS

I.A. Yakhnitsa, A.I. Pilyavsky, and N.V. Bulgokova

A.A. Bogomoletz Institue of Physiology
Academy of Sciences of the Ukrainian SSR, Kiev

Summary

Studies were carried out on dorsal root potential (DRP) and dorsal root antidromic action potentials (AAPs) during two types of locomotion (swimming and stepping), which could be distinguished by the intensity of their afferent inputs (AI).

AAPs were taken to be an index to the level of depolarization of low-threshold afferent fibres and DRP, to that of high-threshold ones. During single swimming or stepping locomotor cycles, two DRP and AAP maxima, were found to be associated with the stance (or retraction) and swing phases. The values of AAPs and DRP and the intensity of AI during locomotion and passive hindlimb movements, respectively, were found to be clearly correlated. No dependence between intensity of depolarization of afferent fibres and efferent output was observed during real locomotion, contrary to what was observed during passive hindlimb movement.

It was concluded that DRP and AAPs during real locomotion are initiated mainly by influences from peripheral afferents.

Introduction

It has been suggested that real locomotion may result from the integration of central motor program and segmental reflex reactions (Grillner, 1975; Shik and Orlovsky, 1976). At the same time, reflex reactions are modified during locomotion (Akakawa et al., 1982 Duysens and Pearson, 1980; Forssberg et al., 1975). This modulation can be effected in several different ways. One is through oscillation of membrane potential in motoneurons and segmental interneurons evoked by the functioning of the spinal locomotor generator (Lennard and Hermanson, 1985; Shefchyk et al., 1984), while another is through modulation of afferent input by presynaptic inhibition that can be evoked both by the activity of the spinal locomotor generator (Bayev, 1980; Bayev and Kostyuk, 1982) and by peripheral influences (Schmidt, 1973).

It has been shown that modulation of segmental reflexes during fictive locomotion (Shefchyk et al., 1984) occurs less than during real locomotion (Akazawa et al., 1982). The main difference between real and fictive locomotion is the presence of phasic afferent inflow. Taking this factor into account, we postulate that rhythmic influences from peripheral receptors can play an important role in the modulation of reflexes during real locomotion and that this action is probably caused by presynaptic mechanisms restricting afferent inflow. In the present study, the reactions reflecting primary afferent polarization during two different types of locomotion (swimming and stepping) differing in the intensity of the phasic afferent inflow were investigated.

Methods

Studies were performed on white rats (200 to 400 g). Two kinds of experiments were carried out: in the first, DRP and in the second, AAPs were recorded in dorsal roots in fixed animals performing stepping or swimming movements.

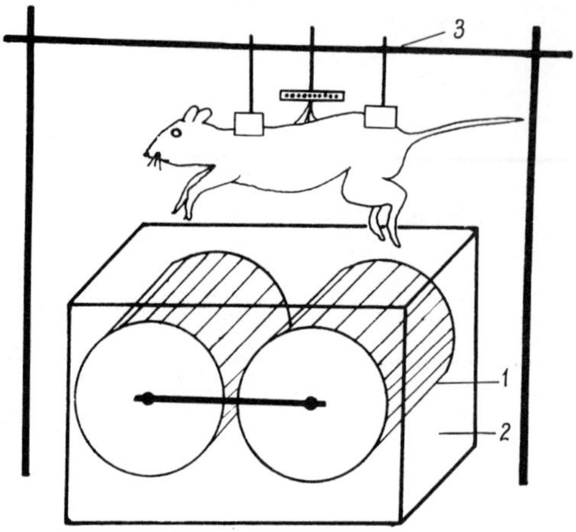

Fig. 1. Experimental arrangement. 1 - treadmill; 2 - water bath; 3 - fixation frame.

A diagram of the experimental arrangement is given in Fig. 1. It included a device for fixing the animal (Bayev and Esipenko, 1985), a water bath (Bayev, 1980) and a treadmill (Akazawa et al., 1982). Depending on the animal's position relative to the water surface and the treadmill two different types of locomotion (swimming and stepping) can be elicited (Yakhnitsa et al., 1985).

Under Hexenal anaesthesia (80 mg/kg), laminectomy (L_1-L_5) and separation of dorsal root L_5 were performed. The animal was placed in the special device (see Fig. 1) and the exposed vertebral column was fixed. One filament of the dorsal root was dissected. Its peripheral branch was used for afferent activity recording and the central one for DRP and AAP recordings. For recording the DRP, Ag-AgCl electrodes were used with an AC amplifier (time constant 2.2 s). The central part of the filament was mounted in a modified sucrose bath (Roberts and Wallis, 1978). AAPs were recorded with silver electrodes; the AC amplifier had a time constant of 0.005 s. The horizontal component of the hindlimb shift was recorded by strain gauge. Recording of AI and EMG of m. gastrocnemius (G) and m. tibialis anterior (TA) was performed using standard methods.

Results

Phasic changes in DRP during real locomotion

Phasic changes in DRP were recorded during fictive (Bayev, 1980; Bayev and Kostyuk, 1982) as well as real locomotion (Bayev and Esipenko, 1985). In the latter case, the shape of negative DRP deflection substantially differed from the DRP obtained during fictive locomotion. These data indicate that phasic afferent influences can play an important role in the generation of primary afferent depolarization during real locomotion.

Comparative analysis of DRP deflections developed during swimming and stepping was carried out taking into account the marked difference in phasic AI between these two types of locomotion.

The typical DRP patterns recorded during swimming and stepping are shown Fig. 2. Two maxima in the DRP appeared during one swimming locomotion cycle (Fig. 2 a, 3 a, b). One occurred during the second half of the swing phase (Sw DRP) and the other coincided with the retraction phase (Rt DRP). The amplitude of Rt DRP was higher than that of Sw DRP and reached 0.3 mV (Fig. 2a). The amplitude of Sw DRP was highly variable; it increased markedly with growing swimming intensity. The amplitude of Rt DRP, on the contrary, was more stable.

Similar changes in DRP were observed during stepping of the rats on the treadmill. In this case, the DRP also exhibited two maxima. One of them corresponded to the stance phase (St DRP) and the other one to the swing phase (Sw DRP). The amplitude of St DRP was as a rule higher than that of Sw DRP but not significantly so (Fig. 2 b). During intensive stepping, the amplitude of St DRP was considerably higher than that of Sw DRP and reached 0.4-0.8 mV.

Comparative analysis of AI during swimming and stepping revealed a similar AI pattern but a considerable increase in AI was observed in the latter case. Fig. 3 gives examples of averaged AI during one normalized swimming (Fig. 3a) and stepping locomotor cycle (Fig. 3b). Two maxima in AI were observed during one locomotor cycle with both types of locomotion. One of them coincided in time with the stance phase and the second one with the swing phase.

In order to elucidate the mechanism underlying generation of DRP during real locomotion, the correlation between DRP and intensity of AI and efferent output (EMG) was calculated. We assumed that if PAD is evoked mainly by AI, the amplitude of DRP must depend on the intensity of AI, and vice versa, if PAD is formed predominantly by the activity of the internal spinal generator, a correlation between amplitude of DRP and intensity of motor output must exist. Fig. 4 a gives averaged patterns of DRP, AI and EMG G, and Fig. 4c, e gives the correlation fields between these parameters. The analysis showed that a high degree of correlation exists between the value of DRP and AI intensity (corr. coef. 0.45-0.62) but no significant correlation between amplitude of DRP and EMG G (corr. coef. 0.06-0.23). In Fig. 4 b, easily identifiable components of DRP are shown and in Fig. 4 d, f correlation fields between these components of DRP and the corresponding AI are given. It was found that the correlation between St DRP and AI was somewhat higher (corr. coef. 0.40-0.71) than between Sw DRP and AI (corr. coef. 0.32-0.5).

The dependence of the DRP value upon AI and EMG G during passive movement was studied. In this situation, the influences from spinal locomotor generator are absent. Two maxima in the DRP were observed during one cycle of passive movements. One of them occurred when the limb was shifted backwards and the second one when it was moved ahead (Fig. 5 a, b). One can see that the AI and DRP patterns generated during passive movement are very similar to those observed during locomotion.

Evaluation of dependence between DRP and AI revealed the existence of a clear correlation between them (corr. coef. 0.6-0.7) (Fig. 5 c). A correlation between DRP and EMG G (corr. coef. 0.3-0.45) was also found to exist (Fig. 5 d). It should be noted that

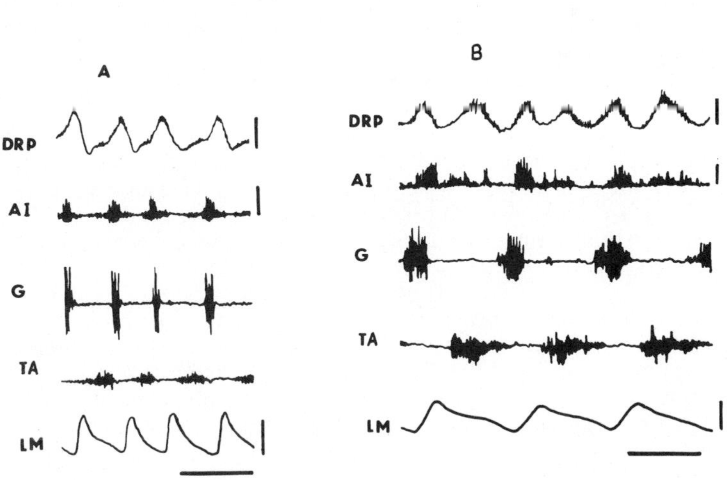

Fig. 2. DRP during swimming (A) and stepping (B). Polygraph records; DRP - dorsal root potential; AI - mass discharge in afferent fibres in distal part of dorsal root; G - EMG from m.Gastrocnemius; TA - EMG from m.Tibialis anterior; LM - horizontal component of hindlimb movement (downward deflection of the line corresponds to the forward limb displacement). Calibrations: 0.5 s; DRP 0.3 mV; AI - 50 μV; LM - 30 mm.

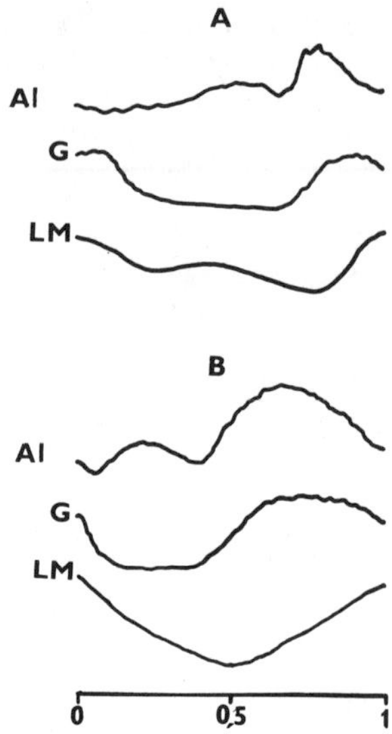

Fig. 3. Integrated AI during one normalized cycle of swimming (A) and stepping (B). EMG is also presented in integrated form. Zero of the normalized cycle corresponds to the beginning of swing phase. Other designations are the same as in Fig. 2.

muscle activity during these movements is caused by stretch reflex and that DRP is therefore reflexly but not centrally evoked.

Phase-dependent changes in dorsal root AAPs during real locomotion

During real locomotion, a low-amplitude spike activity consisting of AAPs can be recorded from the central part of the dorsal root. This activity was visible on the background of the DRP (Fig. 2). No clear coincidence of AAPs with DRP was observed, but the phase-dependent changes in AAPs were prominent.

By discriminating between high-frequency signals, it was possible to record AAPs in net form and to study these potentials in more detail.

During one swimming locomotor cycle, two volleys of AAPs can be observed. One of them coincided with retraction (Rt AAPs), and the second one with swing phase (Sw AAPs) (Fig. 6 a, 7a). The amplitude of Rt AAPs was several times higher than that of Sw AAPs and reached 30 μV. The maximum of AAPs coincided with the maximum of AI.

During stepping, two discharge phases (St AAPs and Sw AAPs) were again found to occur. The amplitude of AAPs during stepping was higher than that during swimming and reached 50 μV (Fig. 6 b). Contrary to what was observed with swimming, the amplitudes of St AAPs and Sw AAPs during stepping were almost equal.

To estimate the role of AI and spinal locomotor generator influences in the forming of AAPs during real locomotion, the correlation between intensity of AAPs and AI or EMG G was assessed. Fig. 7 a gives the averaged pattern of AAPs, AI and EMG G during one normalized swimming cycle, and Fig. 7 b, c shows the dependence between these reactions during 75 sample cycles. One can see the existence of a strong correlation between values of AAPs and AI (corr. coef. 0.52-0.7) but significantly less dependence between AAPs and efferent output (corr. coef. 0.24-0.35).

During passive movements, AAPs were seldom observed, but if they did occur, two volleys of AAPs could be recorded during one cycle (Fig. 8 a, b). One of them coincided with forward movement and the other one with backward movement. During passive movements , an increase in AAP intensity was found to correspond to a rise in AI.

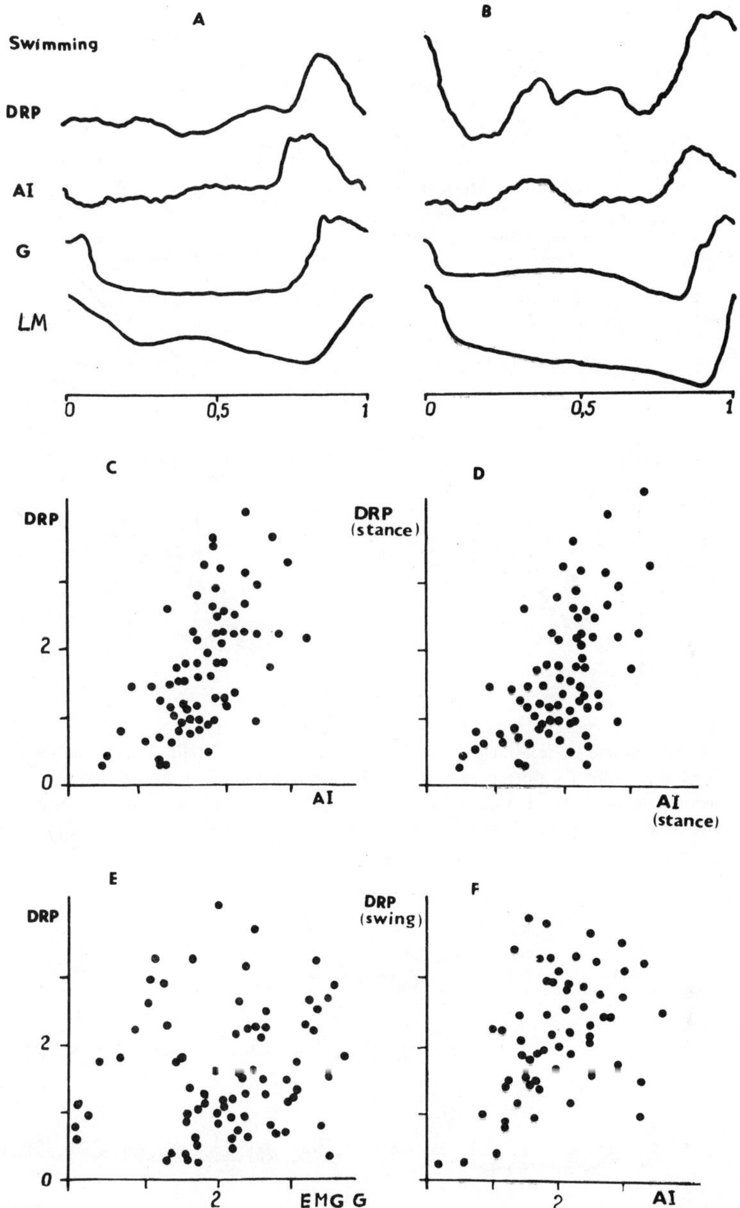

Fig. 4. DRP, AI, EMG G and correlation between them during swimming. A, B - examples of averaged DRPs during single normalized swimming cycle (designations are the same as in Fig. 2); intensity of swimming is higher in B. C, F - correlations between DRP integral (square above zero line) and integrated afferent inflow (C, D, F) and efferent output - integrated EMG from m.Gastrocnemius (E). Correlation between DRP and AI in whole population is given in C, while separate values during stance and swing phases are given in D and F, respectively. Values of DRP, AI and EMG in conventional units. Other designations are the same as in Fig. 2. Tonic and phasic components of AAPs were distinguishable during real locomotion. The relative value of these components depended on the type and intensity of locomotion. The tonic component of AAPs was prominent in the early stage of locomotion and decreased after some time. Whenever the tonic component was weak, the phasic component was stronger.

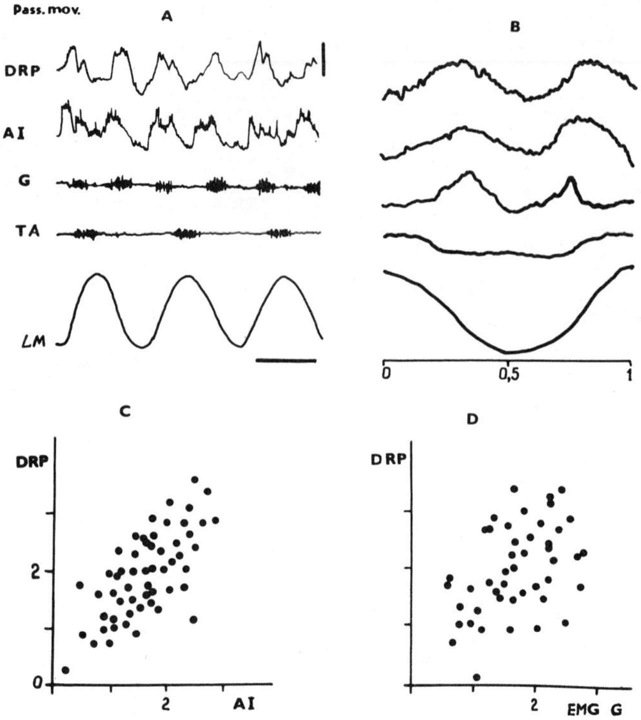

Fig. 5. DRP, AI, EMG G and correlation between them during passive hindlimb movements. A - changes in DRP, AI, EMG G and EMG TA during passive hindlimb movements ; B- an example of averaged DRPs during single normalized cycle (designations are the same as in Fig. 2) C - dependence between DRP values and AI intensity; D - dependence between DRP values and intensity of EMG G. Values of DRP, AI and EMG in conventional units. Other designations are the same as in Fig. 2.

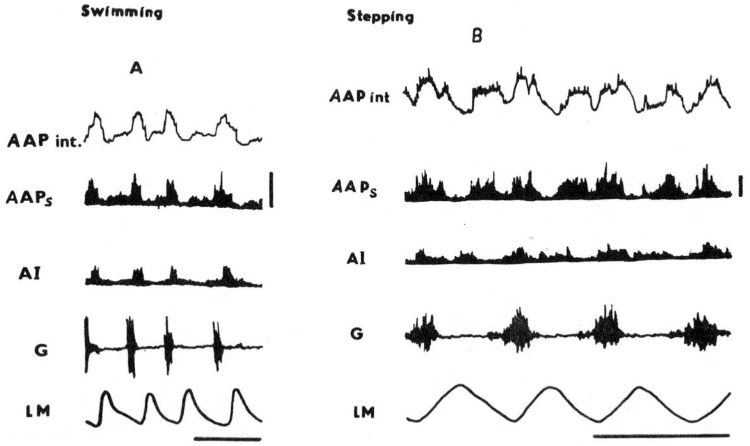

Fig. 6. AAPs during swimming (A) and stepping (B). Polygraph recordings AAPs - antidromic action potentials; AAP int. - integrative form of AAPs; Calibrations: 0.5 s; AAPs - 30 μV. Other designations are the same as in Fig. 2.

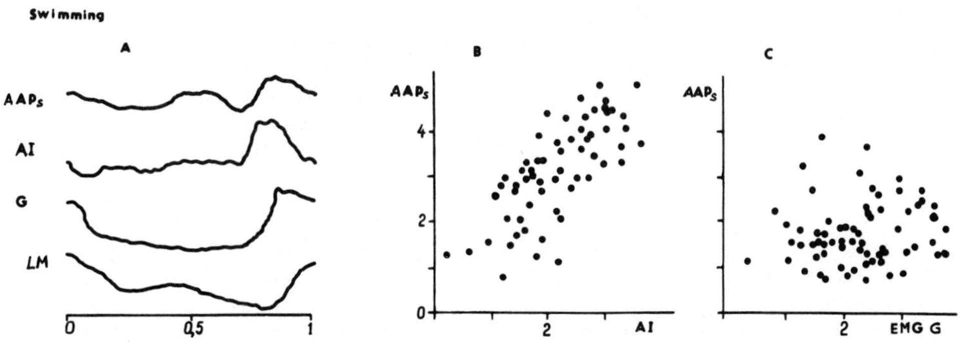

Fig. 7. AAPs, AI, EMG G and correlation between them during swimming. A - example of averaged AAPs during single normalized cycle. B - correlations between AAP integral and integrated AI and C - correlations between AAP integral and integrated EMG from m.Gastrocnemius. Other designations are the same as in Fig. 2 and 4.

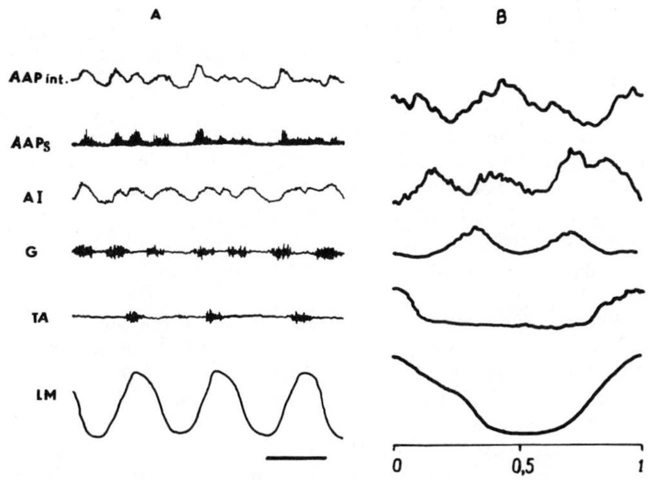

Fig. 8. AAPs during hindlimb passive movement. A - example of AAPs, AI, EMG G and EMG TA and their average during a single normalized cycle. Designations are the same as in Fig. 2 and Fig. 6.

Discussion

PAD is known to be a factor contributing to the restriction of AI in some physiological phenomena (Levy, 1980), in which DRP and AAPs have been studied in detail (Brooks and Koizumi, 1956; Eccles et al., 1967; Eccles et al., 1962; Schmidt, 1973). AAPs have turned out to result from the suprathreshold depolarization of afferent terminals that evoke the action potentials, and their propagation along the afferent fibres in the antidromic direction (Brooks and Koizumi, 1956; Levy, 1980).

The modulation of DRP and AAPs indicates the functional significance of the presynaptic mechanisms in the control of AI during real locomotion. Evidence that afferent influence suppression results from presynaptic inhibition at several levels in the central nervous system shows the importance of these mechanisms in motor control (Chapin and Woodward, 1982; Ghez and Pisa, 1972; Dietz et al., 1985).

It is known that PAD can be evoked in afferent fibres of various groups (Schmidt, 1973). The magnitude of the PAD phenomena varies from one group of afferents to another.

However, AAPs are developed only in low-threshold afferents (Brooks and Koizumi, 1956; Eccles et al., 1961), while subthreshold depolarization expressed as DRP originates mainly in terminals of high-threshold fibres (Eccles et al., 1962; Schmidt, 1973). In view of these differences, AAPs can be used as an index to the restriction of AI in low-threshold fibres and DRP predominantly in high-threshold afferents. These data help to explain why coupling between periods of maxima in AAPs and DRP was sometimes not observed during real locomotion.

It is note worthing that the sequence of DRP changes recorded during fictive locomotion differed substantially from that observed during real locomotion. The decrease in DRP in stance phase and the increase in swing phase are typical of fictive locomotion (Bayev, 1980; Bayev and Kostyuk, 1982). By contrast, during real locomotion two negative deflections of DRP occur during one cycle. These differences may be caused by interactions between central and peripheral mechanisms of DRP generation. But the results of the present study indicate that the value of DRP correlates with the intensity of AI and does not depend on the intensity of the efferent output. A similar correlation between DRP and AI was found to exist during passive movements of the hindlimb, i.e. when the spinal locomotor generator is "silent". These results indicate that oscillations in DRP during real locomotion are evoked predominantly by AI. Actually, if DRP was caused by the internal spinal generator, a strict relationship between DRP and motor output could be observed, as was the case during fictive locomotion (Bayev, 1980). Furthermore, this relationship was found to exist during passive hindlimb movements, when the spinal generator was not active and the muscle activity was of a purely reflex nature.

Nevertheless we found that the correlation between DRP and AI was greater in stance phases than in swing ones. What is the possible reason for this difference? If one compares real with fictive locomotion, DRP had only one deflection developed at swing phase in the latter case (Bayev, 1980; Bayev and Kostyuk, 1982). The fact that slightly less correlation exists here between Sw DRP and AI may therefore indicate that Sw DRP is evoked not only by peripheral but also by central mechanisms.

On the basis of recent literature and our own data, one may conclude that during locomotion AAPs in dorsal roots are initiated by both peripheral and central mechanisms (Bulgakova et al., 1985; Dubuc et al., 1985). Indeed, Dubuc et al., (1985) have shown that rhythmic AAPs occur during fictive locomotion, when phasic AI is absent. On the other hand our results prove that the sources of AAPs are mainly peripheral. It should be noted that: 1/ The intensity of AAPs increases during transition from swimming to stepping, i.e. when AI becomes more intensive. 2/ A clear correlation between AAPS and intensity of AI was observed. 3/ AAPs can be generated during passive hindlimb movements. Our data therefore indicate that the effciency of these two sources of AAP initiation is not equal and that during real locomotion, antidromic discharges are evoked mainly by effects of AI.

It can thus be concluded that DRP and AAPs are indexes to the presynaptic inhibition of high- and low-threshold afferent influences, respectively. During real locomotion, development of DRP and AAPs is determined mainly by AI intensity. It is therefore likely that the peripherally-evoked presynaptic mechanisms underlying AI restriction do in fact operate during various types of real locomotion.

References

Akazawa, K., Aldridge, J. W., Steeves, J. D., and Stein, R. B., 1982, Modulation of stretch reflexes during locomotion in the mesencephalic cat, J. Physiol., 329:553-567.

Bayev, K. V., 1980, Polarization of lumbar primary afferent terminals during fictitious locomotion, Neirofiziologiya, 12: 481-488.

Bayev, K. V., and Esipenko, V. B., 1985, Primary afferent depolarization in lumbosacral spinal cord during real locomotion, Neirofiziologiya. 17: 119-121.

Bayev, K. V., and Kostyuk, P. G., 1982, Polarization of primary afferent terminals of lumbosacral cord elicited by the activity of spinal locomotor generator, Neuroscience, 7: 1401-1409.

Brooks, C. M. C., and Koizumi, K. 1956, Origin of the dorsal root reflex, J. Neurophysiol., 19: 61-74.

Bulgakova, N. V. , Pilyavsky, and A. I., Yakhnitsa, I. A. , 1985, Phase-dependent modulation of antidromic spike potentials in rat dorsal root during locomotion, Neirofiziologiya, 17: 707-710.

Chapin, J. K., and Woodward, D. J., 1982, Somatic sensory transmission to the cortex during movement: Gating of single cell responses to touch, Exp. Neurology, 78: 654-669.

Chez, C., and Pïsa, M., 1972, Inhibition of afferent transmission in cuneate nucleus during voluntary movement in the cat, Brain Res., 40: 145-151.

Dietz, V., Quintern, J., and Berger, W., 1985, Afferent control of human stance and gait: evidence for blocking of group 1 afferents during gait, Exp. Brain Res., 61: 153-163.

Dubuc, R., Cabelguen, J. N., and Rossignol, S., 1985, Rhythmic antidromic discharges of single primary afferents recorded in cat dorsal root filaments during locomotion in the cat, Brain Res., 359: 375-378.

Duysens, J., and Pearson, K. G., 1980, Inhibition of flexor burst generation by loading ankle extensor muscles in walking cats, Brain Res., 187: 321-332.

Eccles, J. C., Kostyuk, P. G., and Schmidt, R. F., 1962, Presynaptic inhibition of the central actions of flexor reflex afferents, J. Physiol., 161: 258-281.

Eccles, J. C., Kozak, W., and Magni, F., 1961, Dorsal root reflexes of muscle group I afferent fibres, J. Physiol., 159: 128-146.

Forssberg, H., Grillner, S., and Rossignol, S., 1975, Phase dependent reflex reversal during walking in chronic spinal cats, Brain Res., 85: 103-107.

Grillner, S., 1975, Locomotion in Vertebrates: Central mechanisms and reflex interaction, Physiol. Rev., 55: 247-304.

Lennard, P. R., and Hermanson, J.W., 1985, Central reflex modulation during locomotion, Trends in Neurosci., 8: 483-486.

Levy, R., 1980, Presynaptic control of input to the central nervous system, Canad. J. of Physiol. a. Pharm., 58: 751-766.

Roberts, H. T., and Wallis, D., 1978, Dorsal and ventral root potentials recorded in vivo by the sucrose-gap method, J. Physiol., 277: 42-43.

Schmidt, R. F., 1973, Control of the access of afferent activity to somatosensory pathways, in: "Handbook of Sensory Physiology", Berlin: Springer. 2: 151-206.

Shik, M. L., and Orlovsky, G. N., 1976, Neurophysioloy of locomotion automatism, Physiol. Rev., 56: 465-501.

Yakhnitsa, I. A., Pilyavsky, A. I., and Bulgakova, N. V., 1985, Investigation of different kinds of locomotor movements in rats, Neirofiziologiya, 17: 183-189.

ROLE OF MOTOR CORTEX IN CONTROL OF LOCOMOTION

I.N. Beloozerova and M.G. Sirota

Institute of Biomedical Problems
Ministry of Public Health
Moscow, USSR

Summary

1. The activity of 252 motor cortex (MC) neurons (including 70 pyramidal tract neurons) was recorded extracellularly in the cat by means of a mobile electrode during free locomotion in a box.

2. The activity of 89% MC neurons was modulated during locomotion. The modulation was related to the stepping movements, since it increased in one stepping phase and decreased in the next.

3. MC neurons were also studied while the animal moved up a flat inclined surface, walking at different speeds, with a load of 85g attached to each forelimb, when the cat had to perform snakelike movements (turns) or walk on the flat surface placed in a horizontal plane. The pattern of MC neuron activity changed little under these conditions in comparison with uncomplicated locomotion.

4. The activity of 68 neurons was recorded in experiments with barriers and involving locomotion on a horizontal ladder which restricted the possible paw positions along the direction of locomotion. These tasks greatly affected the MC activity.

5. Neither bilateral MC lesion nor tetrodotoxin inactivation hampered uphill locomotion, walking along a moving floor, or locomotion involving turns and loaded forelimbs. On the contrary, it proved to be necessary for the MC to be intact for locomotion with space linked stepping limb movements (i.e. with barriers, on a ladder) to be possible.

6. Bilateral destruction of the ventrolateral nucleus of the thalamus (VL) resulted in a decrease in the rhythmical modulation of MC neurons during locomotion. After VL lesion the cat could walk quite well on the horizontal surface and uphill, at various speeds, with the forelimbs loaded; it could perform turns and could walk on the moving floor. The cat proved to be incapable, however, of walking with barriers and on the ladder.

Introduction

Locomotion under natural conditions always includes non-standard, "voluntary" components, which are necessary for overcoming obstacles, changing the direction of movements, and space orientation. In many cases, an animal has to carefully pick out the point at which its paw is to be placed. The system of reception with which the spinal locomotor generator itself is equipped is sufficient to ensure locomotion on a flat surface, but is not capable of adapting movements to a complex environment (Grillner and Rossignol, 1978). Natural locomotion is not possible without the contribution of supraspinal centres to the control of stepping movements. Commands from these centres reach the spinal mechanisms by various descending tracts (Asanuma, 1981; Armstrong, 1986; Arshavsky et al., 1986).

The activity of neurons in the reticulo-, vestibulo- and rubrospinal tracts during locomotion has been studied in decerebrated cats (Orlovsky, 1972 a & b; Orlovsky and Shik, 1976). These neurons were found to be rhythmically modulated, the modulation being related to the stepping rhythm. Similar results have been obtained on pyramidal tract cells in intact cats running on a treadmill band (Schmidt et al., 1976; Durelli et al., 1978; Palmer et al., 1980; Armstrong and Drew, 1984).

It is well known that the activity of motor cortex (MC) neurons changes during the performance of voluntary movements and in some cases, before the beginning of movements (Brooks and Stoney, 1971; Brooks, 1974; Conrad et al., 1974; Porter and Levis, 1975; Evarts and Tanji, 1976).

Lesion of the MC deprives an animal of the capacity to perform some voluntary limb movements, or at least interferes with the performance of these movements. But locomotion on a flat surface is still possible after this kind of lesion (Stepien et al., 1961; Denny-Brown, 1966; Burlachkova and Ioffé, 1978). Even complete removal of the cortex does not prevent locomotion on a treadmill or a flat surface (Orlovsky and Shik, 1976).

In the present study, we recorded the activity of MC neurons (including pyramidal tract neurons, PTNs) in a freely moving cat during locomotion. The external conditions for locomotion were varied as follows: (i) locomotion on a flat horizontal surface with no disturbances (which will be referred to as "normal" locomotion); (ii) the same as in (i) but with additional loading on some muscle groups; (iii) overcoming obstacles of various kinds

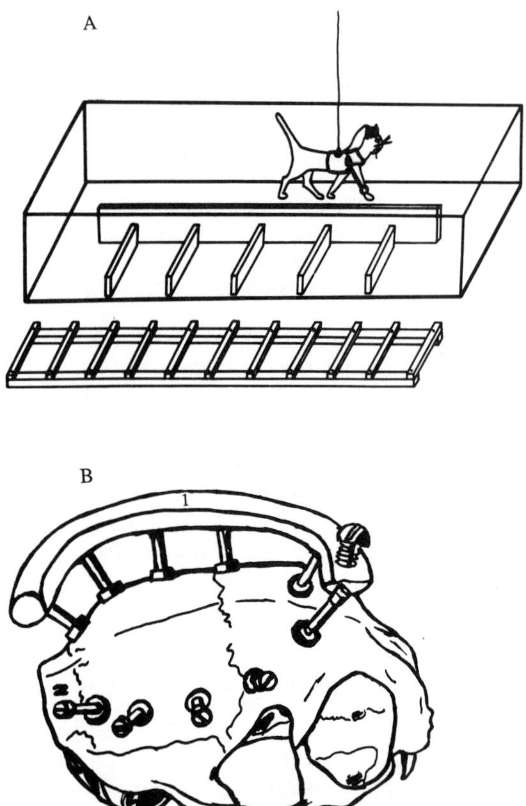

Fig.1. A: The experimental box was divided into two corridors (2.5 m long and 0.5 m wide); barriers or a horizontal ladder could be placed in one of them. The MC neuron recordings were carried out as the cat walked along the corridors. B: the cat's skull with a plastic base (1) fixed on it by means of screws (2). All the equipment for neuron recording was attached to the base.

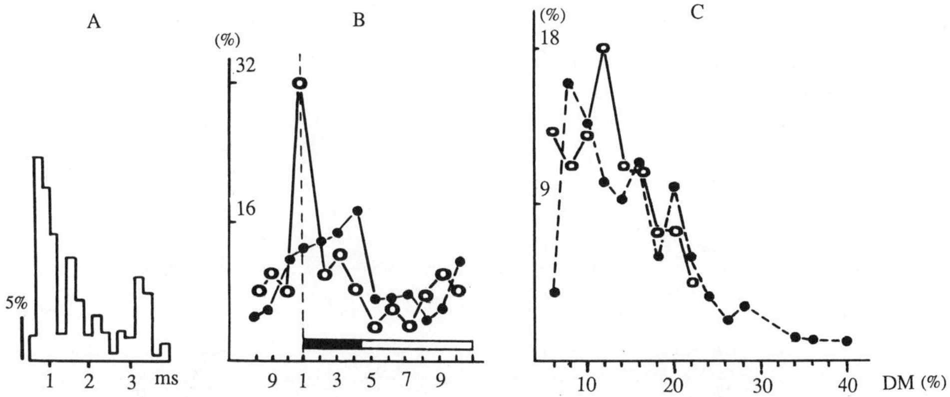

Fig. 2. A: Distribution of latent periods of antidromic responses. B and C: locomotion on the flat horizontal surface; B: distribution of points of maximum activity throughout the step cycle of the contralateral forelimb. The step was divided into 10 intervals, black and white bars - swing and stance phase, respectively. C: Distribution of values of depth of locomotor modulation (DM) open circles - PTNS, blackdots - non-PTNs.

during locomotion. We compared the activity of MC neurons under simple (i) and complicated conditions (ii) and (iii), assuming that the changes in the activity of MC neurons are likely to be related to changes in the task (a supraspinal "voluntary" component of movements) when changes in the movement itself (i.e. in the corresponding afferent inflow) are small.

Method

Seven cats, trained to walk or run inside a box in response to a sound signal, were used. A longitudinal wall divided the box into two corridors, each 2.5 meters long and 0.5 meters wide. On reaching the end of the one corridor, the cat returned taking the other one (Fig. 1A). Once training was completed, the cat underwent surgery, and a plastic base was fixed to its skull (without removing the skin) by means of radially inserted screws (Fig. 1B). A microdriver and an amplifier were then attached to the base. The bone and the dura above the MC (the area corresponding to the anterior and the rostral part of the posterior sigmoid gyrus) were removed, and the aperture was covered with a plastic plate. The plate had about 60-70 small holes, with thin glass tubes glued into them. The recording electrode was subsequently introduced through these tubes into the brain. A bipolar stimulating electrode was also implanted in the medullar pyramid (Snider-Niemer, 1960, coordinates P 10, R 1.5). The activity of MC neurons was recorded extracellularly by a mobile wolfram electrode (tip diameter about 5 μm) insulated with varnish. The movement of the forelimb contralateral to a MC cell was recorded by means of an electric tranducer attached to the limb, with which it was possible to distinguish between the stance and swing phases of the step. In other experiments, the stance phase duration was monitored by a contact attached to the paw. The signals were recorded on magnetic tape. When analyzing the neuronal activity, the step was divided into 10 intervals, and a post-event time histogram of spike distribution throughout the step was obtained for 10-50 successive steps. Besides, the mean value of the activity of a neuron in the step was assessed as well as the activity before locomotion. The difference between the maximum and minimum frequency in the step, divided by the mean frequency, was used as an index to the depth of rhythmic modulation.

Results

Locomotion on the horizontal flat surface

The activity of 252 MC neurons was recorded during locomotion on the flat horizontal surface ("normal locomotion"). Among these there were 70 PTNs which responded to high-frequency antidromic stimulation of the pyramidal tract. The distribution of the latency of antidromic responses is shown in Fig. 2A. The conduction speed of PTN

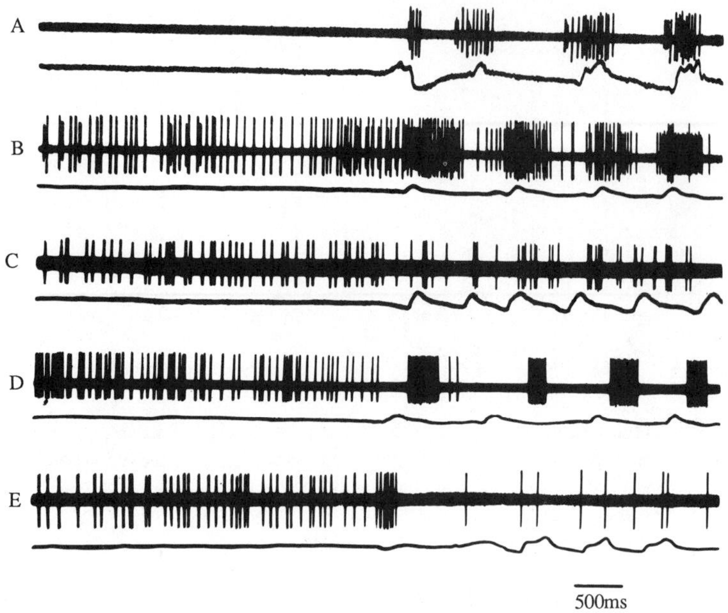

Fig. 3. Examples of activity of MC neurons before and during locomotion (flat horizontal surface). Bottom lines monitor the movement of a contralateral forelimb in rostro-caudal direction. Strongly modulated units (A, B, D), a unit without any clear rhythmical modulation (C) and a unit with decreased firing during locomotion (E) are shown.

axons ranged from 13 to 83 m/sec. Since we found no marked difference between the activity of PTNs and that of non-PTNs, the results of these two groups will be presented together except for a few cases which will be specially indicated. Seventy MC neurons had somatic receptive fields, i.e. they could be activated by tactile stimuli, by muscle squeezing or by passive joint movements.

The activity of 222 (89%) MC neurons was modulated during locomotion in relation to the stepping movements: it increased in one step phase and decreased in the next (Fig. 3). The maximum activity in 65% of "modulated" neurons coincided with the swing phase of the contralateral forelimb, and in 35% of neurons, with the stance phase. Fig. 2B shows the phase distribution of points showing maximum activity, for PTNs and non-PTNs. Most PTNs showed maximum activity at the beginning of the swing phase (black bar), while the maxima of most non-PTNs were distributed throughout the swing phase. The depth of modulation of PTNs and non-PTNs did not differ on the average but the maximum value with PTNs was 25%, while with non-PTNs it could reach 35-40% (Fig. 2C).

In the absence of locomotion, the discharge frequency of MC neurons was 1-30/sec, with an average of 9.8 ± 1 sec (Mean ± SD). The average activity (M ± m) of 34% of the neurons showed no change during the transition from standing to locomotion. In 36% of the neurons, the mean firing-rate increased about two-fold, and in 20% it even increased 3-5-fold. The activity of 10% of the neurons decreased to about half the initial level during locomotion.

We analyzed the activity of modulated neurons just before the first step performed by the contralateral forelimb. For this purpose, we counted the number of spikes in each 100 msec interval during the 2 sec period preceding the swing phase of the first step. No change in the activity of 70% of the neurons (34 out of 54) was observed during this period. The activity of 14 (25%) neurons increased and that of 2 neurons decreased. No rhythmical modulation of activity was observed in MC neurons before the onset of locomotion (Fig. 3).

Locomotion uphill, at various speeds, involving turns, loaded forelimbs, and locomotion on a moving floor

The power developed by the locomotor system of an animal increases in the case of uphill locomotion, provided the speed remains the same. Does any difference in the activity of MC neurons accompany the change in the power developed by limb muscles?

The activity of 37 MC neurons (including 21 PTNs) was recorded when the cat moved up the flat inclined (10°) surface. The average activity over the step cycle and the de of its modulation in 20 (55%) neurons during uphill locomotion were practically the same as in normal locomotion (Fig. 4A). In 10 (27%) neurons the average activity increased by 78 ± 16% during uphill as compared with normal locomotion. An increase in the depth of modulation was observed in 7 (19%) neurons, amounting on the average to 56 ± 19%. The changes in activity during uphill locomotion in all the neurons recorded are shown in the graph in Fig. 7A (see legend for explanation).

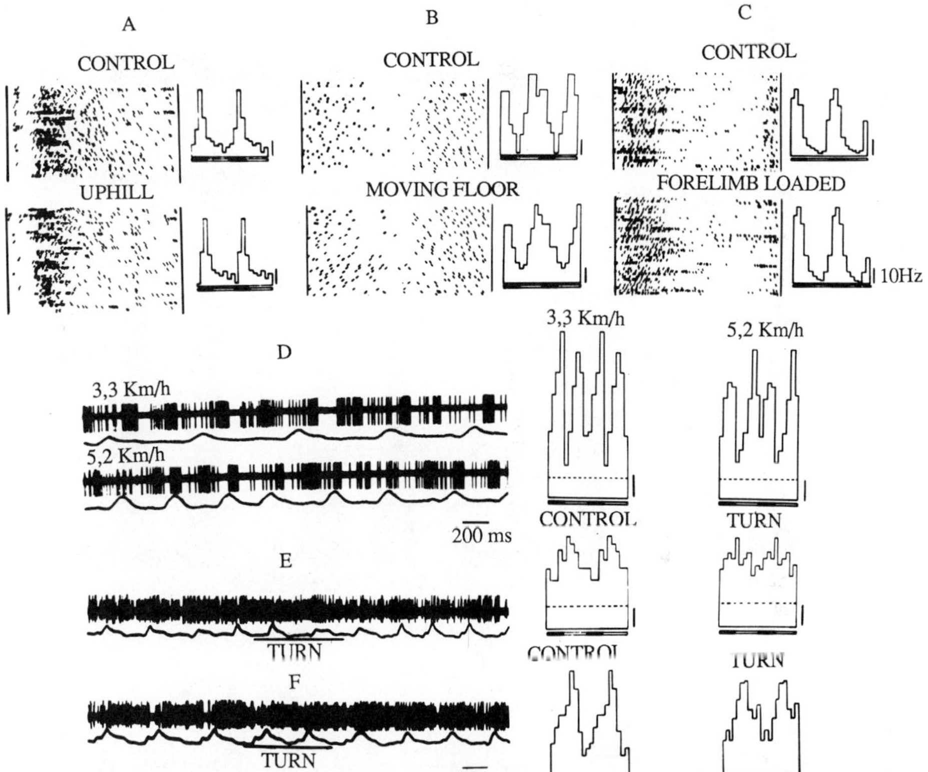

Fig. 4. A - C: examples of activity of MC neurons under various conditions. For each neuron the activity is shown during normal, simple locomotion (Control) and when conditions for locomotion were changed. In A, cat walked uphill (10°); in B it walked on the moving floor (side-to-side movements with frequency of 0.8/sec amplitude of 60 cm); in C with loads (85 g) attached to the elbows. The activity of each neuron is presented in the form of a raster, i.e. sequence of spikes in a cycle, for 30 successive steps. Each spike is represented by a short line: the inclination of the line indicates the phase of the step (right - swing, left - stance). A post-event time histogram of the firing frequency (throughout the step cycle), corresponding to the raster, is also presented. D: activity of MC neuron in a subject walking at two different speeds and corresponding phase distribution of spike activity. E, F: activity of two MC neurons (E and F) when a cat performed the turn (marked by a bar). As a control, in phase distributions, the activity before and after the turn was used. In D - F, the dashed line in the phase distribution indicates the average activity of a neuron before locomotion. A black bar under the phase distribution marks the swing phase of the forelimb contralateral to a neuron.

In our experiments, an animal sometimes spontaneously change its speed, which allowed us to compare the neuron activity at various speeds of locomotion. Fig. 4D illustrates locomotion at speeds of up to roughly twice the lowest speed. The average activity of a neuron, as well as the depth of its modulation, (see the histogram in Fig. 4D), changed little with the increase in the speed. This was also true of the other neurons tested (n = 7).

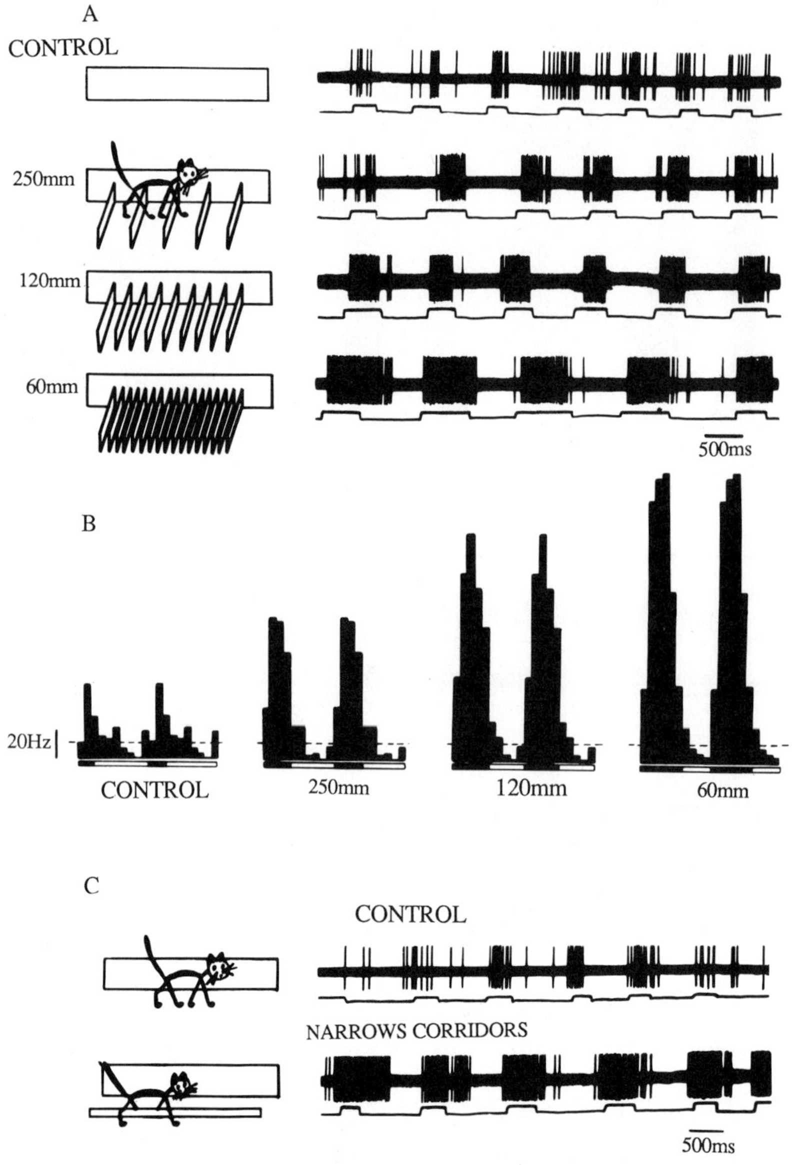

Fig. 5. Changes in the activity of a PTN with restrictions of possible limb positions when stepping. In A, a cat initially walked with no obstacles (Control), and then the barriers (70 mm in height, interbarrier distance as indicated) were arranged in the box. Corresponding phase distributions of the neuron activity are shown in B. C: after locomotion under normal conditions (Control), a longitudinal rod forming two narrow corridors (for left and right limbs) was placed on the floor. In A and C, the bottom trace shows the stance and swing phases of the forelimb contralateral to a neuron. (Deflection down and up, respectively).

Locomotion uphill, as well as locomotion at increased speed, requires greater limb muscle activity (in both flexors and extensors) (Orlovsky et al., 1966). It is therefore surprising that the activity of MC neurons changed so little under these conditions.

In another series of experiments, a load of 85 g was attached to each forelimb at the distal part of the elbow. During locomotion with the load, the average activity in 7 MC neurons (over the 11 tested) increased by 60 ± 10%; in 1 neuron it decreased; in 3 neurons, it remained the same. The depth of modulation in 5 neurons (45%) incrased by 30 ± 1%; in 3 neurons it decreased by 37 ± 7%; and in 2 neurons, it did not change. Fig. 4B shows an example of the neuron in which the pattern of activity persisted in spite of the load. The effects of the load upon all the neurons tested are shown in Fig. 7C.

On reaching the end of a corridor in the box, the cat turned and continued running, taking the other corridor. The beginning and the end of the turn were monitored by means of photodiodes. In a series of special experiments, a series of partitions was placed in the box in such a manner that a cat had to perform snake-like body movements, turning to the right, then to the left, and so on. The step length during the turn differed from that produced while walking straight ahead: it increased in the case of the "outer" limbs and decreased in that of the "inner" ones. If compared to incomplicated locomotion, the average activity during the turns increased by 56 ± 11% in 52 (70%) neurons (out of the 75 modulated MC neurons tested); in 7 neurons it decreased by 30 ± 3%; and in 17 (23%) neurons, it remained the same. The depth of modulation in 24 (32%) neurons increased by 64 ± 9%; in 16 (21%) neurons it decreased by 32 ± 3% and in 35 (47%) neurons, it did not change. Examples of turns are shown in Fig. 4E, F. The data on all the neurons tested in one cat are summarized in Fig. 7B.

In a special series of experiments, the supporting surface was made unstable. For this purpose, the whole box with a walking cat was put on a platform performing sine oscillations in a horizontal plane (the period was 1.3 sec, and the amplitude, 60 cm). The direction of the oscillations was perpendicular to that of locomotion. Under these conditions, the pattern of stepping movements (joint angles, muscle forces, etc.) varied continuously and considerably. Nevertheless, the pattern of MC neuron activity changed little as compared to simple locomotion. In most of the 35 MC neurons tested (including 19 PTNs), neither the average activity nor the depth of modulation changed. In only 7 (20%) neurons was a change in pattern observed. In 3 of these, the average activity increased by 112 ± 36%. The depth of modulation increased in 2 neurons and decreased in 4, the change being 49 ± 10%. Fig. 4C shows the activity of a PTN during locomotion on stable and unstable supporting surfaces, and the histograms of spike activity for these two cases. The persistence of the pattern of activity observed in most MC neurons is also illustrated in Fig. 7D.

Summarizing the results presented in this section, we should like to emphasize that most of the variations and complications of locomotion we used, i.e., different speeds, uphill movement, turns, movement of the floor, and additional loading of the forelimbs, resulted in relatively little change in the pattern of rhythmic activity of MC neurons. This will become more obvious later, when we come to report the great changes in the pattern which occurred under visually controlled locomotion conditions.

Influence of destruction and TTX inactivation of MC upon locomotion on a flat surface

At the end of each experiment, we destroyed the MC bilaterally. Under nembutal anesthesia the MC grey matter was sucked out until the white matter became visible under the holes used previously for the recording electrodes. The capacity of animals for locomotion was tested within two days of the MC ablation. Another method for inactivating the MC consisted of poisoning the cortex by bilateral application of tetrodotoxin (TTX) (5 μl, 10^{-4} mol, Armstrong and Drew, 1984). The drug was injected through a needle which replaced the recording electrode. Recording of MC neurons at a distance of 2 mm from the needle tip showed that their electric activity disappeared within about 10 min of the TTX injection. At that time the capacity of an animal for locomotion was tested.

Animals deprived of the MC, as well as those with the MC inactivated with TTX, seemed to have no difficulty in performing the whole range of our tests: normal locomotion

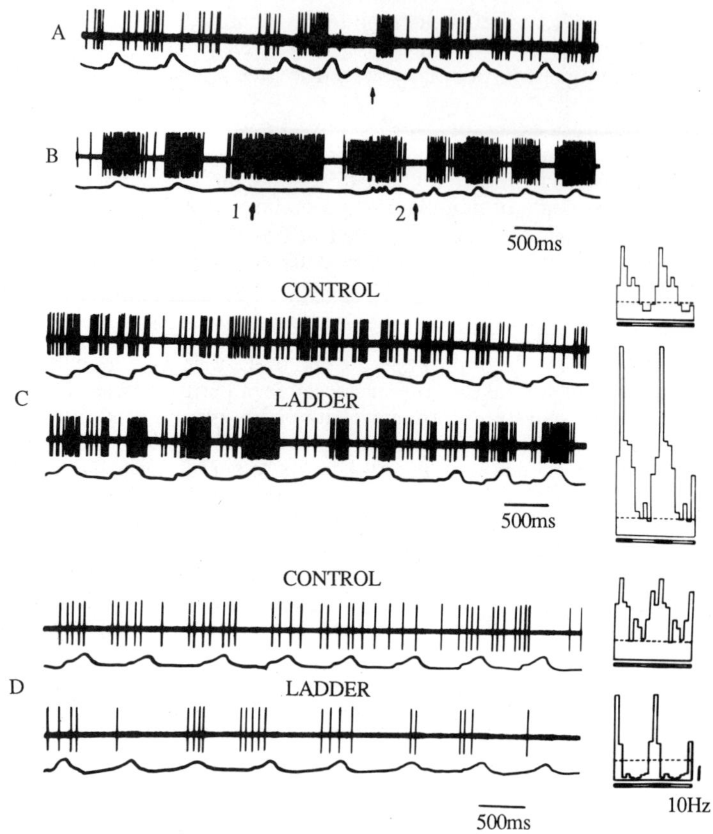

Fig. 6. A, B: activity of various PTNs in a cat overcoming a single barrier (height 70 mm in A and 200 mm in B). The arrow in A indicates the moment at which the contralateral forelimb was above the barrier. The first arrow in B indicates the moment when an animal stopped before the barrier, the second one indicates that when the cat jumped over the barrier. C, D: changes of the activity of two PTNs in a subject walking on the horizontal ladder (interrung distance, 250 mm) as compared to normal locomotion (Control). Corresponding post-event time histogram are presented on the right. In the first unit, the depth of modulation increased due to the growth of a maximum activity, and in the second unit, due to the decrease in activity in between the maximums.

at various speeds, locomotion uphill, locomotion on a moving floor, locomotion with additional loading of the forelimbs, and turns. This is in agreement with the results obtained by other authors (Stepien et al., 1961; Denny-Brown, 1966), who have demonstrated that bilateral MC lesion does not disturb locomotion on a flat surface.

Walking with specially restricted limb positions

Several series of experiments were carried out, in which a cat had to use its visual system for controlling (or correcting) its stepping movements.

1. Walking with barriers. We restricted the possible paw positions on the floor along the X axis (the direction of locomotion) by arranging barriers along the cat's path (the height of the barriers was 70 mm, and the distance between successive barriers 250 mm or less, Fig. 5A). In order to walk along the box, an animal had to step over the barriers and to place its paw in between them.

The activity of 68 MC neurons, including 12 PTNs, was recorded in the experiments with barriers (interbarrier interval, 250 mm). The average activity during locomotion with barriers differed from that in normal locomotion in 54 (79%) MC neurons. In 34 neurons (including 7 PTNs) it increased by 75 ± 15%; in 20 neurons (including 4 PTNs) it decreased by 30 ± 4%. The depth of modulation changed in 57 (84%) MC neurons. It increased in 40 (59%) neurons (including 9 (75%) PTNs). In the case of non-PTNs, the increase was 35 ± 5%; in that of PTNs, 67 ± 32%. In 17 (25%) neurons (including 1 PTN) the depth of modulation decreased by 21 ± 3%. The results obtained in one cat are shown in Fig. 7E.

The deepening of modulation observed in most MC neurons in subjects overcoming barriers was determined by two factors: (i) an increase in the firing rate during the phase of the maximum activity and (ii) an inhibition of the discharge in between the maxima. Usually (in 91% of cases) the phases of maximum activity were the same in both normal and complex locomotion, as in the case of the neuron shown in Fig. 5A.

In a few experiments, we tested the activity of the same MC neuron not only with a 250 mm interbarrier interval but with shorter intervals as well, i.e. such that the animal had to perform stepping with greater accuracy. In all the neurons tested the average activity and depth of modulation increased with the shortening of the interbarrier interval, as in the case shown in Fig. 5 A, B.

In this series of experiments we used not only regularly arranged barriers but, in a few cases, a single barrier, with a view to investigating what strategies an animal used to overcome the obstacle. When the height of a barrier was the same (70-90mm) as in the experiments described above, an animal could overcome it without interrupting a normal sequence of steps, as shown in Fig. 6A, but the amplitude of movement at the joints increased in the step with the obstacle (visual observation). As can be seen in Fig. 6A, a MC neuron burst was more intense in the step with the obstacle, but its phase remained unchanged. When the height of a barrier was 200mm, the animal stopped in front of it and then jumped over it; in this case, the regular modulation of MC neurons was disturbed and then resumed with the beginning of regular stepping (Fig. 6B).

2. Walking on a ladder. We used a horizontal ladder with 50 mm rungs, the distance between successive rungs being 250 mm (i.e. equal to an average cat's step). When walking on the ladder, the possible positions of a limb landing on the supporting surface were farther restricted along the X axis, as compared to walking with barriers. The activity of 108 MC neurons was recorded in a subject walking on the ladder. Among these, there were 24 PTNs and 7 neurons responding antidromically to stimulation of the ipsilateral N. ruber (corticorubral neurons, CRNs). Examples of recordings are shown in Fig. 6 C, D.

In a subject walking on the ladder, as compared to normal locomotion, the average activity changed in 88 (81%) MC neurons. In 77 (71%) neurons it increased by 56 ± 6%; and in 11 (10%) neurons it decreased by 28 ± 4%. The depth of modulation was also altered in 87 (80%) MC neurons (including 19 (80%) PTNs and 6 (88%) CRNs). In 61 (56%) neurons (including 12 (50%) PTNs and 4 (57%) CRNs) it increased by 40 ± 3%; in the case of PTNs the increase being 46 ± 8%; in that of CRNs, 34 ± 18%; and in that of nonidentified MC neurons, 36 ± 5%. The depth of modulation decreased in 26 neurons (including 7 PTNs and 2 CRNs) by 35 ± 4%.

During locomotion on the ladder, the deepening of modulation observed in most MC neurons resulted from the same causes as in walking with barriers, i.e. the increase in firing rate in the phase of maximum activity (Fig. 6C) and/or inhibition of the discharge in the phase of minimum activity (Fig. 6D). The phases of maximum and minimum activity persisted in the overwhelming majority of neurons, as in the examples given in Fig. 6 C, D.

3. Walking along two narrow corridors for left and right limbs. We restricted the possible limb positions along the Y axis (perpendicular to the direction of locomotion) by disposing a long longitudinal rod on the box floor in such a manner that only narrow corridors (50 mm width) between the rod and the box walls were left free for the animal to place its feet (Fig. 5C). Ten neurons were tested in these experiments, and their activity was compared to that recorded during normal locomotion. In 8 neurons, both the average activity and the depth of modulation increased (by 50 ± 3% and 70 ± 5%, respectively) during locomotion along narrow corridors, as shown in the example in Fig. 5C.

Table 1

Percentage of MC neurons in which a change in activity (as compared to that recorded during normal locomotion) accompanied the performance of complex locomotor tasks.

	Uphill	Various speeds	Loaded forelimbs	Turns	Moving floor	Barriers	Ladder	Narrow Corridors
Change in average activity	27	10	60	70	9	79	81	60
Change in depth of modulation	19	8	45	53	17	84	80	72

The main results of this and preceding sections are summarized in Table 1, which shows the percentage of MC neurons in which a change in activity accompanied various complications of the locomotor task. One can see that **Barriers, Ladder** and **Narrow Corridor** are the tasks which most affected the MC activity. In all these tasks, an animal had to link its stepping movements to the coordinates of the outer space.

Influence of destruction and TTX inactivation of MC on locomotion involving space-linked limb movements

Bilateral MC lesion and TTX inactivation hampered neither uphill locomotion, locomotion along the moving floor, nor locomotion with loaded forelimbs (see above). On the contrary, the intact state of the MC proved to be necessary for locomotion involving space-linked stepping limb movements to be possible.

After the MC ablation or TTX inactivation, the animal could no longer step over obstacles (it knocked them over). It could not walk on the ladder either (it missed the rungs and fell down). The cat failed to place its paws at the appropriate points for 5-7 days after the MC lesion, then its attempts became more successful, and within about one week the animal managed to carry out all the locomotor tasks without making any mistakes.

Influence of destruction of N. ventralis lateralis thalami on locomotion and on activity of MC neurons

One of the main inputs to the MC is through the ventrolateral nucleus of the thalamus (VL), which transmits signals from the cerebellum and basal ganglia (Strick and Sterling, 1974). We tried to eliminate this input by destroying the VL bilaterally. For this purpose a macroelectrode was inserted into the VL region stereotaxically and a current of 3 mA was delivered through it for 30 sec. Histological inspection confirmed the destruction of the VL. The results obtained in this animal are presented below.

On the day following the VL lesion the cat could walk quite well on the horizontal surface and uphill, at various speeds, with a loaded forelimb; it could perform turns and walk on the moving floor. On the contrary, the cat proved to be incapable of walking with barriers or on the ladder. The cat's mistakes while stepping were the same as after the MC destruction (see above): it could not step over an obstacle and missed the rungs of the ladder. For about 3 days the animal was completely unable to perform stepping placing its paws accurately. Then its attempts became more successful, and within one week it could overcome the obstacles with only a few mistakes.

In this cat we recorded the activity of 19 MC neurons before the VL lesion and that of 66 MC neurons after the lesion. Comparing the behaviour of MC neurons before and after the VL destruction, one can see an increase (76%) in their average activity (17 ± 1 versus 9.7 ± 1 per sec) and a decrease (36%) in the depth of modulation (8.5 ± 1 versus 11.6 ± 1). The activity of MC neurons was also recorded on the 4th day when the operated cat made the first attempts to walk on the ladder. In this case as well, the average activity increased while the depth of modulation remained small (Fig. 8).

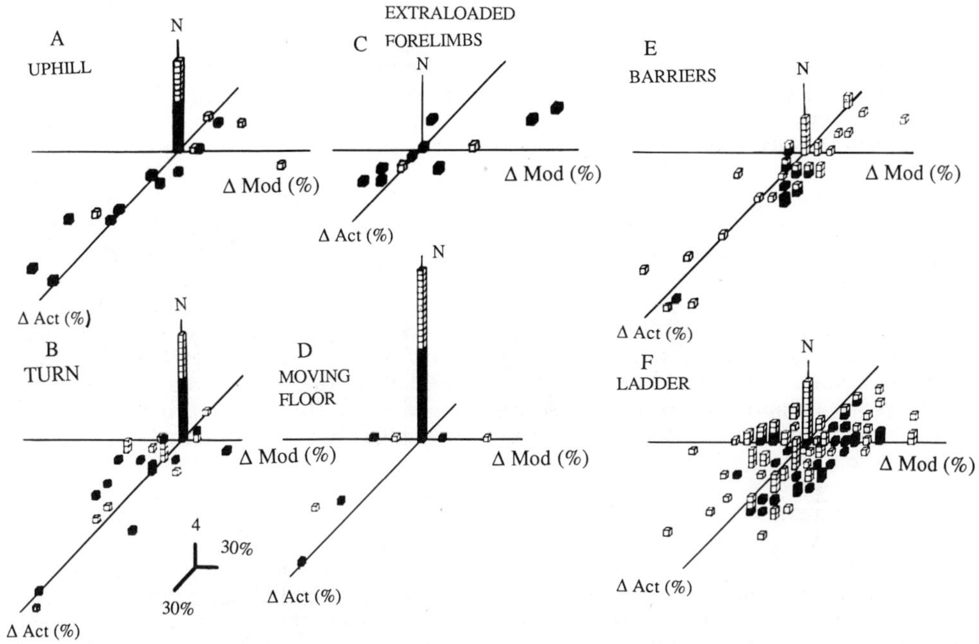

Fig. 7. Changes in the activity of MC neurons with complex locomotor tasks, as compared to normal locomotion. Horizontal axes: change in the average activity (Act) and the depths of modulation (Mod); vertical axis: number of neurons. In the graph each neuron is shown as a cube (black, PTN; white, non-PTN). One can see that in cases A, B, C and D most neurons did not change their pattern of activity, while in E (barriers) and F (ladder) the majority of neurons showed a change in either the average activity, or the depth of modulation, or both these characteristics.

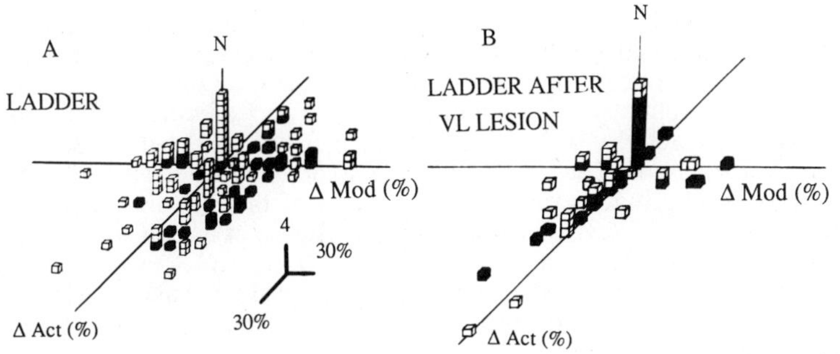

Fig. 8. The effect of VL destruction. A: changes in the activity of MC neurons in subjects walking on the ladder as compared to normal locomotion (repetition of Fig. 7F). B: the same after bilateral VL destruction. One can see that transition from normal locomotion to walking on the ladder, after the VL lesion, is not accompanied by any increase in the depth of modulation but only by an increase in the average activity level.

Discussion

Uncomplicated locomotion. When a cat walked directly on the flat horizontal surface with no obstacles, the activity of most MC cells (both PTNs and non-PTNs) was rhythmically modulated, the modulation being related to the rhythm of the stepping movements. This rhythmic modulation arose only with the onset of locomotion. In some MC neurons, however, a tonic change in activity, could be observed prior to the first step. This change seems to be associated with the postural ajustments involved in the transition from standing to locomotion.

Various MC neurons showed maximum activity in various step phases, but the majority of them were more active in the swing phase of the corresponding limb. This is in agreement with the commonly held view that the pyramidal tract exerts its action mainly on the limb flexors (Brooks and Stoney, 1971; Asanuma, 1981).

Our results, as far as simple locomotion of freely moving cats is concerned, are in complete agreement with those obtained on cats walking on a treadmill band (Armstrong and Drew, 1984a): in both cases, most MC cells were modulated, the maximum overall activity being in the swing phase.

What is the source of the rhythmic signals modulating the activity of MC neurons? Armstrong and Drew have demonstrated (by blocking corresponding nerves with novocaine) that elimination of afferent flow from the receptive field of MC neurons did not abolish the rhythmic modulation of this neuron (Armstrong and Drew, 1984b). This finding suggests that the source of modulatory signals is not at the periphery but rather in the central neuronal mechanisms.

In experiments on decerebrated cats walking on a treadmill band (Asanuma, 1981), it has been demonstrated that neurons of various descending tracts of the brainstem are rhythmically modulated in relation with the stepping cycle. The source of modulatory signals proved to be the spinal neuronal mechanism generating the stepping rhythm. It was also found that spinal influences upon the neurons of descending tracts were mediated by the cerebellum. It seems likely that MC neurons receive the modulatory commands via the same route since their rhythmical modulation was found to decrease considerably after destruction of the VL, the nucleus mediating cerebellar influences upon the MC.

Locomotion with various loads. During locomotion at higher speeds or uphill, limb muscles have to develop more force which is provided by an additional excitation of the appropriate motoneurons. The present study shows that the pattern of MC neuron activity changes little with the increase in the load upon the motor apparatus. In a minority of neurons only, there were changes in the average activity, and in an even smaller number of neurons, in the depth of modulation. These results are in agreement with those obtained in connection with treadmill locomotion, with a cat walking at various speeds or uphill (Armstrong and Drew, 1984a).

We also observed here that the pattern of activity of MC neurons changed little when locomotion was complicated in other ways, namely when a cat performed turns and when it walked on the moving floor. Without doubt, in these cases not only the afferent activity of the spinal mechanisms but also the afferent flow from various limb receptors changed. The fact that the pattern of activity of MC neurons changes little under these conditions means first, that the afferent signals from limb receptors do not play a decisive role in modulating the MC neurons. This provides further confirmation of the idea that modulation of MC neurons originates at central level. Secondly, the persistence of the MC output pattern, while changing the spinal cord afferent activity, suggests that the MC does not participate in the control of stepping movements in the locomotor tasks considered above. This suggestion is further supported by the fact that the destruction of the MC did not hamper the performance of these locomotor tasks.

Locomotion involving space-linked limb movements. The main finding in the present study is that the pattern of activity of most MC neurons changed dramatically when an animal had to perform stepping movements with accuracy, i.e. had to take into account, when stepping, the peculiarities of the surrounding space. Since the information concerning these peculiarities was obtained via the visual system, one may conclude that the activity of MC neurons changes considerably when the visual system is involved in the control of locomotion.

Since destruction of the MC makes an animal incapable of performing space-linked stepping (Barrier and Ladder tests), it seems very likely that the activity of pyramidal tract neurons, involving very pronounced rhythmical modulation, as observed in these experiments, constitutes the cortical commands addressed to the spinal stepping mechanisms. These commands suitably adjust the operation of the spinal mechanisms to external conditions. This is in agreement with Hancock's point of view (1985).

Do the brain mechanisms (those of the MC in particular) replace the spinal mechanisms involved in locomotion when an animal performs stepping movements under visual control? In our opinion, they do not replace the spinal mechanisms, which in all kinds of locomotion determine the sequence of muscle contractions for a given limb and interlimb coordination, as well as the stepping rhythm. This point of view is based on the fact that the phase of activity of most MC neurons in the step cycle does not change when locomotion falls under visual control. Besides, the decrease in the rhythmic modulation of MC neurons observed after the VL lesion indicates that this rhythmicity has an external (as regards the MC) origin. We would like to extrapolate to the MC the hypothesis (Arshavsky et al., 1986) that all the commands addressed from the brain to the spinal mechanisms during locomotion must take the current state of the spinal mechanisms, i.e. the phase of the locomotor cycle, into account. Only on this condition can the highest brain centres correct stepping movements without disturbing the basic locomotor pattern.

References

Armstrong, D.M., and Drew, T., 1984a, Discharges of pyramidal tract and other motor cortical neurons during locomotion in the cat, J. Physiol., 346: 461-470.

Armstrong, D.M., and Drew, T., 1984b, Locomotor related neuronal discharges in cat motor cortex compared with peripheral receptive fields and evoked movements, J. Physiol., 346: 471-496.

Armstrong, D.M., 1986, Supraspinal contribution to the initiation and control of locomotion, Progress in Neurobiol., 26: 273-361.

Arshavsky, Ju.I., Gelfand, I.M., and Orlovsky, G.N., 1986, Cerebellum and rhythmical movements, Springer, Berlin.

Asanuma, H., 1981, Input-output relationship in cat's motor cortex after pyramidal section, J. Neurophysiol, 46: 694-703.

Brooks, V.B. and Stoney, S.D. Jr., 1971, Motor mechanisms: the role of the pyramidal system in motor control, Ann. Rev. Physiol., 33: 337-392.

Brooks, V.B., 1974, Some examples of programmed limb movements: Motor aspects of behaviour and programmed nervous activities, Brain Res., 71: 299-308.

Burlachkova, N.I. and Ioffe, M.E., 1978, About motor cortex nuclear specificity in organization of precision motor reaction, J. visch. nerv. deijt., 28: 475-481 (in russ.).

Conrad, B., Wiesendanger, M., Matsunami, K., and Brooks, V.B., 1974, Precentral unit activity related to control of arm movements, Exp. Brain Res., 71: 299-308.

Denny-Brown, D., 1966, The cerebral control of movement. Thomas, Springfield.

Durelli, L., Schmidt, E.M., McIntosh, J.S. and Bak, M.J., 1978, Single unit chronic recordings from the sensorimotor cortex of unrestrained cats during locomotion, Exp. Neurol., 62: 580-594.

Evarts, E.V., and Tanji, J., 1976, Reflex and independed responses in motor cortex pyramidal tract neurons of monkey, J. Neurophysiol., 39: 1069-1080.

Grillner, S., and Rossignol, S., 1978, On the initiation of the swing phase of locomotion in chronic spinal cats, Brain Res., 146: 269-277.

Hancock, J. , 1985, Motor cortical discharges and locomotion in the cat, J. Physiol., 364, 28P.

Orlovsky, G.N., Severin, F.V., and Shik, M.L., 1966, The influence of speed and load on the coordinated movements during dogs locomotion, Biophisika, 11: 364-366 (in russ.).

Orlovsky, G.N., 1972a, Activity of vestibulospinal neurons during locomotion, Brain Res., 46: 85-98.

Orlovsky, G.N., 1972b, Activity of rubrospinal neurons during locomotion, Brain Res., 46: 99-112.

Orlovsky, G.N., and Shik, M.L., 1976, Control of locomotion: a neurological analysis of the cat locomotor system, Intern. Rev. Physiol., 10: 281-317.

Palmer, C., Marks, W.B., Bak, M.J. and Pedersen, G., 1980, The activity of closely spaced motor cortical pyramidal tract projecting neurons during locomotion, Neurosci. Abstr., 6: 158.

Porter, R.S., and Mc.D. Levis, M., 1975, Relationship of neuronal discharges in the precentral gyrus of monkeys to the performance of arm movements, Brain Res., 98: 21-36.

Schmidt, E.M., Bak, M.J. and McIntosh, J.S., 1976, Long term chronic recording from cortical neurons, Exp. Neurol., 52: 496-506.

Snider, R.S. and Niemer, W.T., 1961, A stereotaxic atlas of the cat brain. The University of Chicago Press, Chicago, 110 p.

Stepien, I., Stepien, L. and Konorski, J., 1961, The effects of unilateral and bilateral ablations of sensorimotor cortex on the instrumental (type II) alimentary conditioned reflexes in dogs, Acta biol. exptl., 21: 121-140.

Strick, P.L. and Sterling, P., 1974, Synaptic termination of the ventrolateral nucleus of the thalamus in the cat motor cortex. A light and electron microscope study, J. Comp. Neurol., 153: 77-106.

ANTICIPATORY POSTURAL ADJUSTMENTS AND DYNAMIC ASYMMETRY OF VOLUNTARY MOVEMENT

S. Bouisset and M. Zattara

Laboratoire de Physiologie du Mouvement
U.A. C.N.R.S. 631
Université de Paris-Sud
Orsay, Paris, France

Summary

Anticipatory postural adjustments (APA) associated with intentional movement were studied in relation to dynamic asymmetry of the movement (bilateral movements, or unilateral ones with and without an added inertia), which integrates the effect of movement on the body structure. Three methods were used simultaneously: local movements were studied by means of electromyography and accelerometry, and general body movements were studied using a force platform.

The results show that APA are specific to the dynamic asymmetry in healthy subjects: they are "preprogrammed" and it can be assumed that they constitute a part of the motor program.

The biomechanical finality of APA may be argued on the basis of a simple biomechanical analysis: they tend to create inertial forces which, when the time comes, will counterbalance the disturbance of balance due to the intentional forthcoming movement.

In Parkinsonian patients, APA no longer exist and the movement velocity decreases.

Introduction

It has been widely accepted since Gelfand et al. (1966) that the programming of a motor act includes a motor and a postural component. The latter starts before the onset of intentional movement and the term anticipatory postural adjustments (APA) is given to these postural activities which precede voluntary movement. Anticipatory postural adjustments show a segmentation into two periods: 1) an earlier one which is slow, the aim of which is to modify the body posture so as to prepare the execution of movement (see Gouny et al., 1977; Brenière et al., 1987); 2) a later one which is dynamic.

This dynamic phase which is at present under consideration has been extensively studied since the pioneering research by Belenkii et al. (1977). Diverse types of movements have been studied and it has shown that these anticipatory postural adjustments are specific to the characteristics of the forthcoming movement. Therefore, it has been said that anticipatory postural adjustments are "preprogrammed".

More precisely it can be assumed that anticipatory postural adjustments depend on two main factors: 1) the constraints of the motor act (body situation, physical environment and **interactions** between them, which are defined by the task instructions: speed, amplitude, precision, ...); 2) the actual functional state of the sensori-motor and musculo-skeletal systems (impairment due to disease or trauma, fatigue, conditioning, etc). Our own interest in the field of anticipatory postural adjustments lies within this context.

Given that the body is a symmetrical anatomical structure in relation to its sagittal plane, it was decided to focus our study on the effect of a factor which integrates the effect of movement to body structure, i.e. the influence of **dynamic asymmetry** on healthy and disabled subjects.

A definition of what we mean by dynamic asymmetry might be appropriate here. Human limb movements can be unilateral or bilateral and inertial loads can be added to the moving limb. It is assumed that movement constitutes a perturbation of balance. And, from a biomechanical view point, raising both upper limbs simultaneously is not the same as raising a mass equivalent to that of the two upper limbs, on the same side: this condition can be fulfilled by adding an inertial load to the upper limb which is to be moved.

Methods

Subjects standing normally were required to perform unilateral upper limb movements (UF), according to a simple reaction time paradigm, at a maximal speed, without (OUF) and with (IUF) an inertial load, and bilateral upper limb movements (BF). The inertial load consisted of a lead bracelet (1 Kg) fastened to the wrist.

This paradigm offers several advantages: in particular, it makes it possible to divide the body into two parts: the upper limb(s) which are voluntarily moved and the rest of the body.

Local movements were studied by means of electromyography (EMG) and accelerometry (ACG). General body movements were studied using a force platform, which gives the resultant forces as well as the resultant torque about the vertical axis.

Contrary to most previous studies, the influence of the experimental factor under consideration was systematically examined in a quantitative manner.

Study of local movements

This study included a systematic description of APA, mostly from a temporal point of view (Bouisset and Zattara, 1981, 1983; Zattara, 1982).

EMG anticipatory activities

The pattern of EMG postural activities was specific to the intentional movement (Fig. 1). Synchrony of homonymous muscle activities was the rule with BF, contrary to UF with which asynchrony increased when an inertial load was added to the upper limb.

As far as APA were concerned, we observed a given sequence of deactivation (-) and activation (+): SOLi(-), TFLc(+) and RFc(+), STi(+) and GMi(+), ESc(+) for UF and both SOL(-), ST(+)/GM(+) and ES(+) for BF. There were no consistent anticipatory EMG modifications in either foot or neck muscles. The APA increased from BF to OUF and from OUF to IUF, i.e. when dynamic asymmetry was increased. Indeed, the mean APA ranged from 25-30 ms with BF, 40-45 ms for OUF and 60-70 ms with IUF: they almost doubled from BF to IUF.

It has been shown that in parkinsonian patients, EMG activities no longer show any anticipation or specificity (Bazalgette et al., 1985).

ACG anticipatory activities

The ACG activities involved all the body segments and their pattern was also specific to the forthcoming movement (Fig. 2). The existence of accelerations indicates that anticipatory activities are movements.

The ACG had the same main features as the EMG: a) identity of APA at a given level was the rule with BF, contrary to UF and b) duration of APA increased from BF to OUF and from OUF to IUF; they almost doubled from BF to IUF, showing values similar to EMGs.

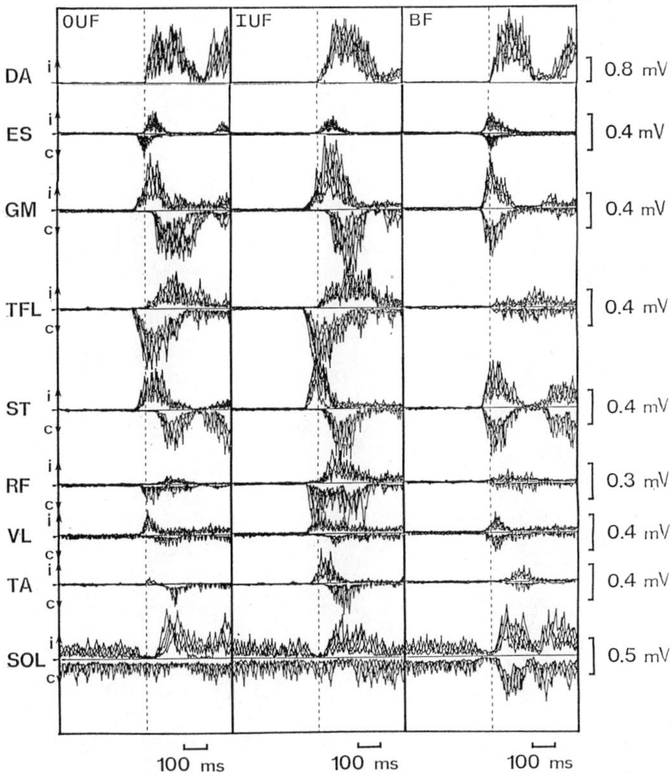

Fig. 1. EMG activities of lower limbs and pelvis recorded under the three experimental conditions. EMG activities are given from one subject performing the three types of movements. With each type of movement, rectified and smoothed EMG activities of five trials were superimposed by synchronizing records on the onset of DA_i activities (dotted line). From left to right: OUF, unilateral flexions with no additional inertia; IUF, unilateral flexions with an additional inertia; BF, bilateral flexion. From top to bottom: activities of anterior portion of Deltoideus (DA), Erectores Spinae (ES), Gluteus Maximus (GM), Tensor Faciae Latae (TFL), Semi-tendinosus (ST), Rectus Femoris (RF), Vastus Lateralis (VL), Tibialis Anterior (TA), Soleus (Sol), Ipsilateral (i) and contralateral (c) activities with respect to the moving limb are displayed in opposite directions.

There was a given order in anticipatory ACG: first contralateral lower limbs and hip, and then ipsilateral ones with IUF, and, both sides simultaneously with BF.

In parkinsonian patients, ACG activities no longer show any anticipatory ACG or specificity. Furthermore, the velocity of intentional movement decreases (Bazalgette et al., 1985).

Moreover, it has been shown that the peak amplitude of anticipatory ACG increases in a quadratic manner with the peak velocity of the intentional movement. Besides, at a given velocity, the peak amplitude is higher with inertial loaded movements (Zattara et Bouisset, 1983).

Interpretation

From these results, it can be seen that APA are specific to the dynamic asymmetry of forthcoming movements in healthy subjects: APA are "preprogrammed". But this is not the case with parkinsonian patients. As the velocity of intentional movement decreases in these patients, the question arises as to whether APA are necessary for the intentional movement to be performed efficiently: without appropriate APA, is intentional movement bound to be slower?

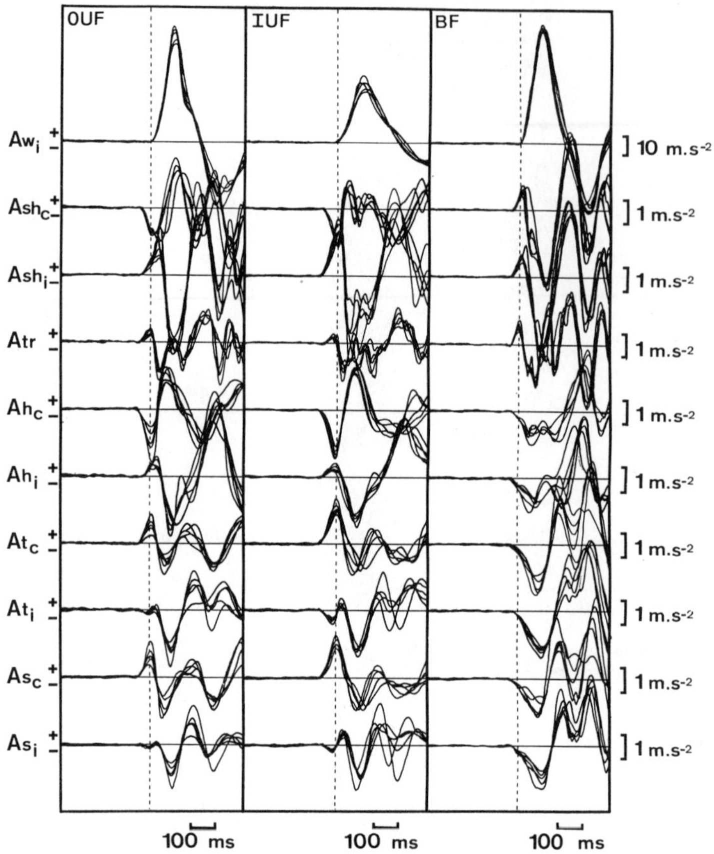

Fig. 2. Local accelerations recorded under the three experimental conditions. Antero-posterior local accelerations are given from one subject performing the three types of movement. With each type of movement accelerations of five trials were superimposed by synchronizing records on the onset of Aw_i (dotted line). From left to right: OUF, unilateral flexions with no additional inertia; IUF, unilateral flexions with an additional inertia; BF, bilateral flexion. From top to bottom: Aw_i, tangential acceleration of the upper limb measured at wrist level (positive sign corresponds to the acceleration phase of the movement); Ash, Atr, Ah, At, and As, antero-posterior accelerations measured at level of shoulders, trunk, hips, thighs and shanks (positive sign corresponds to forward accelerations); i and c, ipsilateral and contralateral accelerations with respect to the moving limb.

Study of general body movements

The general shape of the kinetic curves was reproducible under to a given experimental condition. Upper limb movement was always diphasic: an acceleration was followed by a deceleration phase (Fig. 3). As far as the resultant force or the resultant moment were concerned, the following was observed. During BF, ΔRz and Rx were diphasic, while the variations in Ry and Mz were nil or negligible. On the contrary, with UF, resultant force and moment showed consistent variations with all the variables considered, but these were more complex than with BF (Bouisset and Zattara, in press).

As can be seen in Fig. 3, ΔRz and Rx variations began before the onset of Aw; thes accelerations were positive, i.e. corresponded to upward and forward accelerations of the body center of gravity. Ry and Mz were positive with UF: these accelerations corresponded to a lateral acceleration of the body center of gravity directed toward the contralateral side and induced a moment which was also directed toward the contralateral side.

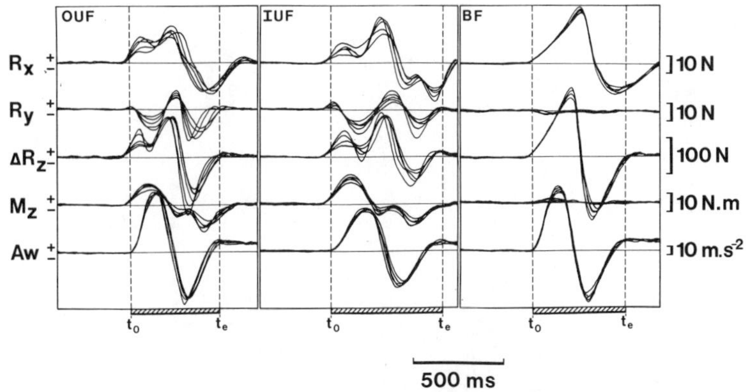

Fig. 3. Force platform data recorded under the three experimental conditions. Force platform data recorded in one subject, performing upper limb(s) movement under the three experimental conditions. With each type of movement, records of five trials were superimposed by synchronizing them on the onset of Aw_i. From left to right: OUF, unilateral flexions with no additional inertia; IUF, unilateral flexions with an additional inertia; BF, bilateral flexion. From top to bottom: Rx, Ry and ΔRz, antero-posterior, lateral and vertical components of the resultant force (ΔRz = Rz-P, with P being the weight of the subject); positive sign corresponds, to forward, right-to-left and upward forces respectively; Mz, resultant moment about the vertical axis (positive sign corresponds to a moment which tends to rotate the body from the right to the left side); Aw_i, tangential acceleration of the upper limb recorded at wrist level. t_0 and t_e correspond to the onset and the end of the upper limb movement.

Durations of APA were longer with ΔRz than with the other variables. Moreover, they increased from BF (29 ms) to OUF (50 ms) and from OUF to IUF (56 ms): they doubled in IUF as compared to BF.

It is tempting to consider that the postural perturbation associated with voluntary movement might result from the arrangement of two factors. These two factors may be defined from a biomechanical standpoint. The one is linear, and the other is rotational. Both may include three components.

Interpretation

The biomechanical finality of APA may be suggested on the basis of an analysis of forces acting at shoulder level and at the body center of gravity (Fig. 4).

Study of reaction time

The above results tend to prove that dynamic asymmetry is a movement parameter which is "preprogrammed". A direct argument was provided by a chronometric study (Zattara et Bouisset, 1984; Zattara and Bouisset, 1986).

This simple reaction time was studied under the same three conditions. The results showed that RT varied from one condition to another (Fig. 5). But when RT was divided into two parts, motor latency (ML) and postural anticipation (PA), it could be seen that ML was constant whereas PA varied. More precisely, PA increased with the dynamic asymmetry, which was, in fact, not surprising because it corresponds to the APA, presented above.

Therefore, ML was equivalent to the "true RT" in that type of motor activity. And it can be assumed that PA was part of the motor program.

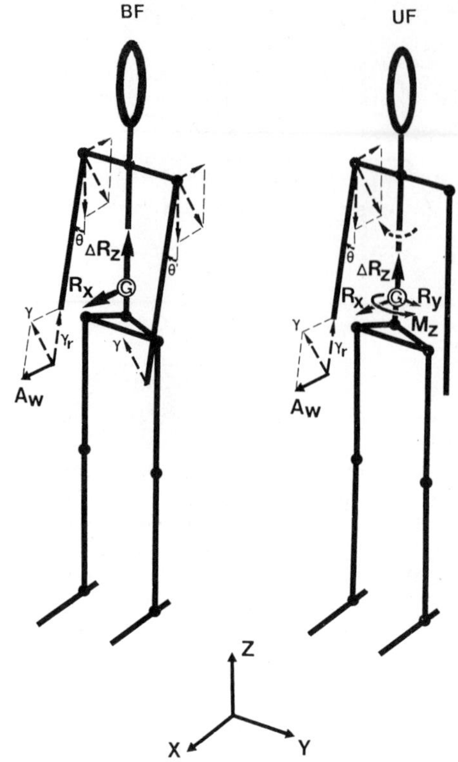

Fig. 4. Interpretation of the finality of anticipatory postural adjustements. The filled arrows correspond to the actual recorded biomechanical data, the interrupted arrows correspond to theoretical parameters. θ, angular displacement of the upper limb(s). Aw, γr, and γ, tangential, radial and total upper limb acceleration. Rx and ΔRz, antero-posterior and vertical acceleration of the body center of gravity, G.Mz, resultant moment about the vertical axis crossing G. From this analysis, it can be assumed that APA tend to create inertial forces which, when the time comes, will counterbalance the disturbance to postural equilibrium due to the intentional forthcoming movement (Bouisset and Zattara, 1984).

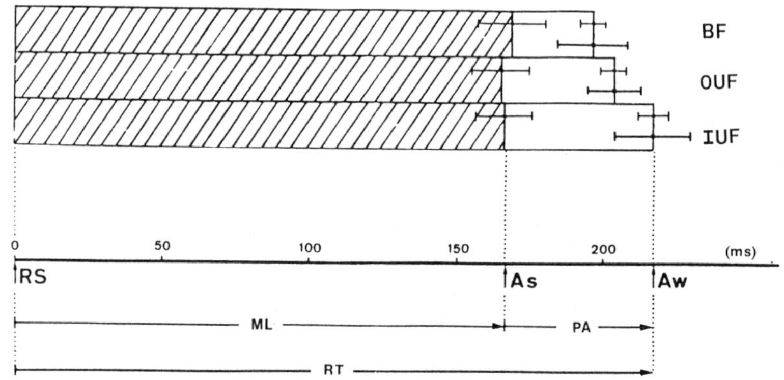

Fig. 5. Chronometric indexes under the three experimental conditions. On the lower part of the figure, the chronometric indexes are illustrated (RT, reaction time; ML, motor latency; PA, postural anticipation). (RS being the response signal, As and Aw indicating the onset of the earlier shank acceleration and the onset of the wrist acceleration, respectively). On the upper part of the figure, mean values and standard deviations of RT, ML (hatched areas) and PA (non-hatched areas) are given for five subjects performing ten movements of each type during two experimental sessions. From top to bottom: bilateral flexions (BF), unilateral flexions without additionnal inertia (OUF) and with an additionnal inertia (IUF).

Conclusion

Dynamic asymmetry has been shown to be an informative experimental factor for the study of APA. Since it integrates the effects of movement parameters on body structure, it allows a more comprehensive approach to APA. It might be interesting to test the hypothesis according to which APA might result from a combination of two biomechanical factors. These factors correspond to the two vectors to which the perturbing force system may be reduced, i.e., its resultant and its moment (in relation to the body center of gravity, for example).

References

Bazalgette, D., Zattara, M., Bathien, N., Bouisset, S., and Rondot, P., 1986, Postural adjustments associated with rapid voluntary arm movements in patients with Parkinson's disease. VIII International Symposium on Parkinson's Disease, New York, 1985, in: Yard, M.D. and Bergmann, K.J. : "Advances in Neurology", 45, Raven Press, 371-374.

Belenkii, Y. Y., Gurfinkel, V., and Paltsev, Y. I., 1967, Element of control of voluntary movements, Biofizika, 12(1), 135-141.

Bernstein, N., 1975, Coordination and regulation of movements, New York: Pergamon Press (Amer. translation 1967), 196pp.

Bouisset, S., and Zattara, M., 1981, A sequence of postural movements precedes voluntary movement, Neuroscience Letters, 22: 263-270.

Bouisset, S., and Zattara, M., 1983, Anticipatory postural movements related to a voluntary movement, International Conference on Space Physiology (C.N.E.S.), Toulouse, in: "Space Physiology", 1 vol. (Cepadues Pubs), 137-141.

Bouisset, S., and Zattara, M., 1987, To which biomechanical necessities do anticipatory postural adjustments related to voluntary movement respond?, Xth Int. Congress of Biomechanics, Umea, 1985. In: Jonsson, B., "Biomechanics X, Human Kinetics Pub.", Champaign, I11.,7p., (in press).

Breniere, Y., Do, M. C., and Bouisset, S., 1987, J. of Motor Behavior, 19, 62-76, 1987.

Gelfand, I. M., Gurfinkel, V. S., Tsetlin, M. L., and Shik, M. L., 1966, Problems in analysis of movements, in: Gelfand, I.M., Gurfinkel, V.S., Fomin, S.V., and Tsetlin, M.L., eds. , "Models of the structural functional organization of certain biological systems" (Amer. translation, 1971), M.I.T. Press, Cambridge, 330-345.

Gouny, M., Breniere, Y., Do, M. C., et Lestiene, F., 1977, Modalité d'intervention de la vision sur les réactions posturales liées à la mobilisation volontaire du bras, C.R. Acad. Sci., Paris, 285, série D, 1115-1118.

Zattara, M., 1982, Ajustements posturaux antecipateurs associés au mouvement du membre supérieur, Thése de 3e cycle, Orsay, 1 vol., 1134pp.

Zattara, M., et Bouisset, S., 1983, Influence de la vitesse d'exécution du mouvement volontaire sur les accélérations anticipatrices, 8ème Congrès de la Société de Biomécanique, Lyon, in résumé des communications, 113-114.

Zattara, M., et Bouisset, S., 1986, Etude chronométrique du programme posturo-cinétique lié au mouvement volontaire, J. Physiol., Paris, 81: 14-16.

Zattara, M., and Bouisset, S., 1986, Chronometric relations between postural program and voluntary motor command, J. of Motor Behavior, 18, 2, 215-224.

BODY SCHEME IN THE CONTROL OF POSTURAL ACTIVITY

V.S. Gurfinkel, Yu.S. Levik, K.E. Popov, B.N. Smetanin, and V.Yu. Shlikov

Institute for Problems of Information Transmission
U.S.S.R. Acad. Sci., Moscow, U.S.S.R.

Summary

The changes in the amplitude and direction of vestibulo-motor responses were studied during illusory body tilt and altered perception of the head position. The results obtained under these conditions were similar to those recorded during the real body movement and real head rotation. These results are discussed in the context of the involvement of the body scheme in the control of involuntary spatially oriented movements.

Introduction

Investigation of the physiological mechanisms underlying postural control dates back to the classical studies by C. Sherrington. In the issue of "Brain" dedicated to Huglin Jackson (Sherrington, 1906), he distinguished between an animal's whole body posture, total posture, and the postures of the individual body segments, segmental postures. Sherrington argued that the segmental postures are based on the stretch reflex while the stretch reflexes in various body segments form a common functional system aimed at counterbalancing the body weight by the muscle forces. Subsequent studies by Magnus (1924) and Rademaker (1931) demonstrated the existence of numerous static and statokinetic reactions, position reflexes, and righting reflexes. When attempting to understand how the reflexes are integrated, Magnus suggested that the central nervous system at any time reflects the body status, its posture, the positions of the limbs, and the contact between the body and external space. The mechanisms underlying this reflection have not yet been elucidated. They are very complex since posture includes not only the maintenance of certain relative positions of the body segments but also fine postural adjustments associated with voluntary movements, stability control during locomotion, and the ability to assess the body's spatial orientation.

The functional system on which postural stabilization is based is unlikely to involve direct interactions between the stretch reflexes in various body segments. It is more likely that some central organization may be responsible for postural control. For this central organization we have used the traditional term "body scheme" but not in its traditional clinical or psychological meaning. We assume that body scheme is an internal body representation or internal body model, the role of which is not limited to conscious perception of the body, but also includes the important functions in the processing of multisensory inputs at the unconscious level. The postural control system organized on the basis of this internal model operating with highly integrated information from various sensory systems could possess a much wider range of functional possibilities than a system organized on the basis of interaction between "single level" reflexes.

In this study, the role of the body scheme in the control of postural activity was investigated using vestibulo-motor responses to electrical stimulation of the labyrinth as an experimental paradigm. These responses have been shown to depend on both the spatial orientation of the head (Lund et Broberg, 1983) and the body movement (Smetanin et al., 1986) during electrical stimulation. We assumed that this dependence is due to a central processing of proprioceptive signals involving the body scheme. To test this assumption, postural responses to vestibular stimulation were studied during the illusory body sway or rotation induced by vibratory stimulation of muscle proprioceptors.

Proceeding on the assumption that the body scheme is involved in the control of postural activity, one might expect the changes in both global postural responses and the sensory illusions that occur at the conscious level to be based on the same changes in the internal body model. In other words, the postural responses might be determined by the illusory posture although most primary afferent signals can be in conflict with the perceived spatial configuration of the body. If one accepts this argumentation, the data presented below confirm the hypothesis that the body scheme is involved in the control of involuntary spatially oriented movements.

Results

Experiment 1

Vibration applied to the lower leg muscles in the upright posture is capable of inducing the illusion of dynamic body inclination sway in the direction of the lengthening of the vibrated muscles (Goodwin et al., 1972; Gurfinkel et al., 1977). We used this phenomenon to induce in the standing subject an illusory forward body sway similar to the actual body movement which occurs when rising slowly on tiptoe. This kind of movement has been shown to enhance the vestibulo-motor response (Smetanin et al., 1986).

During the experiment, the subject with his/her eyes closed maintained the upright posture on the stabilographic platform used to record postural responses. The vestibulo-motor responses were evoked by stimulation of the right labyrinth with DC current lasting 4 sec. To induce the illusory body sway, actual movement of the subject's trunk was prevented as shown in Fig.1. The illusion forward body sway was induced by vibration applied to the Achilles tendons. (1-2 mm, 60-80 Hz vibration produced by hand-made vibrator based on small DC-motor with eccentric mass on the axis). The postural responses were also recorded without trunk fixation during the voluntary body tilt forward as well as during the involuntary body sway forward induced by vibration of the tibialis anterior muscles (the so-called "vibration induced fall", VIF, - Eklund, 1972).

The voluntary forward sway resulted in an increase in the lateral body sway occurring in response to galvanic vestibular stimulation applied during the dynamic phase of

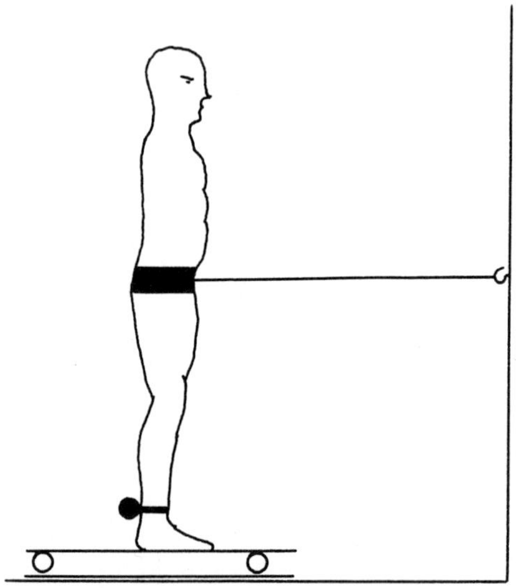

Fig. 1. Scheme of the body fixation for inducing the illusory body tilt forward.

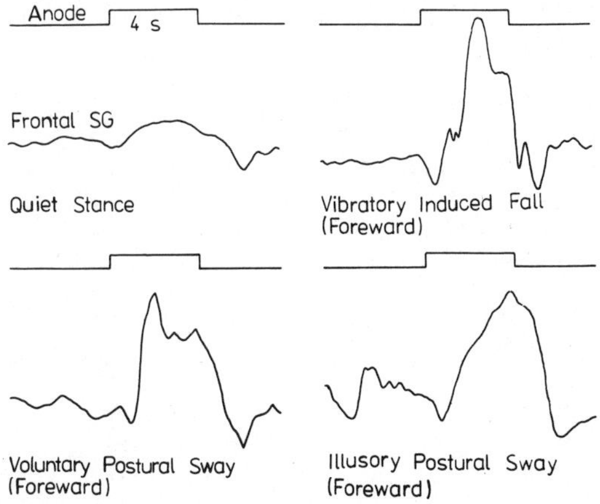

Fig. 2. Frontal stabilographic responses to electrical stimulation of the right labyrinth (1 mA) recorded during actual and illusory body motions. Upward deflection of the stabilogram represents body sway to the right.

the movement (Fig. 2). The increase in the vestibulo-motor response was similar to that observed when rising on tiptoe, squatting, and performing other movements resulting in changes in the total posture (Smetanin et al., 1986).

The involuntary body tilt forward (VIF) induced by vibration of the tibialis anterior muscles resulted in a similar enhancement of postural response to the electrical labyrinth stimulation (Fig. 2).

The illusory body sway forward also resulted in an increase in the lateral stabilographic response to vestibular stimulation (Fig. 2). In some subjects, the stabilographic response increased 2 to 4 times as compared to that recorded during quiet stance, and in others the increase was much larger. The lateral body sway under the illusion was often larger than that recorded during the actual forward body sway. Larger responses correlated with higher velocities of the illusory movement as assessed by the subject. Under

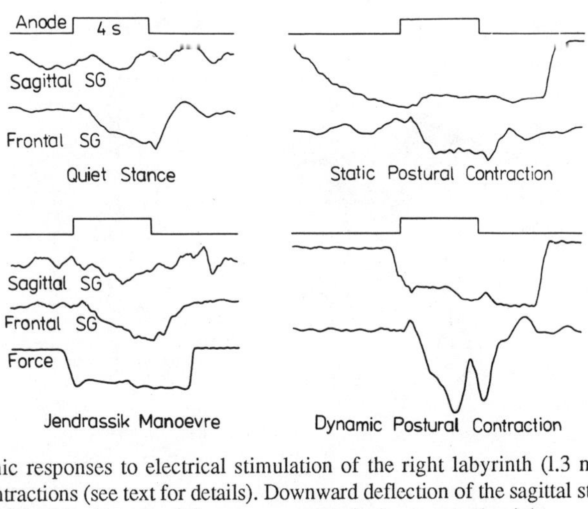

Fig. 3. Stabilographic responses to electrical stimulation of the right labyrinth (1.3 mA) recorded during isometric muscle contractions (see text for details). Downward deflection of the sagittal stabilogram represents body sway forward, that of the frontal stabilogram represents body sway to the right.

the illusion, the body sway in response to the vestibular stimulation was usually slower than during the actual movement.

Attempts to slowly sway the body fixed as shown in Fig. 1 did not result in any enhancement of the vestibular motor response as compared to the quiet stance condition (Fig.3). Similar "static motion" when performed more rapidly resulted in an approximately two-fold increase in the response to the galvanic stimulation applied during the dynamic phase of the force exertion. Dynamic contraction of the arm muscles (the Jendrassik manoeuvre) failed to modify the magnitude of the vestibulo-motor response (Fig. 3).

Thus the vibration induced illusory body sway resulted in an enhancement of the vestibulo-motor respnse which was similar to that observed during the actual dynamic changes of posture.

Experiment 2

The phenomenon whereby the direction of postural response to electrical labyrinth stimulation changes depending upon the spatial orientation (rotation) of the head is well known. It was studied using stabilography (Lund et Broberg, 1983; Hlavacka et Njiokiktjien, 1985) as well as electromyography (Nashner et Wolfson, 1974). This phenomenon is thought to be due to interaction between the vestibular and neck afferent inputs at the level of motoneurons and/or vestibulospinal neurons.

However this interaction seems to be more complex since similar dependence of the direction of the postural response upon the spatial orientation of the head has also been observed when rotation of the head relative to the feet resulted from rotation of the trunk (Lund et Broberg, 1983), i.e. the neck receptors were not stimulated. One can therefore assume that this interaction involves higher level integrative mechanisms. To test this assumption, we studied the vestibulo-motor response under the experimentally induced conflict between subjectively perceived and actual spatial orientations of the head.

The illusion of head and trunk rotation was induced by vibratory stimulation of the gluteus maximum muscle while any actual body rotation was prevented mechanically (Lackner et Levine, 1979). We modified this method to allow both sagittal and frontal postural sways to occur in response to the vestibular stimulation. Rotation of the trunk was

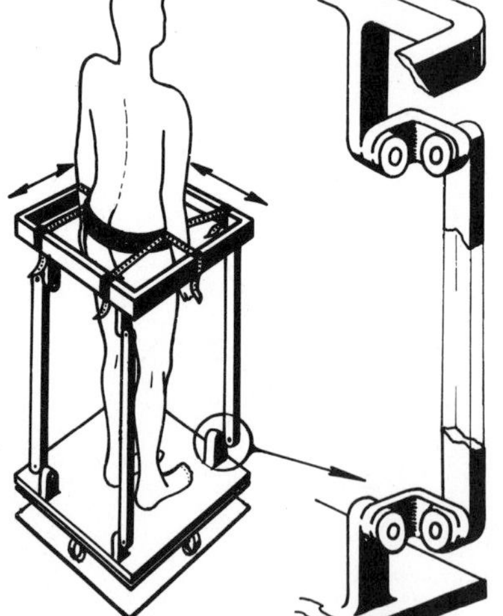

Fig. 4. Experimental set-up for inducing illusory head and trunk rotation.

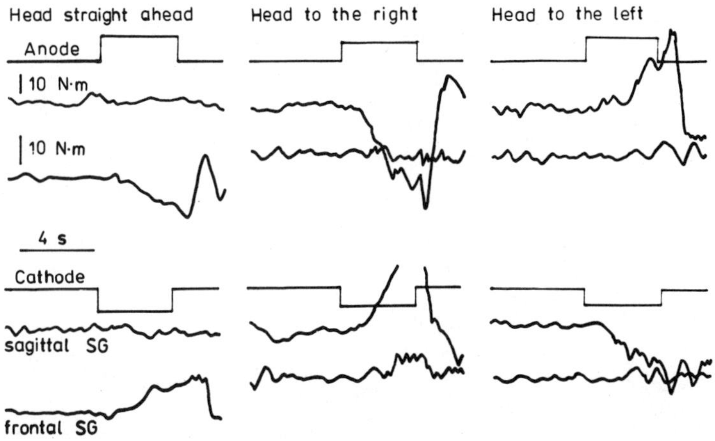

Fig. 5. Stabilographic responses to electrical stimulation of the right labyrinth (2 mA) recorded in a standing subject with his head held in three different positions. Upward deflection of the sagittal stabilogram represents body sway forward, that of the frontal stabilogram represents body sway to the left.

prevented by means of a mechanical set-up shown in Fig. 4. The set-up was placed over the stabilograph platform used to record vestibulo-motor responses.

Experiments were performed in a dark room. The subject with his (her) eyes closed maintained erect posture on the platform of the set-up as shown in Fig.4. The subject's pelvis was fixed to prevent it from rotating. Under this condition, vibration of the right gluteus maximum muscle was capable of inducing illusory head and trunk rotation to the right. In the 5 subjects who were most susceptible to vibration (selected from among 15 healthy persons), the illusory rotation consistently approached 90°. In these 5 subjects, directions of postural response to electrical stimulation of the right labyrinth were compared during actual and illusory head rotation.

Fig. 5 shows stabilographic responses of one subject to stimulation of the right labyrinth by DC current of both polarities with the head held in a normal position as well as with the head turned either to the right or to the left. With the head held in the normal position, the resulting body sway was in the frontal plane while with the head turned 90° to either side it was in the sagittal plane. With a positive stimulating current polarity, the body swayed in the direction of the mastoid under stimulation, whereas with negative polarity it swayed in the opposite direction.

Fig. 6 shows stabilographic responses of two subjects to vestibular stimulation delivered after the subject has signalled the presence of the vibration induced illusion of rotation. This signal was switched on when the subject perceived his (her) head (together with the trunk) as having rotated approximately 90° to the right.

Electrical stimulation of the right labyrinth applied during the illusory head rotation to the right resulted in body sway in the posterior direction when the polarity of the current was positive and in sway in the anterior direction when the polarity of the current was negative. The frontal component of the postural response was usually small or absent but a diagonal response pattern, i.e. simultaneous changes in the sagittal and frontal components of the stabilogram, was occasionally observed (Fig. 6, right column).

Thus the postural responses to electrical labyrinth stimulation applied during illusory head rotation were similar to those recorded during actual head rotation. This result can be seen from Fig. 7 where records of the sagittal and frontal stabilograms similar to those shown in Fig. 5 and 6 are presented in a two-dimensional display.

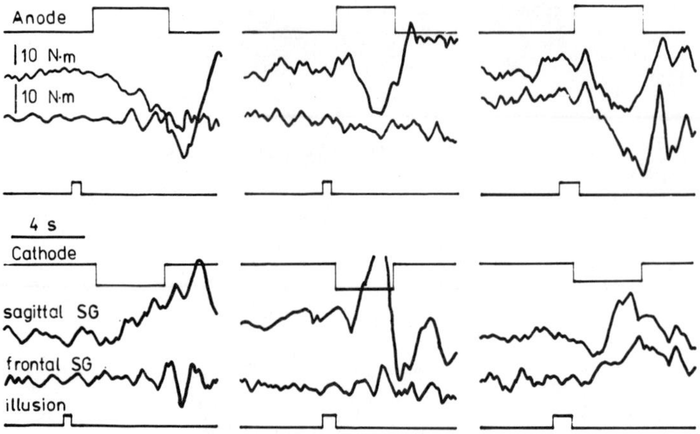

Fig. 6. Stabilographic responses to electrical stimulation of the right labyrinth (2 to 2.5 mA) recorded in standing subjects under the illusion of head and trunk rotation to the right. Upward deflection of the sagittal stabilogram represents body sway forward, that of the frontal stabilogram represents body sway to the left.

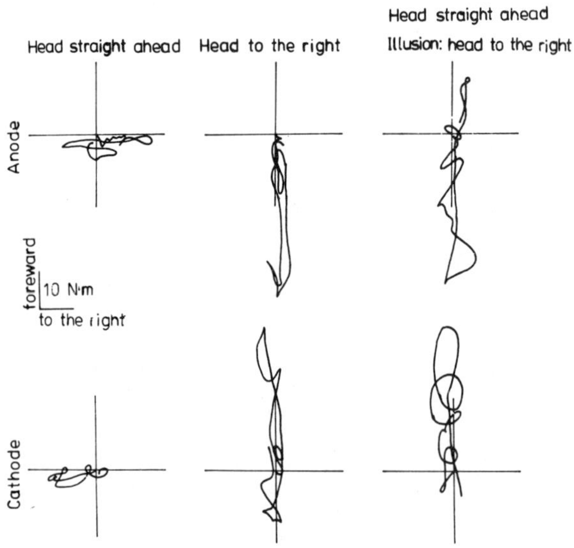

Fig. 7. Two-dimensional display of stabilographic responses to electrical stimulation of the right labyrinth recorded in a standing subject with his head held in two different positions as well as under the illusion of head and trunk rotation. The stimulus duration was 4 sec, the trajectory recording 8 sec.

Fig. 8 shows the averaged data (mean directions and standard deviations) on the direction of the postural responses obtained with 4 subjects. These data suggest that the mean direction of the postural sway was always close to an either strictly frontal or strictly sagittal one. The direction of the postural response during the illusory rotation was similar to that observed during the actual head rotation. It should be noted that the variability of the response was considerably higher under the illusion than during the actual rotation. This difference can be easily explained by the observation that the illusory head rotation was felt by the subject less distinctly and was more variable than the actual one. The results of experiment 2 have been published elsewhere (Popov et al., 1986).

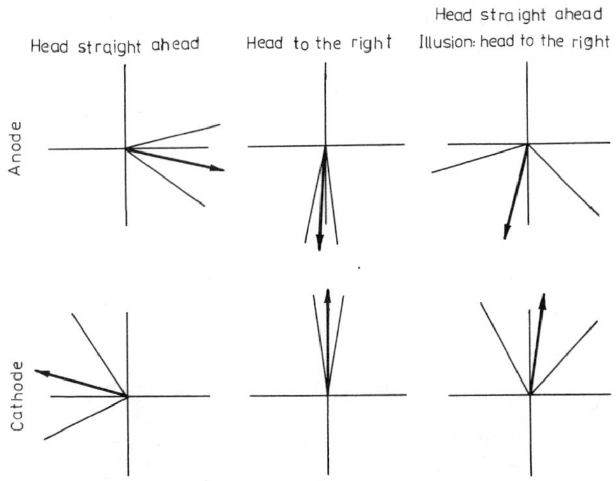

Fig. 8. Mean vectors (and standard deviations) representing the directions of stabilographic responses to the electrical stimulation of the right labyrinth recorded in a standing subject with his head held in two different positions as well as under the illusion of head and trunk rotation.

Discussion

The changes in the vestibulo-motor responses observed under illusory body tilt (Experiment 1) and altered perception of the head position (Experiment 2) were similar to those recorded during the actual body movement and actual head rotation, respectively. The present findings suggest that proprioceptive input modifies the vestibulo-motor response not by "re-addressing" the lower level reflex responses but by inducing some modifications into a common central organization that provides a basis for both conscious perception of the body posture and control of involuntary postural responses. Conscious perception of the body configuration is traditionally thought to be based on the body scheme. Therefore, it seems to be reasonable to use this term for the whole hypothetical central organization although this interpretation of the term "body scheme" is wider than that accepted in clinical literature, where the body scheme relates to phenomena such as the amputee phantom limb and inadequate perception of the body or external space under some pathological conditions. These situations show the effects of body scheme deterioration but they provide little information for understanding body scheme function under normal conditions. Observations on healthy persons under the so called "altered states of consciousness" (Dittrich et al., 1985) may be more useful in this respect. No matter what the origin of these states was, the subjects consistently reported a number of phenomena such as blurring of borders between the body and environment, feeling the body parts as not belonging to the subject, sensation of changed dimensions of the body parts and external objects, etc. (see Table 1).

Analyzing these and other related observations one can conclude that the normal functions of the body scheme include: perception of the borders between the body and extrapersonal space; knowledge of the body dimensions, lengths of the body segments and the sequence of their linkage, their mass-inertia properties; formation of a stationary reference system common for the body and the external space; a proper adjustment of the levels of muscle activity required to maintain a given posture; and postural adaptations to predictable disturbances. The implementation of these functions requires adequate sensory inputs. The control of postural muscle activity is, however, to a considerable extent based on parameters that cannot be directly signalled by specific receptors. These parameters include the segment length, the sequence of segment linkage, the position of the gravity center, the configuration and dimensions of the support surface, and the muscle torques.

Table 1: Etiology-Independent phenomena connected with body scheme functioning during altered states of consciousness.

N° in the APZ questionnaire	ITEM
1	I had the feeling everything around me was somehow unreal.
7	I felt as though I were floating.
13	The boundary between myself and my surroundings seemed to blur.
19	The ground I was standing on seemed to be swaying.
28	Parts of my body seemed no longer to belong to me.
39	I had the feeling my limbs were larger than usual.
43	Things around me appeared to be bigger than usual.
68	It seemed to me as though I did not have a body anymore.
129	It seemed to me that my environnement and I were one.
146	I had the feeling that I could no longer control the movement of my body.

Another important point should be emphasized. Even if a receptor is capable of directly measuring certain parameters, its afferent message will have a local nature because the measured parameters will be processed by a reference system coupled with the segment where the receptor or its receptive field is located rather than by the reference system used in programming whole-body motions. The interpretation of signals from receptors located at distal segments in a multi-link chain requires knowledge of the state of the intermediate segments. Since the receptors are influenced by both external stimuli and the organism's activity, the interpretation of primary afferent signals requires detailed knowledge of the current postural task.

From the above considerations, it can be concluded that direct usage of the receptor primary signals in postural control is unlikely. These signals require an extremely complex processing by a system provided with information on the whole body configuration as well as on the external forces at work. In this connction, it is interesting to note that Head (1920) compared the body scheme to the taximeter that does not give distance in miles but transforms it directly into shillings and pennies, as this is what the passenger wants to know. This image implies that one of the most important functions of the body scheme is transformation of sensory signals into a form suitable for direct usage. The question is, who is the passenger, or, in other words, what is the purpose of the information processing. We suggest that the body scheme is involved in postural control and postural adjustments associated with motor activities, including both conscious and unconscious components of motor programming. To fulfil these functions, the body scheme should not be a kind of "inventory" of the body segments but rather a certain scheme that reflects, though in a simplified manner, the basic biomechanical, geometrical and regulatory properties of the organism, i.e. it should constitute a true model.

At first sight it seems that maintaining the usual posture does not require such a complex central organization. It might be sufficient simply to exert appropriate muscle torques to counterbalance the gravitational torque in every joint. However a study of human posture under weightless conditions has shown (Clement et al., 1984) that the elimination of the gravitational torques failed to result in any dramatic changes in the usual posture although the pattern of muscle activity required for maintaining the posture was quite different from that observed under normal gravity conditions.

Our previous experiments with the so called "cutaneous reading" have shown (Gurfinkel et al., 1985) that the perception of complex tactile stimuli is largely dependent upon the posture (i.e. the orientation of the receptive field). Another study (Gurfinkel et al., 1986) has shown that when the experimental phantom of the upper extremity was induced by a pressure cuff ischaemia, the voluntary movements of the arm were programmed in accordance with the altered perception of the arm position. These observations seem to support the idea that the body scheme is involved in both spatial perception and motor programming. The present findings concerning changes in the vestibulo-motor response due to illusory body tilt or rotation can be better understood within the framework of the internal body model concept than on the basis of simple reflex interactions.

References

Clément, G., Gurfinkel, V.S., Lestienne, F., Lipshits, M.I., Popov, K.E., 1984, Adaptation of postural control to weightlessness, Exp. Brain Res., 57: 61-72.

Dittrich, A., von Arx, S., Staub, S., 1985, International study of altered states of consciousness (ISASC), Summary of the Results, The German Journal of Psychology, 9: 319-333.

Eklund, G., 1972, General features of vibration-induced effects on balance, Uppsala J. Med. Sci., 77: 112-124.

Goodwin, G.M., McCloskey, D.I., Matthews, P.B.C., 1972, The contribution of muscle afferents to kinaesthesia shown by vibration induced illusions of movement and by the effects of paralysing joint afferents, Brain, 95: 705-748.

Gurfinkel, V.S., Lipshits, M.I., Popov, K.E., 1977, Investigation of the system for the vertical posture regulation using vibratory stimulation of the muscle spindles, Human Physiology, Translated by Plenum Publ. C° from Fiziologiya Cheloveka, 3: 635-643.

Gurfinkel, V.S., Debreva, E.E., Levik, Yu.S., 1985, Dependence of tactile stimulus interpretation on receptive field orientation, Human Physiology, Translated by Plenum Publ. C° from Fiziologiya Cheloveka, 11: 3-6.

Gurfinkel, V.S., Debreva, E.E., Levik, Yu.S., 1986, The role of internal model in the position perception and planning of arm movement, Fiziologiya Cheloveka, 12: 769-776 (in Russian).

Head, H., 1920, Studies in Neurology, v.2, Hodder and Stoughton, London.

Hlavacka, F., Njiokiktjien, C., 1985, Postural responses evoked by sinusoidal galvanic stimulation of the labyrinth, Acta Otolaryngol. (Stockholm), 99: 107-112.

Lackner, J.R., Levine, M.S., 1979, Changes in apparent body orientation and sensory localisation induced by vibration of postural muscles: vibratory myesthetic illusions, Aviat., Space Environ. Med., 50: 346-354.

Lund, S., Broberg, C., 1983, Effects of different head positions on postural sway in man induced by a reproducible vestibular error signal, Acta Physiol. Scand., 117: 307-309.

Magnus, R., 1924, Körperstellung, Springer, Berlin.

Nashner, L.M., Wolfson, P., 1974, Influence of head position and proprioceptive cues on short latency postural reflexes evoked by galvanic stimulation of the human labyrinth, Brain Res., 67: 255-268.

Popov, K.E., Smetanin, B.N., Gurfinkel, V.S., Kudinova, P.P., Shlikov, V.Yu., 1986, Spatial perception and vestibulo-motor response in man, Neurofiziologia, 18: 779-787 (In Russian).

Rademaker, C.G., 1931, Das Stehen: Statische Reaktionen, Gleichgewichtsreaktionen und Muskeltonus unter besonderer Berücksichtigung ihres Verhaltens bei Kleinhirnlosen Tieren, Springer, Berlin.

Sherrington, C.S., 1906, The Integrative Action of the Nervous System, Univ. Press, Cambridge.

Smetanin, B.N., Shlikov, V.Yu., Kudinova, M.P., 1986, Facilitatory effect of voluntary movement on the vestibulo-motor response, Fiziologiya Cheloveka, 12: 133-140 (In Russian).

FORWARD AND BACKWARD AXIAL MOVEMENTS: TWO MODES OF CENTRAL CONTROL

P. Crenna *, C. Frigo **, J. Massion ***, A. Pedotti **, and A. Deat ***

* Instituto di Fisiologia Umana II, Milano, Itaty
** Centro di Bioingegneria, Milano, Italy
*** Unité de Neurosciences Intégratives, LNF 3 CNRS, Marseille, France

Summary

Axial upper trunk movements are accompanied by hip and knee movements in the opposite direction, resulting in a close adjustement of the projection onto the ground of the center of gravity (CG). Two patterns of muscle activation were observed during axial movements. The first is characterized by the sequential activation of first the leg muscles and secondly the prime mover. This pattern can be observed with forward axial upper trunk movements. The second pattern of muscle activation consists of the nearly synchronous onset of both the leg muscles and the prime mover EMG. This pattern can be observed with backward upper trunk movements. The first hypothesis concerning the two different patterns is that they may result from the asymmetry of the knee joint, which is able to flex in association with backward movements, but not to hyperextend in association with forward movements. This hypothesis was tested by casting the knee joints in two subjects. It was found that in a subject performing backward upper trunk movements no change in the pattern of EMG activation appeared after casting. It is suggested that the synchronous mode observed with backward movements is less efficient in terms of motor performance whereas the sequential mode observed with forward movements is more efficient. The shift from the synchronous to the sequential mode might result from training which occurs with forward and not with backward movements: forward movements are performed every day from childhood onwards; whereas backward movements are little used in everyday life.

Introduction

It is generally agreed that movement and posture are two opposite aspects of motor control. Movement control governs the displacement of a given segment from an initial to a final position or target; the central program responsible for the trajectory of the displaced segment has been attracting much attention recently (see Bizzi et al., 1984; Feldman, 1986). On the other hand, posture means the holding of a given position by one or several body segments or by the whole body. This position is controled on the basis of an internal frame of reference and a series of reflex actions which stabilize the position as dictated by the internal frame of reference. The sensory inputs responsible for maintaining a position are various, depending on the position in question. The proprioceptive inputs from the protagonist muscles controlling a single joint are the main inputs involved in maintening position of this joint. In the control of whole-body posture, the reflex regulation relies on proprioceptive, visual and labyrinthine cues, acting in a continuous or discontinuous way depending on the frequency of the body oscillation (Diener et al., 1986).

When a voluntary movement is performed, posture and equilibrium are disturbed. One reason for this is that the performance of a movement changes the general geometry of the body and therefore the position of the center of gravity (CG). A second reason is that the moving segment, for example the arm, is supported by the rest of the body; the dynamic forces associated with movement performance act on the body, according to the law that action equals reaction, and thus disturb its posture and equilibrium.

It has frequently been observed that voluntary movements in a standing subject are accompanied by an adjustment of posture resulting in a coordination between posture and movement. The postural adjustments associated with a voluntary movement have been mainly investigated while the subject is performing arm movements (Belenkiy et al., 1967; Bouisset and Zattara, 1981; Cordo and Nashner, 1982; Friedli et al., 1984). By comparing the EMG activity of leg muscles with the electromyographic (EMG) activity of the prime mover, it was first observed by Belenkiy et al., (1967) that the changes in leg muscle activity usually precede by some 50-100 ms the prime mover activation. This has been confirmed by many other workers (see Massion, 1984). One of the main characteristics of these adjustments, then, is the fact that they are "anticipatory" i.e. they precede the onset of the postural perturbation caused by performance of the movement and thus minimize the effects of the perturbation on posture and equilibrium.

The coordination between posture and movement raises specific problems when the voluntary movement involves axial segments such as the trunk. First, the segments participating in the movement are the same as those involved in body support. Secondly, the mass of the moving segment is such that it can be compensated for only by displacements of other body segments which are large enough to keep the projection of the CG onto the ground in the same place. Babinski (1899) was the first to conduct a quantitative analysis of axial movements. He noted that backward and forward head and trunk movements are accompanied by displacements of the hip and knee in the opposite direction, resulting in a stabilization of the CG during the movement. He called the whole set of upper trunk and associated limb movements "synergies". Similar axial synergies have been reported and analysed by Gurfinkel and Elner (1973) during respiratory movements. These authors noticed that hip movements opposite to the trunk respiratory movements occur, preventing CG displacement.

Several questions concerning posture and equilibrium control during axial movements were raised in a previous study (Crenna et al., 1987). Some of them will be mentioned here. First, do the axial synergies efficiently regulate the center of gravity position ? Secondly, is the control of the axial synergies centrally organized, or is it partly of reflex origin ? Thirdly, is the central organization of the synergies, if present, comparable to the coordination previously described with arm movements in the standing subject, where an anticipatory postural adjustment precedes the onset of the arm movement ? Subsidiarily, is the central organization comparable between forward and backward movements ? It will be shown that two modes of control are to be observed: the first is to be found with forward axial movements, where a set of "postural" muscles are put into action before the prime mover, and the second with backward axial movements, where no distinction can be made between the prime mover and the "postural" muscles, the whole set of muscles being activated almost synchronously. The reasons for the existence of a dual mode of control between forward and backward movements will be analysed in the discussion.

Methods

The subjects (n = 8) were asked to stand quietly on a force platform and relax with their hands on their hips and their eyes open. At a given signal (tone) they were asked to bend both their head and trunk forward or backward until reaching a pre-estimated displacement of the head of about 30 cm in the sagittal plane and to move back to the initial position. The subject was required to perform 3 to 5 movements in each direction at different speeds. The results reported here concern the movements performed at the fastest speed (400-500 ms). With two subjects, fast backward upper trunk movements were performed after casting of the knees.

The parameters recorded were as follows (Fig. 1) :

1. Ground reaction forces: the vertical and horizontal forces were recorded through a force platform. The horizontal displacement of the center of pressure and the center of gravity were computed.

2. Kinematic analysis of the movement: seven markers were placed on the main reference points on the subject: two markers on the head, one on the shoulder, one on the

superior iliac crest, one on the trochanter, one on the knee, and one on the ankle. Their position was analysed by means of an automatic TV image processor (ELITE system, Ferrigno and Pedotti, 1985). The sampling rate used was 50 Hz and the accuracy of the system 1/2560.

3. Electromyographic recording: 4 pairs of antagonist muscles were recorded through bipolar surface electrodes and displayed on a miror galvanometer. The following muscles were selected: neck extensor (extensor capitis: Ex C) and flexor (sternocleidomastoideus, Scm); trunk extensor (erector spinae, Er.S.) and flexor (rectus abdominalis, R. Abd.), knee extensor (vastus medialis, VM) and flexor (semimembranosus, Sm or biceps femoris, BF), ankle extensor (soleus, SOL or gastrocnemius medialis, GM) and ankle flexor (tibialis anterior, TA).

Results

1. Efficiency of the coordination

Figure 1 shows that forward upper trunk movements are accompanied by backward displacement of the hip and knee. Backward upper trunk movements are associated on the contrary with forward hip and knee displacement. As can be seen in Figure 1, there is a marked difference between the hip and knee displacements associated with forward and backward trunk movements. With forward movements, the hip displacement is more pronounced than the knee displacement, because hyperextension of the knee joint is not possible for mechanical reasons. With backward movements, the knee displacement is larger than the hip displacement due to the fact that in addition to the hip flexion there is a knee joint flexion. Asymmetry of the hip and knee displacement when forward and backward trunk movements are performed is therefore due to mechanical constraints imposed on the knee joint.

The axial synergies associated with forward and backward movements by regulate the center of gravity position very efficiently. The projection of the center of gravity onto the ground moves less than 1.5 cm, whereas it has been estimated that without the associated hip

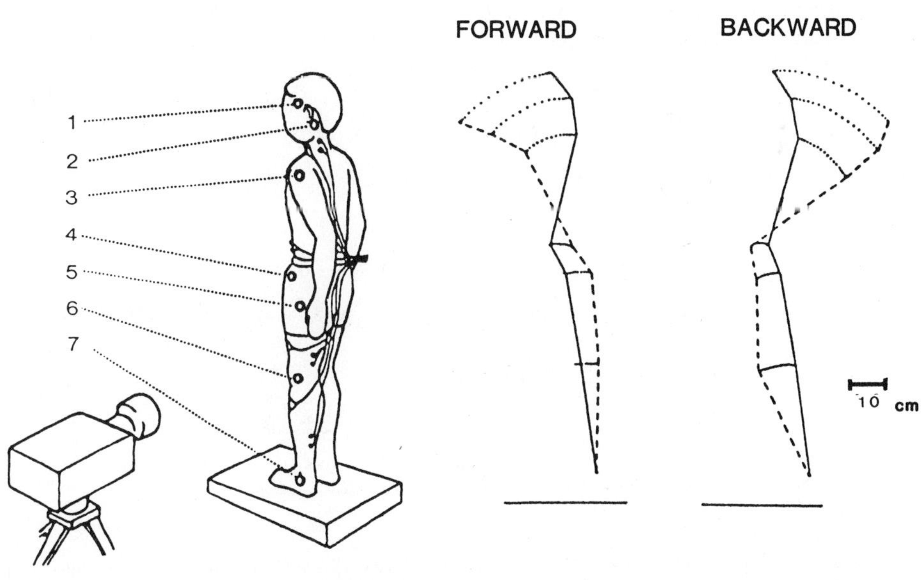

Fig. 1. Experimental set up (on the left) and kinematic analysis of the movement (on the right). Each marker is represented by a dot; sampling rate of the dot position: 50 Hz. A continuous strip links together the markers at the movement onset and a dashed line at the maximal amplitude. Note the displacement in opposite directions of the upper and lower markers.

and knee movements the displacement would amount to 12 cm (for discussion, see Crenna et al., 1987).

2. Central versus reflex control

Analysis of the kinematic characteristics of the axial synergies reveals that the onset of the displacement of the various markers is nearly synchronous. This also applies to the velocity curve measurements. This suggests that the displacement of the various segments involved in the axial synergies is centrally controlled.

The EMG analysis during forward and backward upper trunk movements confirms that the synergies are centrally organized (Fig. 2). With the forward trunk movements, a set of muscles located in front of the body are activated before the onset of movement, as shown by the first change in the velocity curve of the upper head marker. These muscles are the neck flexor (StC1), trunk flexor (prime mover, R. Abd.), hip flexor and knee extensor (VM) and ankle flexor (TA). With the backward trunk movements, a set of muscles is also activated before the onset of movement; these muscles are all located in the back of the body: neck extensors, trunk extensor and prime mover (Er.Sp.), hip extensor and knee flexor (Hamstring), ankle extensor (GM or SOL).

3. Mode of central control

The onset of the postural adjustments associated with arm movements precede the onset of movement (see Massion, 1984). This sequential control of first the postural adjustment and secondly the prime mover is generally interpreted as indicating the existence of a feedforward command of the postural muscles, which minimizes the perturbation of posture and equilibrium that the forthcoming movement would otherwise cause.

Analysis of the EMG activity associated with forward upper trunk movement reveals that among the muscles which are activated before the onset of the movement, two sets can be observed with different activation onset times. The first set of "postural" leg muscles (TA, VM) was usually activated before the prime mover about 100 ms before the movement onset; moreover, the onset of their activation was highly correlated and synchronous (regression line approximately equal to 45°). On the contrary, the onset of the prime mover activation (R. Abd.) was not significantly correlated with that of the leg postural muscles and occurred quite consistently about 50 ms before the movement onset. With the forward axial movements, the central control was thus comparable to that observed with arm

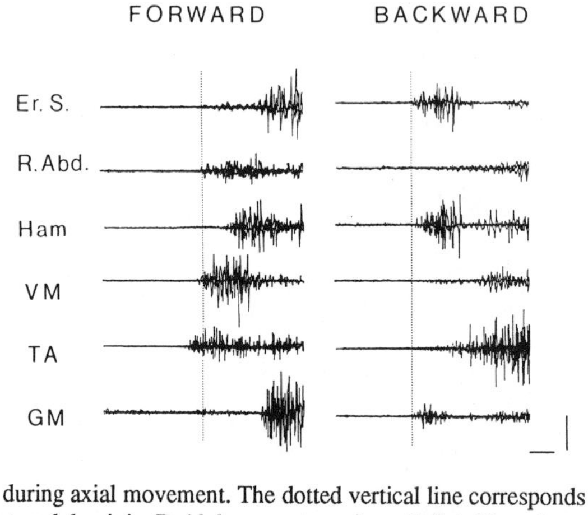

Fig. 2. EMG activation during axial movement. The dotted vertical line corresponds to the onset of activation of the prime mover (rectus abdominis: R-Abd or erector spinae: ErSp). Note that with forward movements, activation of tibialis anterior (TA) and vastus medialis (VM) precedes the activation of the prime mover, whereas with backward movement, activation of gastrocnemius (GM), semi-membranosus (Ham) and ErSp are simultaneous. Further comments in the text. Calibrations: 100 ms; 300 μA.

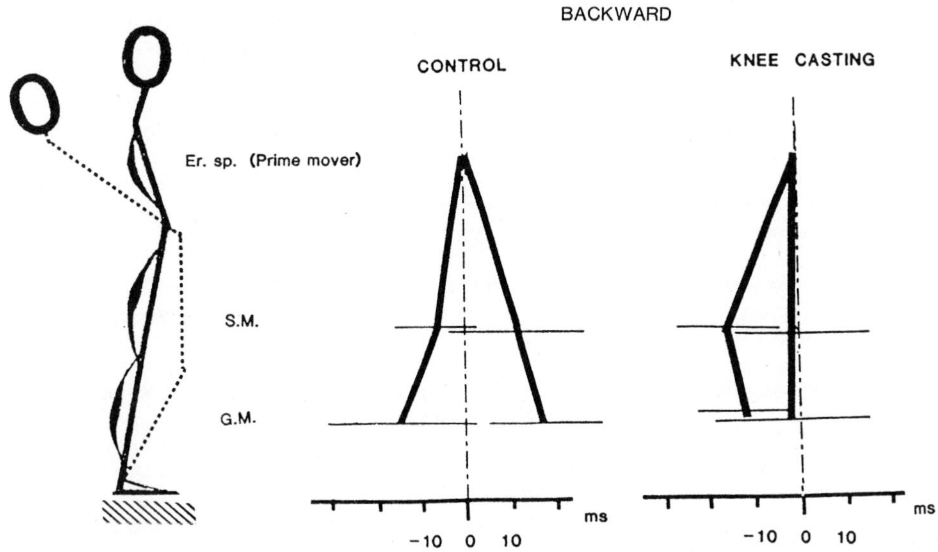

Fig. 3. Patterns of muscle activation during backward axial movements. The onset of activation of the Er.Sp. is taken as time reference (O). The onset of activation of the other muscles is calculated with respect to this reference. The horizontal bars correspond to the standard deviation. Control: two untrained subjects. Each subject underwent 20 trials. Note that no sequential activation comparable to that observed with forward movements is to be seen. Knee casting: the same two subjects were checked after knee casting; no marked change was observed after casting.

movements: a dual control could be seen to operate, involving first an adjustment of posture, and secondly the onset of movement (Fig. 3).

On analysing the EMG activity associated with backward upper trunk movements, a quite different mode of muscle activation was observed. The onset of activation of the prime mover (ErSp) and the leg postural muscles (Sm, Triceps Suralis) was almost synchronous and highly correlated; their activation preceded the onset of movement by some 60 ms. Thus, no clear cut sequential mode of control was observed here and it looks as if a single central command puts the whole synergy at once into action.

4. Why are the forward and backward axial synergies differently organized?

Two different modes of control were thus observed with forward and backward uppertrunk movements. With forward movements, a sequential control first of the postural muscles and secondly of the prime mover was observed: this type of control is comparable to that found with arm movements. With backward movements, a synchronous activation of the prime mover and the leg postural muscles takes place.

Two hypotheses can be proposed in order to explain these differences between the central organization of forward and backward upper trunk movements.

The first hypothesis claims that the difference is linked to anatomical constraints. Actually, we have seen that in forward and backward upper trunk movements, displacements of the hip and knee are in the opposite direction to that of the upper trunk, with an axis of rotation situated somewhere at the lumbar level (see Fig. 1); the ability to rotate is limited by the fact that the feet are fixed on the ground. The displacement of the body can therefore be represented at best by the analogy with an elastic beam which is hinged both at the ankle and at a point located at lumbar level. However, one main difference exists between forward and backward upper trunk movements, namely the ability of the knee to flex but not to hyperextend. As a result, with forward movements, the effect of the rotation is stopped at the knees, and allows only a rather limited displacements of the legs. By contrast, with upper trunk backward bending, the effect of the rotation on the lower segments is larger because of the flexion of the knee.

If the possibility of flexing the knee with backward upper trunk movement explains why a single mode of control serves instead of the dual mode observed with forward movements, then it should be expected that subjects with a cast knee joint preventing flexion of the knee will shift from the single mode of control observed with backward movements to a dual mode comparable to that observed with forward trunk movements.

Two subjects were given a knee cast and asked to perform a series of fast backward movements. The onset times of ErSp, Sm and Gm muscle activation were compared before and after casting.

Figure 3 gives the results obtained with these two subjects. In each of these subjects, the onset of activation of the trunk and leg muscles was not far from being synchronous before casting, with a slight lag by the dorsal muscles in the first subject and by the prime mover in the second subject. After casting, no clearcut change in the pattern of control was to be seen. It can be concluded that the sequential mode of activation (first the postural leg muscles and secondly the primer mover) observed with forward movements does not result from mechanical constraints arising from the knee joint characteristics.

Another possible explanation for the difference between the modes of control of upper trunk forward and backward movements is the fact that people have much more practice with forward than backward movements.

It should be remembered that due to the frontal location of the eyes, arm movements are generally oriented towards the front of the body. Intensive training thus occurs from early life in many types of actions toward the front of the body; whereas, natural backward movements are very rare and most subjects have very little experience of movements oriented towards the back.

In conclusion, the dual and sequential mode of control which accompanies forward voluntary movements of the arms and the trunk seems to be a more elaborate and efficient means of controlling motor activities. The early activation of the "postural" muscles ensures a better minimization of the postural perturbation due to the movement; it might also contribute towards speeding up the onset of movement or allowing it to be performed with less energy expenditure (see Crenna et al., 1987). The single mode of control of the backward axial movements seems to be a more primitive, less efficient mode of control which is probably inborn, and which does not tend to improve due to lack of training.

References

Babinski, J., 1899, De l'asynergie cérébelleuse, Rev. Neurol., 7: 806-816.

Belenkiy, V. E., Gurfinkel, V. S., and Paltsev, E. I., 1967, On elements of control of voluntary movements, Biofizika (in Russian), 12: 135-141.

Bizzi, E., Accornero, N., Whapple, W., and Hogan, N., 1984, Posture control and trajectory formation during arm movement, J. Neurosci., 4: 2738-2744.

Bouisset, S., Zattara, M., 1981, A sequence of postural movements precedes voluntary movements, Neurosci. Lett., 22: 263-270.

Cordo, P. I., and Nashner, L. M., 1982, Properties of postural adjustments associated with rapid arm movements, J. Neurophysiol., 47: 287-302.

Crenna, P., Frigo, C., Massion, J., and Pedotti, A., 1987, Forward and backward axial synergies in man, Exp. Brain Res., 65: 538-548.

Diener, H. C., Dichgans, J., Guschlbauer, B., and Bacher, M., 1986, Role of visual and static vestibular influences on dynamic posture control, Human Neurobiol., 5: 105-113.

Feldman, A. G., 1986, Functional tuning of the nervous system during control of movement or maintenance of a steady posture, III. Mechanographic analysis of the execution by man of the simplest motor tasks, Biophysics, 11: 766-775.

Ferrigno, G., and Pedotti, A., 1985, Elite: a digital dedicated hardware system for movement analysis via real time TV-signal processing, IEEE Trans, Biomed. Eng., 32: 943-950.

Friedli, W. G., Hallett, M., and Simon, S. R., 1984, Postural adjustments associated with rapid voluntary arm movements, 1. Electromyographic data, J. Neurol. Neurosurg. Psych., 47: 611-622.

Gurfinkel, V. S., and Elner, A. M., 1973, On two types of static disturbances in patients with local lesions of the brain, Agressologie, 14D: 65-72.

Massion J., 1984, Postural changes accompanying voluntary movements, Normal and pathological aspects, Human Neurobiol., 2: 261-267.

THE ROLE OF NIGROSTRIATAL AND MESOLIMBIC DOPAMINERGIC BRAIN SYSTEMS IN THE CONTROL OF VOLUNTARY MOVEMENTS AND POSTURAL ADJUSTMENT IN DOGS

K.B. Shapovalova and A.F. Yakimovsky

Pavlov Institute of Sciences, Leningrad, USSR

Summary

Microinjections of 3 g dopamine (DA) into the caudate nucleus head (HNC) and the nucleus accumbens (NAC) produced unidirectional effects on the motor components of avoidance instrumental reaction (IR) and posture adjustment in two groups of dogs (intact dogs and dogs with varying degrees of caudate pathology). However, each of these dopaminergic systems can mainly regulate certain components of the IR motor program, i.e. nigrostriatal system - the initiation of both the conditioned postural adjustment and the voluntary movement and mesolimbic system - the IR amplitude and stability of performance. The effects of the ipsi- and contralateral DA microinjections into HNC were, in principle, similar. The same was true of the microinjections into left and right NAC. The influence of bilateral microinjections into HNC and into NAC was stronger than that of unilateral ones. The greatest effects was obtained after combined "four-cannulae" microinjection of DA bilaterally into both structures in the dog with an akinetic form of caudate pathology. DA microinjections into the HNC of "parkinsonian" animals were followed by an improvement in all IR components and, consequently by an increase in the percentage of correct IR in the experiment. On the other hand, the normal animals with the absolutely correct IR and 100% score on the task solution apparently had a well balanced level of neostriatal mediatory system interaction. Therefore, DA microinjections into the HNC and NAC of these animals, which change this balance, disrupt the instrumental behaviour by increasing the number of interstimulus limb raising and phasic movements, overlapping the tonic instrumental reaction.

Introduction

Recent neurochemical, neurophysiological and neuromorphological investigations on the neostriatum have contribued greatly to our understanding of the role of this structure in the regulation of voluntary movements as well as its participation in the onset of some forms of motor pathology (Pasik et al., 1979; Kitai, 1981; Beninger, 1983; Groves, 1983; Nieoullon et al., 1983; Lidsky et al., 1985; Garcia-Rill, 1986). Among the most important findings in this field have been those concerning the control by brain dopaminergic systems (including the nigrostriatal and mesolimbic systems) of locomotion and postural adjustment (Steg and Johnels, 1979; Beninger, 1983; Groves, 1983; Mogenson and Nielsen, 1983; Shapovalova, 1985). Other important data have also resulted from observations on patients subjected to long-term treatment by neuroleptics (Hornykiewicz, 1979; Bunney et al., 1980; Johnels and Steg, 1980; De Long and Georgopoulos, 1981). According to recent data, striato-limbic integration may play an important part in the initiation and regulation of voluntary movements (Mogenson et al., 1980). One of the mesolimbic dopaminergic structures, the Nucleus Accumbens (NAC) (Heimer and Van Hoesen, 1979), which receives a direct input from the ventral tegmental area - A10 (Ungerstedt, 1971) is regarded as a "bridge" in the transmission of the influence of the limbic system to the neostriatum (Mogenson et al., 1980). The NAC can intensify the striatal dopaminergic mechanisms controlling initiation and performance of motor behaviour on the basis of its afferent and efferent organization (Shapovalova, 1985). On the other hand, NAC has a direct influence of locomotor activity via the descending pathways to the Substantia Innominata and further to the pedunculopontine nucleus, the main components of the mesencephalic locomotor region of the brain (Pijnenburg et al., 1975; Mogenson et al., 1983; Garcia-Rill, 1986).

Investigation were conducted on the role of the dopaminergic brain system in the initiation and performance of instrumental defensive reactions and the prior conditioned postural adjustments in a series of our experiments on dogs using the dopamine (DA) receptor antagonist haloperidol and the DA precursor L-Dopa (Shapovalova, 1984; Shapovalova et al., 1984). In connection with the above, we decided that further neuropharmacological analysis of the role of neostriatal (Head of Nucleus Caudatus - HNC) and mesolimbic (NAC) DA-receptive structures in the control of voluntary movements and postural adjustments might be very interesting. Comparative studies on the involvement of these two dopaminergic structures in the regulation of voluntary movements in normal dogs and in dogs with varying degrees of caudate pathology were conducted in order to shed further light on this problem. One of the main techniques used for this purpose was the observation of effects resulting from DA microinjection through microcannulae implanted into HNC and NAC.

Methods

We applied this technique in chronic experiments on the model of instrumental defensive reaction associated with the maintenance of a certain posture (known as the Petropavlovsky method). Experiments were carried out on 5 adult mongrel dogs: on normal animals (NN 1, 3, 7) and on dogs with various degrees of caudate pathology caused by electrical stimulation of the HNC performed in various experimental series during several years (NN 4 and 5). The animals were trained, after presentation of a conditioned signal (Metronome, M 130/s), to avoid the electrical shock applied to the left hind limb, by raising this leg to a certain height (8 cm), thus breaking the contacts, and keeping this posture during the whole duration of the conditioning signal (10 s). In the absence of the appropriate motor response, the electrical shock was switched on at the 5th s of the conditioned signal and accompanied with the conditioning signal for the remaining 5 s (see for details Shapovalova et al.,1984).

Stainless steel cannulae (d=0.6 mm) were stereotaxically (Lim et al., 1960) implanted into HNC and NAC under Hexenal anesthesia under sterile conditions. During microinjections flexible microinjectors were used (see for details Suvorov and Yakimovski, 1982). DA (3 g) dissolved in double-distilled water was injected into HNC and NAC. The microinjections were conducted under aseptic conditions. In all, 42 DA microinjections into HNC and NAC were performed.

The following IR components were recorded: 1. The tensogram of the working limb, in which the following parameters were analysed: L_1 - the latency of the beginning of postural adjustment; L_3 - the time necessary for "unloading" of the working limb to occur; L_4 - the latency of total posture adjustment. 2. The electromyogram (EMG) of the musculus rectus femoris and semitendinosus of the working (left) and supporting (right) hind limbs. 3. The mechanogram of the working limb, in which the following components were analysed: L_5 - the latency of the working limb raising, H - the amplitude of the IR, and L_6 - the latency of the instrumental task solution (the time necessary for the animal to hit a "safety zone").

Table I: Mean values (throughout a month) of IR components (in s) in the normal dogs (NN 1,3) and in the dogs with caudate pathology (NN 4,5) prior to the experimental session with DA microinjections.

Dogs NN	IR components				
	L_1	L_3	L_4	L_5	L_6
1	0.4±0.06	0.3±0.04	0.7±0.05	0.8±0.04	1.0±0.1
3	0.8±0.07	0.4±0.05	1.2±0.2	1.4±0.1	1.8±0.2
4	1.3±0.4	1.6±0.5	3.2±0.5	3.4±0.6	3.7±0.7
5	2.5±1.1	1.9±0.7	4.5±1.2	4.6±1.3	-

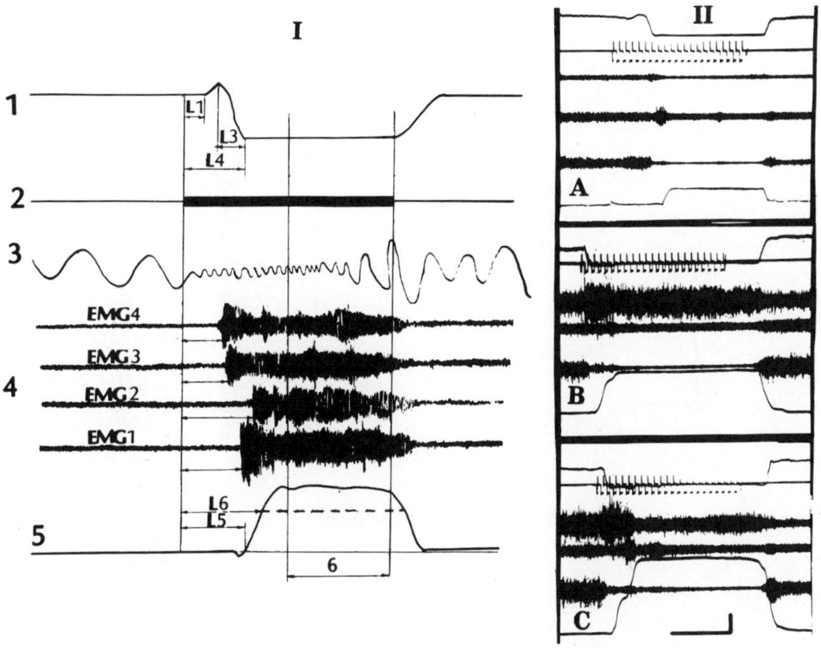

Fig. 1. I. Schema of the data recording and processing. (1) Recording of pressure on the support of the working limb: L_1, latency of the postural adjustment; L_3, time taken to unload the working limb; L_4, latency of the total unloading of the working limb. (2) The CS mark (metronome, 130/min). (3) Respiratory recording. (4) EMG records on the working (left, hind) and the supporting (right, hind) limbs: EMG1 latency of m. rectus femoris activity; EMG2, latency of m. semitendinosus activity of the working limb; EMG3, EMG4, the same for the supporting limb. (5) Mechanogram of the working limb: L_5, latency of the instrumental movement; L_6, latency of the instrumental task performance (time necessary for the animal to hit a "safety zone"). (6) Duration of the electrical current passed. II. Effects of DA microinjections into dopaminergic structures on the performance of the IR by the dog with caudate pathology (dog N4). A typical IR recordings prior to the experimental session with DA microinjections. B. IR performance 20 min after DA microinjection into HNC (ipsilateral, 3 μg). C. IR performance 20 min after DA microinjection into NAC (ipsilateral 3 g). In another experimental session. From top to bottom: recording of the tensogram of the working limb (left, hind), the onset of the CS, EMG of m. rectus femoris of the supporting (right, hind) limb, the same of m. semitendinous and m. rectus femoris of the working limb, mechanogram. Horizontal bar, 5 s. Vertical bar, 8 cm.

The data obtained were processed statistically. Their signifiance was determined by means of Student's t-test.

Results

As mentioned above, we studied these problems in two groups of dogs - normal dogs and dogs with various degrees of caudate pathology. The components of IR can be compared in the pathological (NN 4 and 5) and normal dogs (NN 1,3) prior to the experimental session with DA microinjections (Table I).

At the beginning of the present experimental session, the latency of the postural adjustment (L_1) in dog N4 was more than 1 s (the mean value 1.3±0.4 s) and latency of the instrumental task solution (L_6) was 3.7±0.7 s (Table I). The time of the total period of postural adjustment (L_4) also increased significantly (3.2±0.5 s) in comparison with the normal dog (1.2±0.1 s). The amplitude of the IR in dog N4 was only 8-9 cm (i.e., the minimal level for the animal to hit a "safety zone") and the instrumental responses were characterized by an absolute lack of phasic components in the IR and interstimulus phasic responses (Fig. 1.A). Fig. 1.B shows two instrumental reactions recorded 20 min after DA

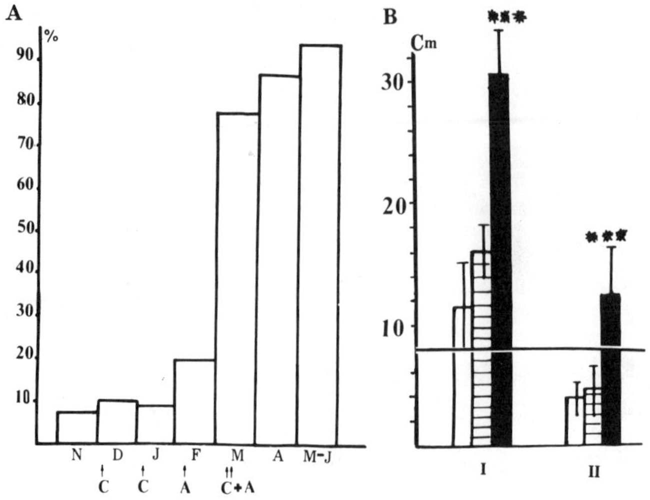

Fig. 2. A. Mean percentages (throughout a month) of the correct task performances in the dog with severe form of caudate pathology (dog N5) prior (November) and after DA microinjections: into HNC (contralateral) - December (C); into HNC (bilaterally) - January (C); into NAC (bilaterally - February (A); simultaneously into HNC and NAC (bilaterally) - 3th March (C + A). B. Effects of DA microinjections into dopaminergic structures on the amplitude (cm) of the IR in normal dog (dog N1) - I, and in a dog with caudate pathology (dog N5) - II. Each column - the mean data on 30 IR tasks. Empty white columns - background (immediately prior to DA microinjection), hatched columns - after DA microinjections into HNC, filled black columns - after DA microinjections contacts (corresponds to 8 cm for the animal). Stars: significantly different from the background ($p<0.001$).

microinjection into the ipsilateral HNC. Considerable shortening of L_1 (mean value 0.5±0.04 s) and the total period of postural adjustment (1.7±0.5 s) are shown in this figure. The amplitude of the instrumental response also increased markedly. After DA microinjections into the NAC performed in another experimental session, similar results were obtained. Fig. 1.C. shows the IR recorded 20 min after DA microinjection into the ipsilateral NAC. It is interesting to note that the amplitude of instrumental reaction increased sharply (more than three times) in comparison with the control (Fig. 1.A).

The most effective DA microinjection was in the dog with the severe form of caudate pathology (dog N 5). Before the beginning of the experimental session with DA microinjections, this dog was akinetic. Over a month the mean level of correct instrumental task solution was only 8-9 % (Fig. 2.A). This was due to the very low amplitude of the movements (4-5 cm), the great duration of the postural adjustment components and the latency of movement initiation (Fig. 3.B). For example, the mean value of component L_1 was in this period 2.5±1.1 s (Table I). In most of the animals' performances, only escape reactions were observed after onset of the electrical current (Fig. 3.A). The avoidance reaction after presentation of the conditioning signal was usually absent. The DA microinjections were performed as a rule once, but also occasionally twice weekly. These microinjections were accompanied by a sharp decrease in the time taken by the total postural adjustment (component L_4). This occurred first of all at the expense of more than two-fold shortening of the latency of the postural adjustment (L_1) - Table II. However , the amplitude of instrumental movements was lower than the correct level, and this appeared to be the main cause of the difficulty subjects had in solving the instrumental task (11 %). The amplitude of instrumental movements increased sharply after bilateral DA microinjections into the NAC (Fig. 3.E). This facilitatory effect was observed only on the day of the microinjections and 1-2 days after it, so that the mean monthly level of IR amplitude was lower than the critical amplitude level necessary for breaking the contacts. This was one reason for the relatively low number (20 %) of correct instrumental tasks performed in this period, (Fig. 2.A). However, the single DA microinjection bilateral into HNC and bilateral into NAC (i.e. simultaneously into four microcannulae) produced very prolonged effects, which persisted for five months (Fig. 2.A). Besides the stable shortening of the L1, L4 and L5 components described above, this microinjection was accompanied by a stable increase in the IR amplitude (Fig. 3.G) and an incrase in the number of correct tasks performed (80-90%). These effects were observed during the following 5 months (Fig. 2.A). It is interesting to note that the duration of the component L3 -the time taken to unload the working limb and redistribute its weight to the supporting legs - did not change during any of the modulations in DA level occurring in the DA-ergic brain structures in comparison with the initial data (Shapovalova, 1984, 1985). The mean value of the component L6 after "combined" microinjection into HNC and was 4.0±0.5 s (Table II).

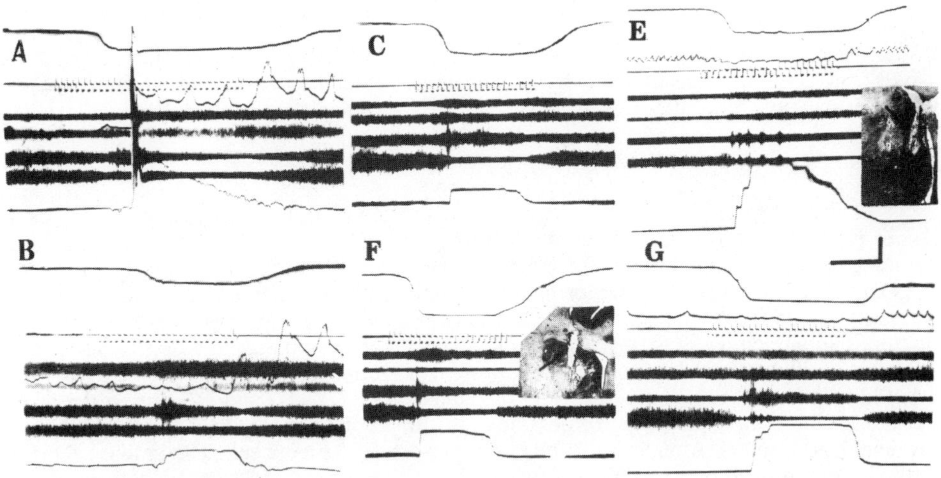

Fig. 3. Effects of DA microinjections into dopaminergic structures on the IR performances in the dog with severe form of caudate pathology (dog N5). A, B. Prior to the DA microinjections: C. 20 min after unilateral DA microinjection into HNC. F. 20 min after bilateral microinjection of DA into HNC. E. 20 min after bilateral DA microinjection into NAC. G. 15 days after simultaneous DA microinjection into HNC and NAC (bilaterally). From top to bottom: recording of the tensogram of the working limb, recording of respiration (in A, B, E and G), the onset of the conditioned signal (metronome, 130/min), EMG activity of mm. rectus femoris and semitendinosus of supporting and working limbs, the mechanogram of the instrumental movement. Arrows show the localization of the microcannulae into HNC (in F) and into NAC (in E). Horizontal bar, 5s. Vertical bar, 8 cm.

Table II

		IR components					
Month	DA injections	L1	L3	L4	L5	L6	H
Nov.	-	2.5±1.1	1.9±0.7	4.5±1.2	4.6±1.3	-	4.5±0.4
Jan.	HNC, bilaterally	1.0±0.3	2.0±0.4	3.0±0.1	3.0±0.3	-	5.5±0.6
Feb.	NAC, bilaterally	1.0±0.5	2.1±0.3	3.0±0.4	3.3±0.5	4.5±0.8	7.2±0.2
Mar.	HNC, bilat. + NAC bilat.	1.3±0.3	2.0±0.3	3.3±0.2	3.4±0.3	3.9±0.6	9.5±0.6
Apr.	-	1.2±0.3	1.9±0.6	3.1±0.2	3.2±0.2	3.8±0.6	8.5±0.5
Jun.	-	1.1±0.2	2.2±0.4	3.4±0.3	3.5±0.2	4.3±0.4	9.0±0.3
Jul.							

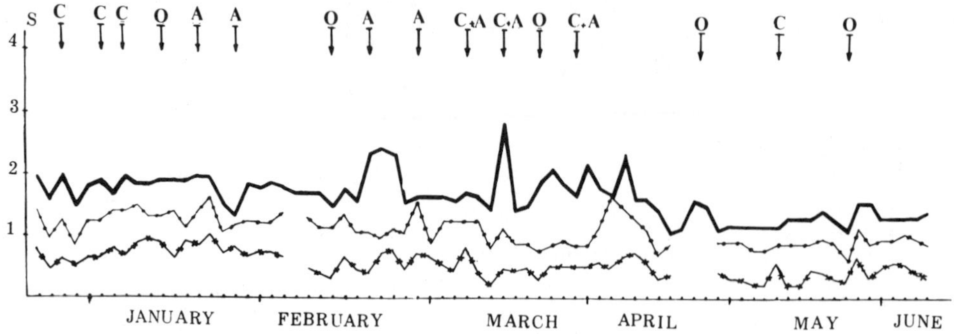

Fig. 4. The effect of DA microinjections into dopaminergic structures on the dynamics of the latent periods (in s) of IR components during several months. The normal dog (dog N3). From top to bottom: L_6, L_4, L_1. DA microinjections into HNC (C). DA microinjections into NAC (A), simultaneous DA microinjections into HNC and NAC (C+A), control microinjections of the double-distilled water into the HNC and NAC (O). Each point - the mean daily data.

The effect of DA microinjection into the HNC and NAC in normal dogs was somewhat different as compared to the results in animals with varying degrees of caudate pathology. The common direction of DA action in normal dogs remained the same, however. It is important to note that in normal dogs, before the experimental session with DA microinjections the latency of all the IR motor components, including the latency of the postural adjustment (L_1) was very small (Table I). For example, in the dog N 3 the mean values of L_1, L_4 and L_6 components were respectively: 0.8±0.07 s, 1.2±0.1 s and 1.8±0.2 s. In Fig. 4, the dynamics of the IR components (L_1, L_4 and L_6) during the whole period of contralateral DA microinjections into the HNC and NAC (December-June) are presented. As a rule, DA microinjections were performed once a week. The gradual decrease in the L_1 component is shown in this figure. After 5 months of DA microinjections the L_1 component had diminished more than twice, compared with the background (December). The mean value of L_1 in May-June was only 0.3±0.05 s ($p < 0.01$).

The DA microinjections into NAC in normal dogs, as in the dogs with the striatal pathology, resulted in a significant increase in the IR amplitude, while similar microinjections into the HNC were accompanied by a small increase in the IR amplitude (Fig. 2.B). After DA microinjections into HNC and NAC in normal animals, the number of interstimulus phasic movements and phasic components of IR intensification were recorded, which led to difficulties in performing the correct instrumental task on the part of the animals. In this connection, the most noteworthy data were obtained in experiments on normal dogs without complete shaping of the IR motor program (dog N7). In Fig. 5 A four typical IR recordings prior to the experimental session with DA microinjections are shown. In all the examples shown in this figure, the dog solved the instrumental task (the electrical reinforcement was avoided), although the pattern of EMG-activity and movements was not the same in each case. The data processing shows that the IR motor components varied greatly from day to day. The absence of any stable postural adjustment pattern prior to the IR performance should be specially emphasized (Fig. 5.A). The level of instrumental task success at the beginning of the session with DA microinjections was about 65-70%, although the conditioning signal was used 250 times. Simultaneous DA microinjections into the HNC and NAC (contralaterally, 3 μg) resulted in disturbances of the conditioned postural adjustment and in total elimination of the IR tonic form (Fig. 5.B). When responding to the conditioning stimulus, the animal performed only phasic leg raising, and even escape from the electrical current was not observed to have the ordinary duration (Fig. 5.B, downward part). Between the conditioning stimulus tests the dog was extraordinarily excited and interstimulus phasic leg raising was constantly observed. It should be mentioned that the time of postural adjustement initiation (component L_1) after DA microinjections shortened in dog N7, as in the other animals. Recovery of the appropriate conditioned behaviour was achieved only three days after DA microinjection, in spite of daily use of the conditioning signal with the electric reinforcement.

Discussion

It is well known that the most typical manifestation of nigrostriatal dopaminergic pathology is akinesia, i.e. impaired movement initiation. The akinesia produced by nigro-caudate pathway damage mainly affects voluntary movements (De Long and Georgopoulos, 1981). This form of akinesia can be produced in experiments on animals by all kinds of influences resulting in a decrease in DA levels in the HNC (Ranje and Ungerstedt, 1971), such as: neuroleptic treatment (Fielding and al., 1978; Shapovalova, 1984), 6-OHDA injection into Substantia Nigra (Brook and Iversen, 1975) and after prolonged bilateral electric stimulation of the HNC as in our experiments (Shapovalova, 1984, 1985). On the other hand, one of the possible ways of eliminating the akinesia is DA microinjection through cannulae into the HNC and NAC (Iversen, 1977; Shapovalova, 1985). The most effective "treatment" of akinetic animals obtained in our experiments consisted of simultaneous DA microinjections administered bilaterally into the HNC and NAC. These results are in agreement with the fact that the strongest deficit in the animals' performance of active avoidance tasks can be due only to the decrease in DA level in both the dopaminergic (i.e. the nigrostriatal and mesolimbic) systems (Koob et al., 1984).

One of the manifestations of the akinetic form of parkinsonism, according to our data, may be the disturbance of postural adjustment initiation preceeding the IR performance. This conclusion is confirmed by the fact that DA microinjection into the HNC resulted in a

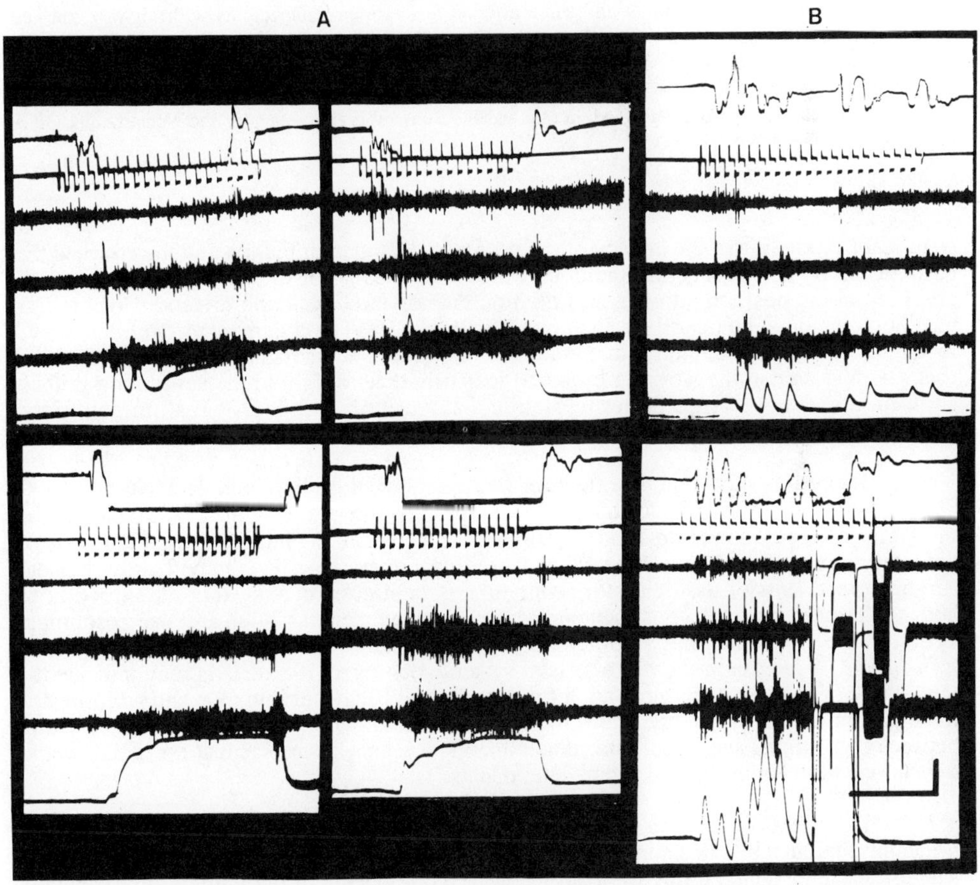

Fig. 5. Disturbances of IR performance after DA microinjection into HNC and NAC in the normal dog (dog N7) without complete shaping of the IR motor program. A. Examples of IR-performances prior to the experimental session with DA microinjections. B. Two IR performances 20 min after contralateral microinjections into HNC and NAC. Other designations as Fig. 1. Horizontal bar, 5s. Vertical bar, 8cm.

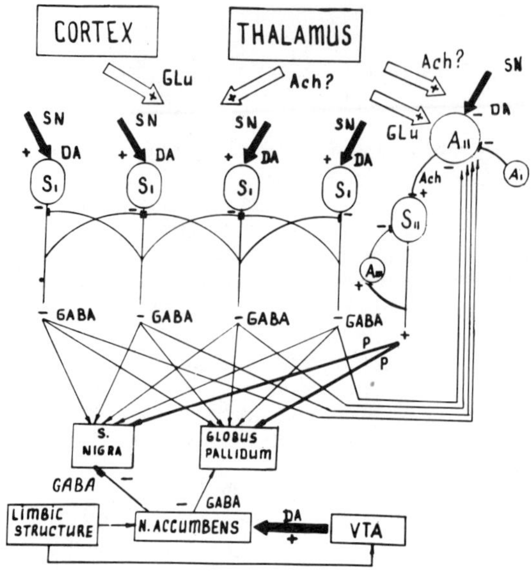

Fig. 6. Schematic diagram summarizing the possible organization of neostriatal efferent systems (according to Groves, 1983, with some additions) and some afferent and efferent connections with nigrostriatal and mesolimbic dopaminergic systems. SN - Substantia Nigra; VTA - Ventral Tegmental Area; AII - the large cholinergic interneuron of the Neostriatum; AI and AIII - the small interneurons of the Neostriatum; SI and SII - the efferent neurons of the Neostriatum; Glu - Glutaminergic, DA - Dopaminergic and Ach - Cholinergic inputs of the Neostriatum; GABA - GABAergic and P - Subst. P-ergic outputs of the Neostriatum. Other explanations in the text.

significant shortening of the latency of the postural adjustment initiation and thus formed the basis of the IR performance in the dogs with caudate pathology. The significant increase observed in the postural adjustment initiation latency after systemic injections of the DA receptor antagonist haloperidol (Shapovalova, 1984, 1985) is in agreement with the above idea. On the other hand, the basic component of postural adjustment (component L_3) - the time taken to unload the working limb and redistribute its weight to the supporting limbs - showed no change after either systemic haloperidol and L-Dopa administrations (Shapovalova, 1984, 1985) or DA microinjections into the HNC and NAC.

The results show that the effects of DA microinjections into both the HNC and NAC have the same direction. We can conclude that these two structures jointly participate in the initiation and regulation of postural adjustments as well as in the initiation and regulation of voluntary movements. However, each of these dopaminergic systems may mainly regulate certain components of the IR motor program, i.e. the nigrostriatal system-the initiation of both conditioned postural adjustment and voluntary movement initiation and the mesolimbic system - the amplitude and stability of performance. It is known that NAC activation influences DA release in the nigro-striatal system (Beninger, 1983). This may indicate the importance of an increase in motivation for instrumental behaviour initiation and performance in dogs with striatal pathology. Thus, the strio-limbic integration initially due to interaction between nigrostriatal and mesolimbic dopaminergic mechanisms is essential for initiation and regulation of the voluntary movement components.

The effects of the ipsi- and contralateral DA microinjections into the HNC were, in principle, similar. The same applies to the effects of microinjections into left and right NAC. The influence of bilateral microinjections on the components of the voluntary movements was stronger than that of unilateral ones, especially in the case of the "four-cannulae" DA microinjection administered bilaterally into the HNC and NAC. Contrary to all the others, this combined microinjection had prolonged effects, especially in the dog with a severe form of striatal pathology, in which the effect lasted for several months. The mechanism of this prolonged effect is still not clear. It is possible that it may be complemented by some additional neuromediatory factors, perhaps of a peptidic nature, that intensify and protract the action of DA.

Recent evidence (Ungerstedt, 1971; Pasik et al., 1979; Dray, 1980; Kitai, 1981) has led to the idea (Groves, 1983) that two efferent systems of the neostriatum may be involved, differing in their morphological, neurophysiological and neurochemical organization and in the effects they evoke on their major targets, the Globus Pallidus and pars reticulata of the Substantia Nigra. It appears that the forms of voluntary movement and their motivation--dependent manifestations are controlled by the balance of the Subst. P-ergic and GABA-ergic neostriatal efferents (Fig. 6).

Neostriatal DA, which has an excitatory influence on the axo-spine synapses of S1 cells and both direct and indirect inhibiting effects on the neostriatal cholinergic cells (Groves, 1983; Nieoullon et al., 1983), plays a key role in the regulation of the excitatory and inhibitory efferent influences on major neostriatal targets. DA microinjections into HNC and NAC of "parkinsonian" animals are accompanied by an improvement in all the IR components, and consequently by an increase in the percentage of correct IR performed in experiments. On the other hand, the normal animals which produced absolutely correct IR performances and 100 % task solution level apparently have a well-balanced level of neostriatal mediatory system interaction. Therefore, DA microinjections into the HNC and NAC of these animals, which change this balance, disrupt the instrumental behaviour by increasing the number of interstimulus limb raising and phasic movements, overlapping the tonic instrumental reaction. This results in the animals' getting into the "danger zone" associated with pain reinforcement and consequently in a reduction in the percentage of correct IR solutions. With the cessation of injections, any DA-dependent imbalance is rapidly compensated for. The most interesting results were obtained with DA microinjections administered to hyper-mobile dogs without a complete shaping of the IR motor program. In this case, the absolute disappearence of the tonic form of IR and its replacement by phasic conditioned responses were observed. Similar data have been obtained in experiments on cats (Albertin, 1985). In conclusion, the available data suggest that phasic and tonic components of learned movements and postural adjustments are probably controlled by different neostriatal transmitter systems.

References

Albertin, S.W., 1985, Participation of the caudate nucleus DA-reactive system in regulation of instrumental conditioning of different degrees of complexity, Fiziol. Zh. im. I.M. Sechenova, 71: 87.

Beninger, B.J., 1983, The role of dopamine in locomotor activity and learning, Brain Res. Rev., 6: 173.

Brook, C., and Iversen, S.D., 1975, Changed eating and locomotor behaviour in the rat after 6-OHDA lesions to the Substantia Nigra, Neuropharmacology, 14: 95

Bunney, B.S., Grace, A.A., and Hommer, D.W., 1980, Changing concept of nigral dopamine system functions within the basal ganglia: relevance to extrapyramidal disorders, J. Neurol. Transmit. Suppl., 16: 17.

Chevalier, G., Thierry, A., Shibazaki, T., and Feger, J., 1981, Evidence for a GABA-ergic inhibitory nigrotectal pathway, Neurosci. Lett., 21: 6.

De Long, M.R., and Georgopoulos, A., 1981, Motor functions of the basal ganglia, in: "Handbook of Physiology", sect.I. The Nervous System, vol. II, part 2:1017-1061.

Dray, A.I., 1979, The striatum and substantia nigra: a commentary on their relationship, Neurosci., 4: 1407.

Fielding, S., and Lal, H., 1978, Behavioral actions of neuroleptics, in: "Handbook of Psychopharmacology", (L. Iversen, S. Iversen, S. Snyder, eds), pp. 91-120, Plenum Press: New-York.

Garcia-Rill, F., 1986, The Basal Ganglia and the locomotor regions, Brain Res. Rev., 11: 47.

Groves, Ph. M., 1983, A theory of the functional organization of the Neostriatum and the Neostriatal control of voluntary movement, Brain Res., 5: 109.

Heimer, L., and Van Hoesen, G., 1979, Ventral striatum, in: "The Neostriatum", (I. Divac and G. Oberg, eds), pp. 147-160, Pergamon Press: Oxford, New-York.

Hornykiewicz, O., 1979, Brain Dopamine in Parkinson's disease and other neurological disturbances, in: "The Neurobiology of Dopamine", (A.S. Horn, J. Korf and B.H.C. Westernik, eds), pp. 633-654, Acad. Press, London.
Iversen, S.D., 1977, Brain Dopamine systems and behavior, in: "Handbook of Psychopharmacology", (L. Iversen, S. Iversen and S. Snyder, eds), vol. 8, pp. 333-384, Plenum Press, New York.
Johnels, B., and Steg, G., 1980, The corpus striatum and the regulation of posture and locomotion, Neurosci. Lett., 19, Suppl., 5: 339.
Kitai, S.T., 1981, Electrophysiology of the Corpus Striatum and brain stem integrating systems, in: "Handbook of Physiology", (J. Broomhart and V. Mouncastle, eds), Sect. I, The Nervous System, v.II, part 2, pp. 997-1016, Bethesda.
Koob, G., Simon, H., Herman, J., and Le Moal, M., 1984, Neuroleptic-like destruction of the conditioned avoidance response requires destruction of both the mesolimbic and nigrostriatal systems. Brain Res., 303: 319.
Lidsky, T., Manetto, C. and Schneider, S., 1985, A consideration of sensory factors involved in motor functions of the Basal Ganglia, Brain Res. Rev., 9: 133.
Lim, R., Liu, Ch., and Moffit, R., 1960, "A stereotaxic atlas of the dog's brain", Springerfield, Illinois, USA
Mogenson, G., Johnes, D;, and Yim, Ch., 1980, From motivation to action functional interface between the limbic system and the motor system, Progr. in Neurobiol., 14: 69.
Mogenson, G., and Nielsen, M., 1984, Neuropharmacological evidence to suggest that the Nucleus Accumbens and Subpallidal region contribute to exploratory locomotion, Behav. and Neural. Bul., 42:52.
Mogenson, G., Swanson, L., and Wu, M., 1983, Neural projections from Nucleus Accumbens to Globus Pallidus, Substantia Innominata and lateral preoptic-lateral hypothalamic area: an anatomical and electrophysiological investigation in the rat, Neurosci., 3: 189.
Nieoullon, A., Kerkerian, L., and Dusticier, N., 1983, Presynaptic controls in the Neostriatum: reciprocal interactions between the nigrostriatal dopaminergic neurons and the cortico-striatal glutaminergic pathway, in: "Neural coding of motor performance", (J. Massion, ed.), pp. 54-65, Springer-Verlag: New York, Berlin, Heidelberg.
Pasik, P., Pasik, T., and Di figlia, M., 1979, The internal organization of Neostriatum in mammals, in: "The Neostriatum", (I. Divac and G. Oberg, eds), pp. 5-36, Pergamon Press: Oxford, New-York.
Pijnenburg, A., Honig, W., and Van Rossum, J., 1975, Inhibition of D-amphetamine induced locomotor activity by injection of haloperidol into Nucleus Accumbens of the rat, Psychopharmacol., 41: 87.
Ranje, C., and Ungerstedt, U., 1977, High correlations between number of dopamine cells, dopamine levels and motor performance, Brain Res., 134: 83.
Shapovalova, K.B., 1984, Conditioned postural adjustment in dogs during systemic administrations of agonists and blocators of the dopaminergic brain system, Zh. Wyssh. Nervn. Deyat. im. I.P. Pavlova (In Russian), 34: 1057.
Shapovalova, K.B., 1985, Possible neurophysiological and neurochemical mechanisms of striatum involvement in the regulation of the voluntary movement, Fiziol. Zh. im. I.M. Sechenova (In Russian), 71: 537.
Shapovalova, K.B., Yakunin, I.V., and Boyko, M.I., 1984, Participation of dogs caudate nucleus head in the mechanisms of conditioned posture reorganization, Zh. Vyssh. Nervn. Deyat. im. I.P. Pavlova (In Russian), 34: 669.
Steg, G., and Johnels, B., 1979, Motor functions of the Striatum, in: "Neostriatum", (I. Divac and G. Oberg, eds), pp. 231-239, Pergamon Press: Oxford, New-York.
Suvorov, N.F., and Yakimovski, A.F., 1982, The role of enkephalin-reactive system of the caudate nucleus in the performance of an alimentary conditioned reflex in dogs, Zh. Vyssh. Nervn. Deyat. im. I.P. Pavlova (In Russian), 32: 840.
Ungerstedt, U., 1971, Stereotaxic mapping of the monoamine pathways in the rat brain, Acta physiol. scand. suppl. 367, 82: 36.

ON THE ROLE OF MOTOR CORTEX IN THE LEARNED REARRANGEMENT OF POSTURAL COORDINATIONS

M.E. Ioffe, N.G. Ivanova, A.A. Frolov, E.V. Birjukova, and N.V. Kiseljova

Institute of Higher Nervous Activity and Neurophysiology
USSR Academy of Sciences, Moscow, USSR

Summary

Two kinds of postural patterns opposite to inborn ones were elaborated in dogs. Some dogs were trained to avoid electrical stimulation of left forelimb by lifting the limb and keeping it lifted for 5s. This reaction had to be accompanied by decreasing the pressure on the support of the ipsilateral hindlimb, so the inborn "diagonal" postural pattern (unloading the limb diagonally opposite to the lifted one and loading the other pair of limbs) was rearranged into the "ipsilateral" one. The other dogs were trained to escape electrical stimulation of one forelimb by increasing the support pressure of the limb (this reaction is opposite to the inborn flexor reflex). Ablation of the motor cortex contralateral to the performing limb resulted in temporary disturbances of the rearranged postural patterns. After bilateral ablation of the motor cortex the elaborated postural patterns disappeared and mainly inborn postural coordinations were performed. The program of the elaborated reaction was maintained and mainly the executive mechanisms were disturbed (inhibition of the inborn reactions). The motor cortex probably controls the rearrangement of postural coordinations modulating functions of the appropriate brain stem structures.

Introduction

Motor learning is the elaboration of new movements. The structural basis of the process is very intriguing and has not yet been properly elucidated. It is known that the motor cortex is a structure involved not only in the control of fine discrete movements (Fulton, 1949; Lawrence and Kuypers, 1968; Phillips and Porter, 1977; Wiesendanger, 1981; Evarts et al., 1984) but also in the rearrangement of motor coordinations as the result of learning. This has been demonstrated in the case of some limb movements (Ioffe, 1973; Burlachkova and Ioffe, 1979). But most of the movements performed by a standing animal or human being are accompanied by postural adjustments which ensure centre of gravity displacement and equilibrium maintenance during limb movement. Postural adjustment is mainly controlled by structures other than the motor cortex (Shumilina, 1949; Korjakin, 1958; Ioffe, 1975) although the motor cortex plays an important role in the integration of postural displacement and limb movement parameters (Massion, 1979).

In view of the fact that motor cortex is, on the one hand, a structure specifically involved in the learned rearrangement of motor coordinations and, on the other hand, that it is not the basic structure controlling the postural displacements accompanying limb movement, it seemed interesting to study the role of the motor cortex in the learned rearrangement of postural displacement patterns. Our study deals with this question.

Methods

The study was performed on 7 adult dogs weighing 11-19 kg. The dogs were trained to stay on 4 independent platforms with strain-gauges (Egiazaryan et al., 1980). The vertical forces of limb pressure on the support and left forelimb lifting were recorded on an ink-writer and disc of an on-line Mera-60 computer. Off-line data analysis was partly performed on a Nord-100 computer.

Two experimental procedures were used to rearrange inborn postural patterns.

The first procedure (Experiment 1) was as follows.

An avoidance conditioned reflex was elaborated. In response to the sound CS (200 Hz, duration 5s, time from the onset of the conditioned stimulus CS to the beginning of the unconditioned stimulus: US O.5 s) the dog had to lift the left forelimb above a certain level (usually 5-7 cm) and keep the limb lifted for 4.5 s to avoid electrical stimulation (US1) of the skin of the limb. Usually limb lifting was accompanied by the so-called diagonal postural adjustment: the limb diagonally opposite to the lifted one was unloaded and the other pair of limbs was loaded (Ioffe, 1975; Gahéry et al., 1980). After elaboration of this conditioned reflex accompanied by the diagonal pattern of postural adjustment, the dogs were trained to perform another one, the so-called unilateral postural adjustment: while decreasing the pressure on the support of the lifted left forelimb to decrease the support pressure of the ipsilateral (left) hindlimb and to increase the pressure on the other pair of limbs. This was obtained by the additional application of weak electrical stimulation (US2) on the hindlimb ipsilateral to the lifted one. The dog could avoid the US2 by decreasing the pressure on the hindlimb by 10% of the initial value and keeping the decreased pressure level during 4.5 s of stimulation. Two electrical stimulations (US1 and US2) were therefore used simultaneously to elaborate 2 simultaneous avoidance reflexes (lifting the left forelimb and decreasing the pressure on the left hindlimb) and thus to rearrange the inborn diagonal pattern of postural adjustment into a unilateral one.

The coefficient of diagonality (D) (Gahéry et al., 1980) was calculated in each trial using the formula

$$D = 1 - |\Delta w_1 - \Delta w_4| + |\Delta w_2 - \Delta w_3| / |\Delta w_1| + |\Delta w_2| + |\Delta w_3| + |\Delta w_4|$$

where Δw_1 is the difference between the final and initial values of the left forelimb support pressure, and Δw_2, Δw_3 and Δw_4 are the same for the right forelimb, left and right hindlimbs, respectively.

D was calculated every 10 ms during 5 s of trial (in 500 points per trial). The time-course of D could thus be established in each trial and averaged for a whole session.

Horizontal trajectories of the centre of pressure (CP) and the centre of gravity (CG) under dynamic conditions were also computed (Ioffe et al., 1982) according to the following equation system

$$z_c m \ddot{x}_c = x_c \sum_{i=1}^{4} w_i - a(w_2 - w_1 - w_3 + w_4)$$

$$z_c m \ddot{y}_c = y_c \sum_{i=1}^{4} w_i - b(w_1 + w_2 - w_3 - w_4)$$

$$m \ddot{z}_c = \sum_{i=1}^{4} w_i - mg$$

where m is the mass of the animal; g = acceleration of gravity; w_i are experimentally measured values of the vertical pressure forces; x_c, y_c, z_c = lateral, antero-posterior and vertical coordinates of CG, respectively; 2a = distance between fore (hind) limbs; 2b = distance between fore and hindlimbs.

After elaboration of the new postural pattern, the motor cortex of the right hemisphere was ablated using the map of the dog motor cortex by Gorska (1974) and the state of the learned movement and pattern of the postural adjustment were tested for 6-18 months. Then the left motor cortex was also ablated and the rearranged postural coordination was studied.

The second procedure for inducing postural pattern rearrangement (Experiment II) was as follows.

The dogs were trained to escape electrical stimulation of the right forelimb by increasing the pressure on this limb to above 5 kg. Duration of electrical stimulation was 10 s. In this case the dog had to inhibit the inborn flexor reflex during electrical shock and to shift the CG to the stimulated limb to increase the pressure of that limb on the support. This postural displacement was opposite to the inborn one, which consisted of a CG displacement from the stimulated limb to ensure body equilibrium during limb flexion.

The percentage of inborn flexor reactions and that of the so-called successful reactions, values of pressure increase on the stimulated limb and some other parameters of the reactions were calculated.

By successful reaction we mean one when the dog escaped electrical stimulation. After elaboration of the reflex to satisfy the criterion 90% of successful reactions during three successive sessions, we ablated the motor cortex contralaterally to the stimulated limb and after two-four months of testing and retraining, the motor cortex of the other hemisphere was also ablated.

After the experiments, the animals were sacrified by nembutal solution injection and a morphological analysis of the brain was carried out to check the completeness of the ablations. A representative reconstruction of one brain is shown in Fig. 1.

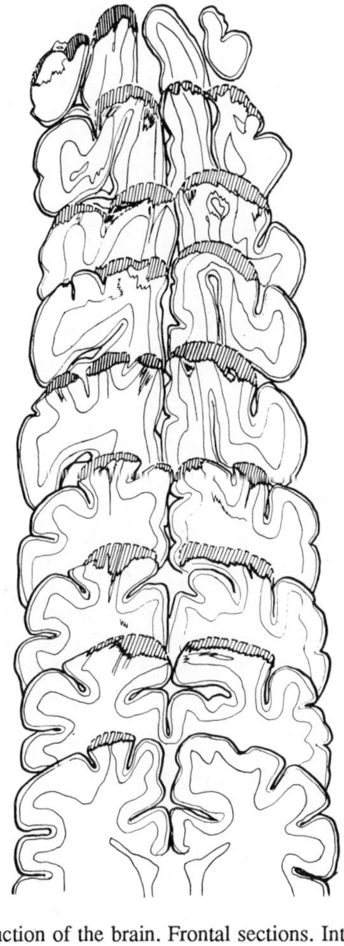

Fig. 1. A representative reconstruction of the brain. Frontal sections. Interval between sections is 1.3 mm. Ablated area is shaded. 1st lesion is on the right side, 2nd lesion is on the left side.

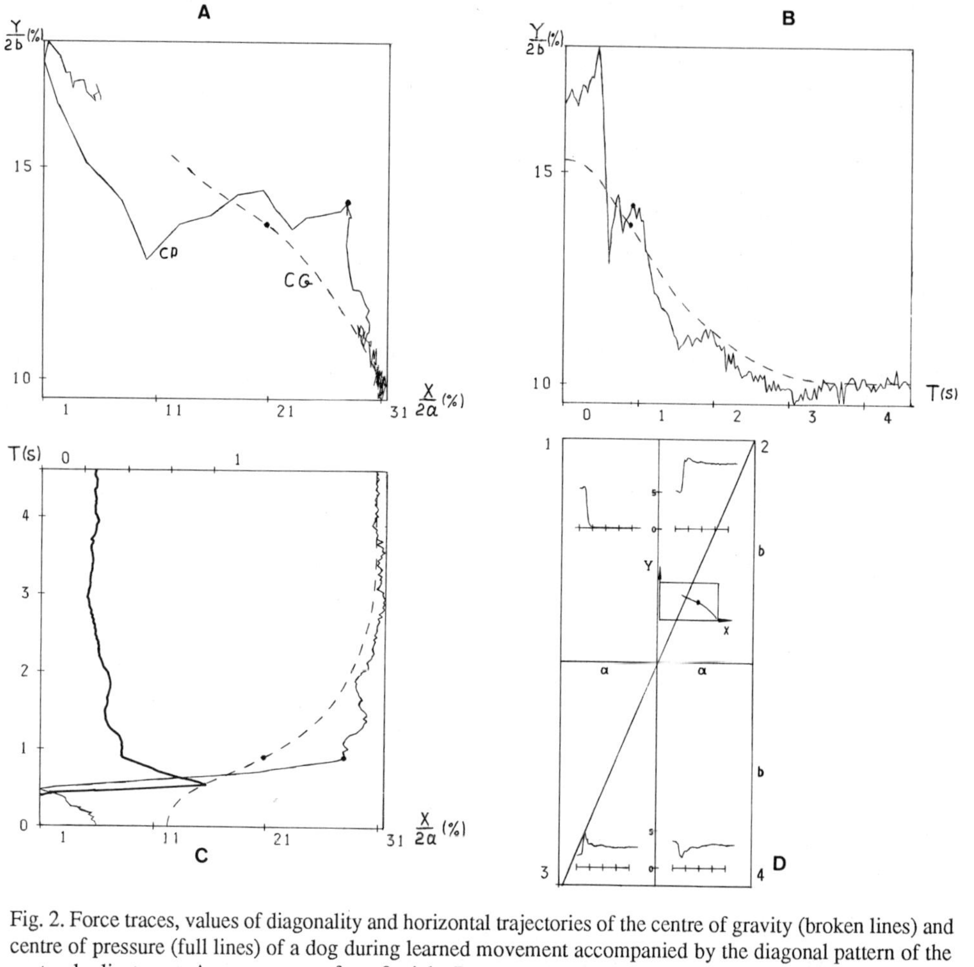

Fig. 2. Force traces, values of diagonality and horizontal trajectories of the centre of gravity (broken lines) and centre of pressure (full lines) of a dog during learned movement accompanied by the diagonal pattern of the postural adjustment. Average curves from 9 trials. D, rectangle corresponds to the area of support (1 and 2, points of forelimb support, 3 and 4, the same for hindlimbs). Force traces for the corresponding limbs are at the corners of the rectangle. Diagonal 2-3 separates the support triangle after the left forelimb is lifted. 2a and 2b are the interlimb distances. Small rectangle inside the rectangle D corresponds to the area in which centre of gravity and centre of pressure moved. X-lateral coordinate, Y-antero-posterior one. The same rectangle is represented as a square in A. Full circles at the centre of gravity and centre of pressure trajectories correspond to the onset of limb lifting (zero level of the support pressure). B, time course of the centre of gravity and the centre of pressure position changes in the antero-posterior direction. C, the same in the lateral direction. On the left side of C the time course of the diagonal coefficient is represented by a thick line. In A: abscissa, lateral coordinates, ordinate, antero-posterior ones. In B: abscissa, time of analysis (seconds), ordinate, antero-posterior coordinates. In C: abscissa, lateral coordinates (below) and values of diagonal coefficient (above), ordinate, time of analysis. Coordinates are indicated as a percentage of the distances between the appropriate limbs. Zero coordinates correspond to the middle of these distances. The time mark is 1s.

Results

Experiment I

The peculiarities of the diagonal and unilateral patterns of postural adjustment have been outlined elsewhere (Ioffe et al., 1985). After elaboration of the avoidance reflex accompanied by the diagonal postural adjustment, we observed maximal diagonality (D = 0.7-0.9) in the dynamic phase of postural adjustment. While the limb were kept lifted, diagonality was reduced (D-0.2-0.4). Postural displacement usually began with a short-lasting increase in the lifting limb pressure on the support (pushing off), which then was followed by a decrease in the limb pressure. Accordingly, the CP first shifted towards the lifting limb. It produced initial acceleration of the CG in the opposite direction, i.e. to the so-called support triangle formed by the other limbs. The CP then followed the CG, left it behind and in the static phase the CP and the CG merged (Ioffe et al., 1982) (Fig. 2). Thus, one can distinguish 5 phases in the CP trajectory: 1) initial CP displacement to the performing limb; 2) fast jump into the support triangle; 3) small loop or direction change of the trajectory during limb lifting; 4) slower displacement into a zone of postural stability; 5) stabilization of CP position in this area while the limb was maintained lifted.

The limb lifting began in the end of the second (fast) phase. The CG was at that moment already in the support triangle and continued to move in the same direction during lifting of the limb.

It is interesting to note that the CP and CG trajectories differed somewhat from one dog to another but each dog had its own pattern of trajectories.

Rearrangement of the postural pattern was very difficult for the animals and took hundreds of trials (Fig. 3). Two reactions - forelimb lifting and ipsilateral hindlimb support pressure decrease - interfered with each other. In the first stage of the training, both reactions occurred alternately in the same trial. This interference between two reactions is a typical stage in the rearrangement of different movements during motor learning (Ioffe, 1973).

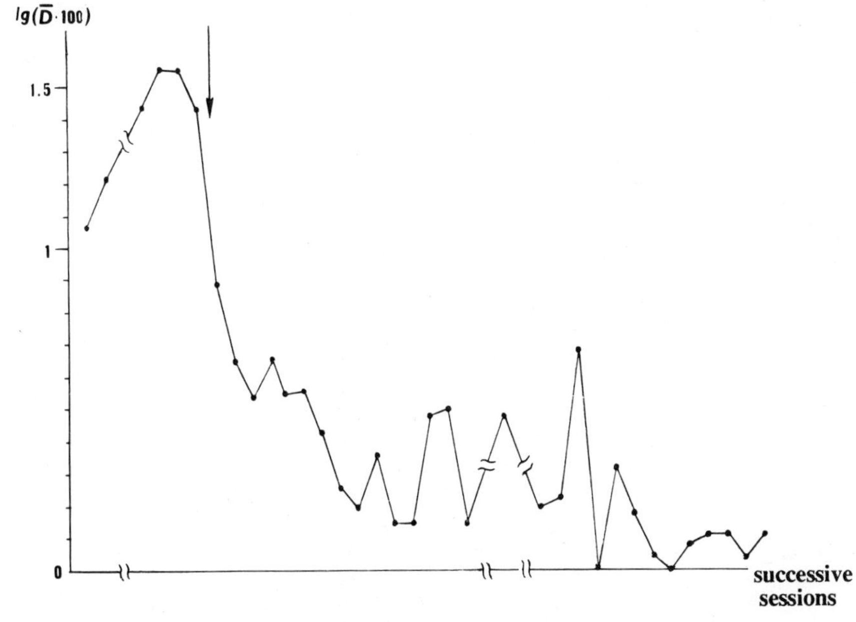

Fig. 3. Time-course of diagonality changes during the rearrangement of diagonal postural pattern into unilateral one. Abscissa, successive sessions, ordinate, logarithm of mean value of diagonality (D) for a session multiplied by 100 (to calculate D the sum of all the 500 values of D obtained in every trial (see Method) or after averaging for a session was divided by 500). Arrow corresponds to the beginning of training to rearrange diagonal postural pattern into unilateral one.

Achieving the right correlation between the US1 and US2 intensities was very important, so that one reaction should not inhibit the other.

The unilateral pattern of postural adjustment was elaborated after training (Fig. 4). D was equal to zero except for a short burst of about 30-60 ms duration in the dynamic phase of postural displacement. This burst emerged about 150 ms after the onset of the postural displacement. It was found to be very difficult to eliminate this burst for biomechemical reasons: the CP crosses a zone in the support square where the pattern must be diagonal (the "zone of diagonality") to reach the "zone of unilateral pattern" (see Frolov et al., this issue).

The CP and CG trajectories were somewhat different from those observed before the postural rearrangement. Namely, the CG moved more laterally than before (Fig. 4). But, as before, each dog had its own pattern of trajectories.

After ablation of the motor cortex contralateral to the lifting limb, the postural displacement and forelimb movement observed in response to the CS in some earlier trials were found to be absent. Reaction to the US depended on the ratio of the US1 to the US2 intensity. If the US1 was much more effective, flexion of the forelimb (flexor reflex) could occur without any postural preparation and the dog began to fall. Then the animal restored equilibrium by increasing the forelimb pressure which was followed by diagonal postural displacement.

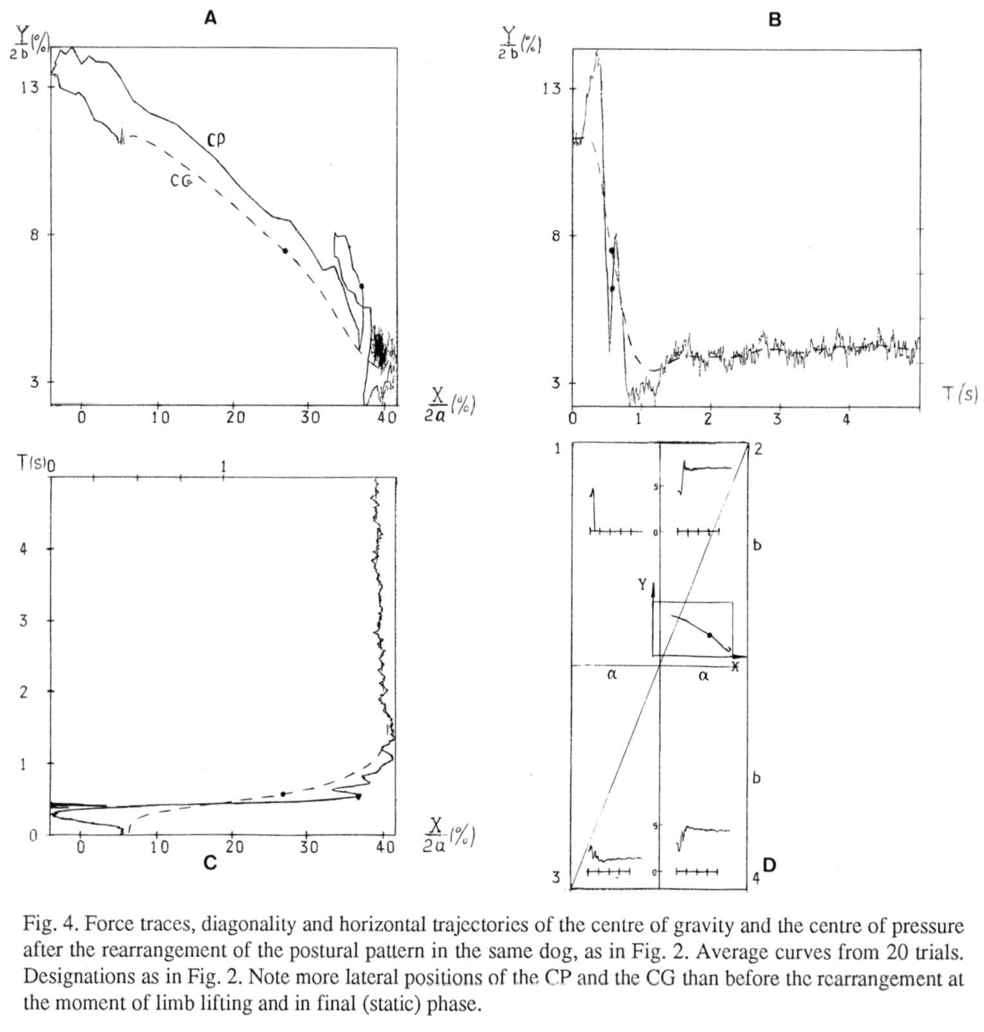

Fig. 4. Force traces, diagonality and horizontal trajectories of the centre of gravity and the centre of pressure after the rearrangement of the postural pattern in the same dog, as in Fig. 2. Average curves from 20 trials. Designations as in Fig. 2. Note more lateral positions of the CP and the CG than before the rearrangement at the moment of limb lifting and in final (static) phase.

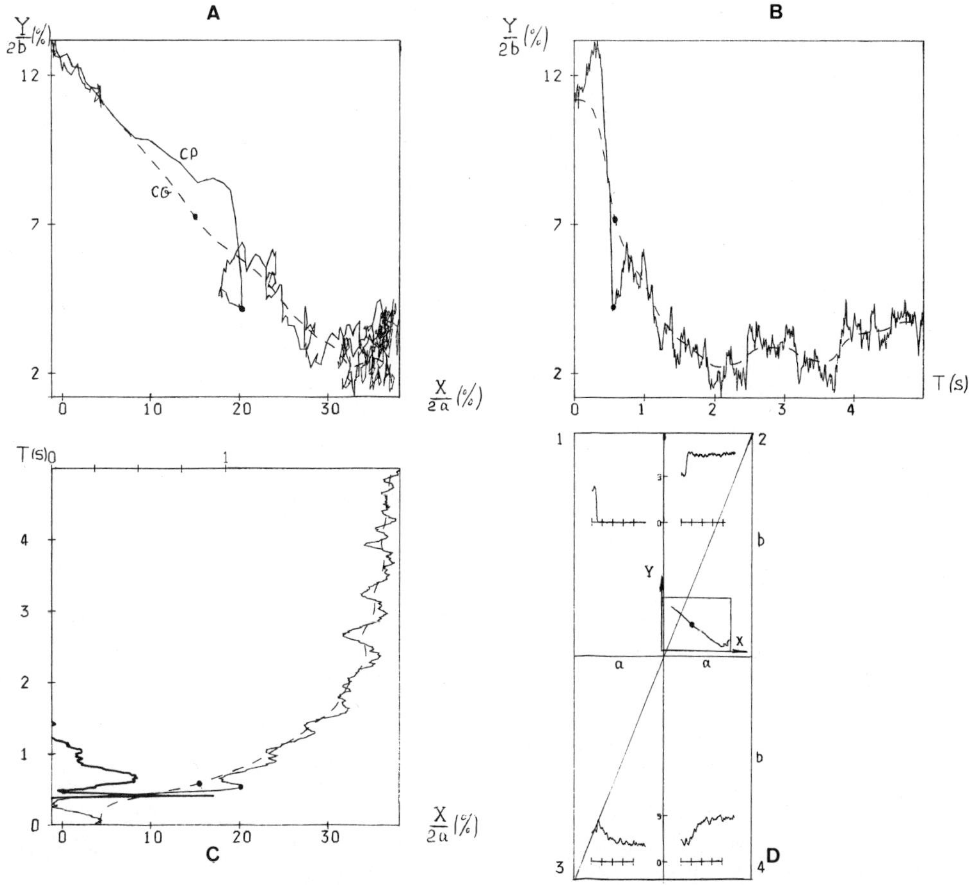

Fig. 5. Force traces, diagonality and horizontal trajectories of the centre of gravity and the centre of pressure 18 days after ablation of the motor cortex of the right hemisphere in the same dog. Average curves from 9 trials. Designations as in Fig. 2. Note decrease on the speed of the CP trajectory, position of the CG near the support triangle border at the moment of the limb lifting start and postural instability in the static phase.

Strongly pronounced interference between the fore and hindlimb reactions lasted 2-3 weeks. It was then possible to obtain some suppression of diagonality, mainly in the static phase of movement (maintaining the limb lifted).

Some peculiarities of the CP and CG trajectories were observed in this period (Fig. 5).

The second phase of the CP trajectory (fast jump into support triangle) decreased in amplitude, mainly in the lateral direction (from 30-40% to 20-25% of the interlimb distance). The lateral component of the CP maximal velocity during the jump also somewhat decreased. As a result the forelimb began to lift when the CG had just crossed the border of the support triangle (Fig. 5D). Then the CP moved into the support triangle step-by-step. The CP fluctuations increased during this phase and during the static phase of movement, stabilization of posture was disturbed.

Some disturbances of the rearranged postural pattern could be observed for some months (Fig. 6). But eventually, after retraining the unilateral pattern of the postural adjustment did not differ from the pre-operative one except for some instability of posture during the static phase of movement (Fig. 7).

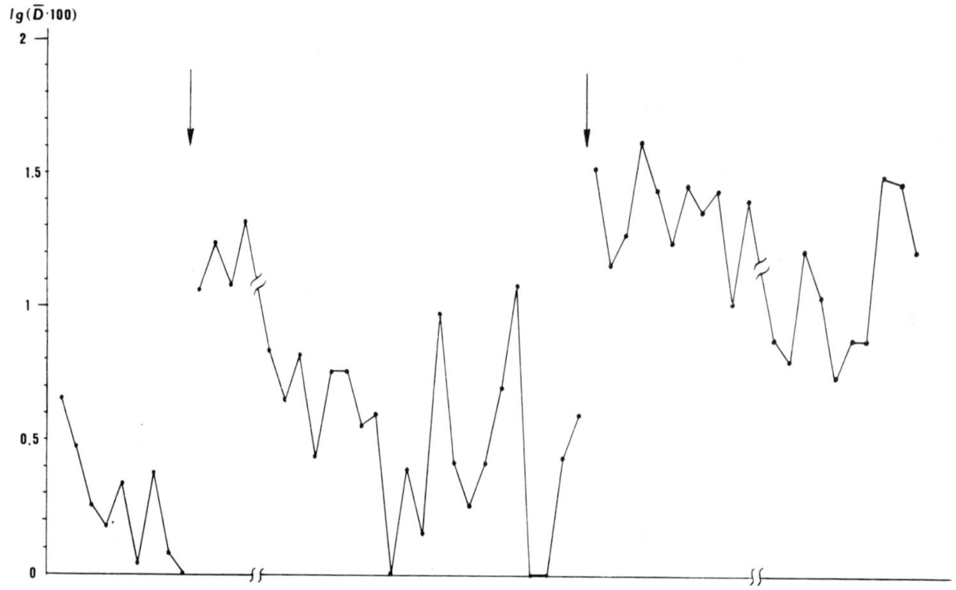

Fig. 6. Changes in diagonality after ablation of the motor cortex. Designations as in Fig. 3. Arrows correspond to the first and second operations.

The next step in the study was the ablation of the motor cortex of the other hemisphere. Unfortunately, only one dog survived the operation.

The dog was extremely excited in the experimental situation during 1 1/2 months after the surgery. The postural displacement and the limb lifting in response to the CS were abolished for about 2 months. When trying to lift the forelimb in response to the US, the dog often fell down to the right side. The pattern of the postural adjustment was diagonal during 5 1/2 months after the operation (Fig. 6). Then successful reactions began to appear. The percentage of such reactions was not high (33% of 36 trials). An example of the reaction is shown in Fig. 8. The duration of the diagonality peak was about 200 ms. Quite large fluctuations of the CP and CG were observed in the static phase of movement.

Experiment II

The rearrangement of the inborn flexor reaction into one of the opposite type was also very difficult to achieve. Active increases in the stimulated limb support pressure interfering with flexor reaction appeared after 50-100 trials. After 200-500 trials the escape reaction antagonistic to a flexor one was elaborated. In response to the US the CG shifted to the stimulated limb and the support pressure on the limb increased.

Ablation of the motor cortex contralateral to the stimulated limb resulted in some changes in the elaborated postural reaction. Inborn flexor reactions often appeared interfering with the elaborated reaction. The percentage of successful reactions somewhat decreased (Fig. 9). The latency of the reactions grew.

After 2-4 months of retraining the percentage of successful reactions was restored in two dogs out of the 3 taking part in this series, but the increase in latency persisted.

The consequences of the motor cortex ablation on the second side were much more severe. During the first month after the operation flexor reactions (decreasing the pressure or lifting the limb) dominated. The percentage of successful reactions sharply decreased (Fig. 9). But the latencies increased in only one dog.

In the next period (about 3 months after the second operation) flexor reactions again dominated. Corrective reactions appeared (some lowering of the lifted limb or an increase in the pressure) but usually they were not intensive enough for the dog to avoid electrical stimulation. The percentage of successful reactions thus continued to decrease.

No recovery was observed and in the final period (six-ten months after the second operation) successful reactions were performed in only 8-18 % of trials (Fig. 9). The escape reaction was practically completely destroyed. But even in the last period the postural reactions were very diverse. Purely flexor reactions were recorded in only 30-32 % of the trials. This means that the majority of trials had a rearranged component (increasing of support pressure to the stimulated limb) which became insufficient to break the electrical stimulation. It can be seen from Fig. 10 that, in fact, after the first operation the mean values of the pressure increase did not decrease but that after the second operation they decreased significantly. The animals thus tried to perform the rearranged postural reaction but its amplitude was usually not large enough. This was obviously the main reason for the successful reaction percentage decrease after bilateral ablations of the motor cortex.

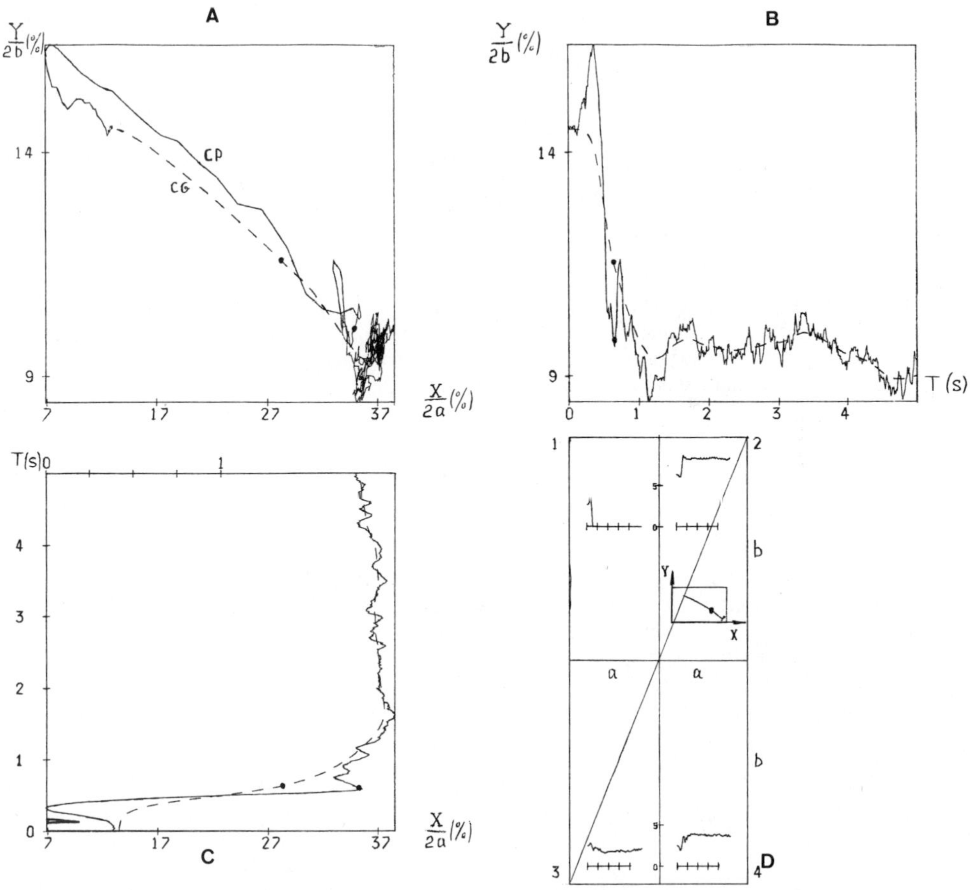

Fig. 7. Force traces, diagonality and horizontal trajectories of the centre of gravity and centre of pressure 1 year after ablation of the right motor cortex in the same dog. Average curves from 10 trials. Designations as in Fig. 2.

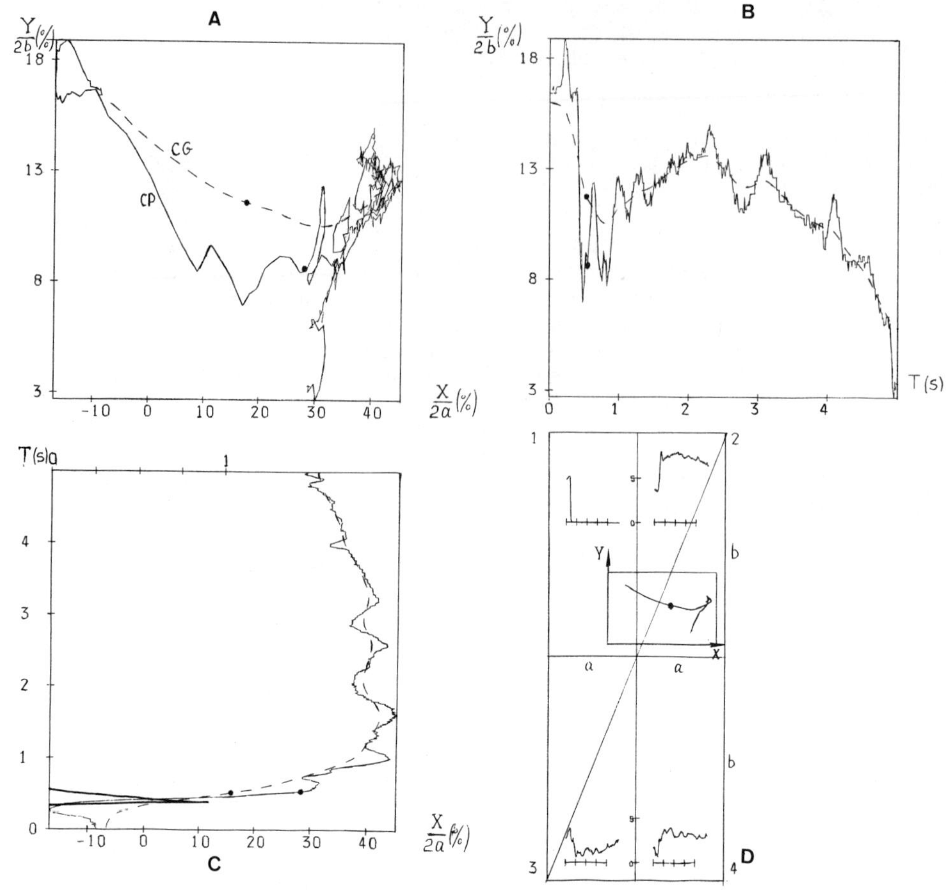

Fig. 8. Force traces, diagonality and horizontal trajectories of the centre of gravity and centre of pressure 6 months after ablation of the motor cortex of the left hemisphere in the same dog. Individual trial. Designations as in Fig. 2. Note position of the CG just near the border of the support triangle at the moment of limb lifting and postural instability in the static phase.

Discussion

The above data show that the motor cortex is specifically involved in the learned rearrangement not only of local motor reactions but also of postural coordinations. A difference was observed, however, between the contralateral and bilateral ablation results. The contralateral ablations resulted in temporary disappearance of the rearranged postural pattern, which could be recovered. The bilateral extirpations were followed by serious disturbances of the new postural coordination which could not be recovered during 6-10 months of retraining.

The reason for this difference is probably a bilateral arrangement of postural adjustment. Central command must be precisely and is obviously bilaterally organized to provide such very well integrated bilateral postural coordination.

The structures controlling postural adjustment are not yet precisely known. They may be the basal ganglia (Martin, 1967), cerebellum (Massion, 1984) or the ponto-bulbar reticular formation (Shumilina, 1949). These structures are probably influenced by the motor cortex modulating postural control (Massion, 1979). As postural coordination is arranged

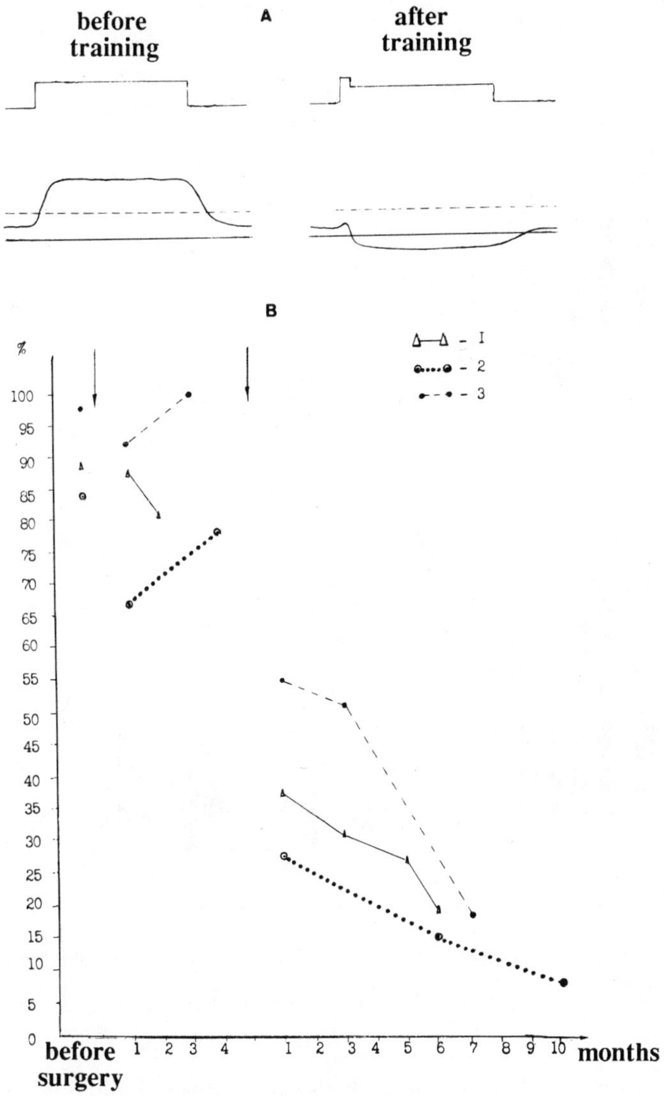

Fig. 9. Scheme of Experiment II (A) and percentage of the successful reactions after motor cortex ablation (B). In A: mark of electrical stimulation (above), curve of support pressure and limb lifting (thick line), zero level of support pressure (broken line), level of 5 kg pressure corresponding to switching off of the electrical stimulation (thin line). Decrease in the line of electrical stimulation corresponds to the moment of the stimulation switch-off. In B: abscissa, months of training, ordinate, percentage of successful reactions. 1, 2, 3 are numbers of dogs. Arrows correspond to the first and second operations.

bilaterally, the cortical influences are also bilateral. After unilateral motor cortex ablations these influences still remained and only bilateral ablations removed them.

What is the mode of cortical influences during the rearrangement of the postural adjustment pattern? Let us remember that even after bilateral motor cortex ablation the dog tried to perform the elaborated coordination, sometimes successfully. The program of the rearranged postural reaction was probably preserved (the dog "knew" what to do) but the performance was disturbed.

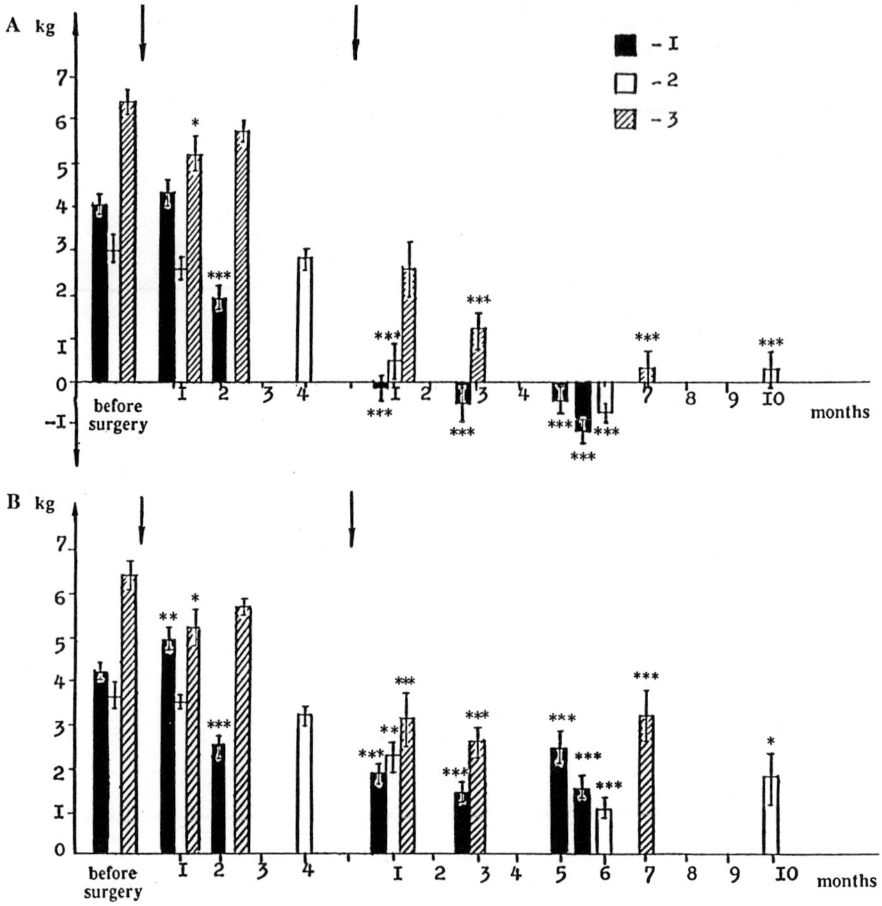

Fig. 10. Values of support pressure increased after motor cortex ablation. A, all the reactions, B, reactions with true increase in the initial presure. Abscissa, months of training, ordinate, values of pressure increase, kg, 1, 2, 3 are numbers of dogs. The other designations as in Fig. 9.

During learned rearrangement of local movements the motor cortex probably inhibits motor reactions interfering with the performance of a newly elaborated movement (Ioffe, 1975). Similar mechanisms may have intervened in Experiment II. After the bilateral motor cortex ablation the corticofugal influences are insufficient to inhibit the inborn flexor reaction. To interpret the results of Experiment I in terms of inborn coordination inhibition it is necessary to decide what coordination is to be inhibited.

To perform the unilateral pattern it is necessary to move the CP into the "zone of unilateral pattern" in the support square (Frolov et al., this issue). In fact, this consists mainly of stronger lateral displacement of the CP. Biomechanically, the task is not to inhibit the diagonal pattern but only to transfer the CP somewhat more laterally. But physiologically, we should probably accept that brain structures controlling postural adjustment have a special built-in program of diagonal interlimb coordination designed to minimize the CG displacement (Ioffe, 1975) and it is necessary to inhibit this program execution to obtain the unilateral pattern. How can this be done? It has been shown by Tsukahara (1986) that synaptic reorganization in the central nervous system may occur not only after structural but also after some functional changes. This might also take place during rearrangement of the postural pattern.

One might therefore suppose that during training to rearrange the diagonal postural pattern into the unilateral one, a program of new coordination is being formed (perhaps, in associative cortical regions) (Balezina et al., 1984). The program is carried out through the motor cortex, which rearranges its influences on the structures controlling postural adjustment to inhibit mechanisms of the diagonal pattern and to elicit performance of the unilateral one.

An alternative hypothesis might be considered here. The functional reorganization does not necessary need an inhibitory mechanism in order to supress the diagonal pattern. It is perhaps a mechanism in the brain which selects either the diagonal or the unilateral pattern, depending on the behavioral context. But in view of the interference between two reactions during elaboration of the unilateral pattern and after motor cortex ablation, inhibition of the diagonal pattern seems to be more plausible.

Reorganization of coordination patterns by learning is usually very difficult to achieve (Zhuravin and Bures, 1986). This is also true in cases of postural pattern reorganization. Sometimes, however, the diagonal postural pattern is replaced by another one spontaneously. This takes place, for example, in the cat while learning to lift one limb (Dufossé et al., 1982). A study of the role of the motor cortex in this period of spontaneous postural pattern rearrangement would be of great interest.

The authors wish to thank Dr. V. N. Mats for histological analysis and brain damages reconstructions, Drs. M.I. Konnov, A.V. Alexandrov, N.O. Matchinsky for help in data processing and Drs. E.I. Mikljaeva and S.V. Chromova for technical assistance.

References

Balezina, N.P., Ioffe, M.E., Mats, V.N., and Pavlova, O.G., 1984, On the role of cortical association systems in performance of learned motor coordination (on the model of alimentary instrumental reflex in dogs), Int. J. Neurosci., 22: 198.

Burlachkova, N.I., and Ioffe, M.E., 1979, The analysis of the postural adjustment accompanying a local movement, Agressology, 20, B: 141-142.

Dufossé, M., Macpherson, J., and Massion, J., 1982, Biomechanical and electrophysiological comparison of two postural supporting mechanisms in the cat, Exp. Brain Res., 45: 38-44.

Egiazaryan, E.L., Ioffe, M.E., Paramonov, A.A., Krovyakov, Yu.V., Vyatkin, A.V., Geletsyan, F.G., and Nikolaev, G.V., 1980, Device for posture studies in dogs (in Russian), Zh. Vyssh. Nervn. Deyat., 30: 205-207.

Evarts, E.V., Shinoda, Y., and Wise, S.P., 1984, "Neurophysiological Approaches to Higher Brain Functions", A Neuroscience Institute Publication, John Wiley & Sons, New York.

Frolov, A.A., Birjukova, E.V., and Ioffe, M.E., 1988, The influence of motion kinematics on a support pressure pattern in quadruped, in "Stance and Motion: Facts and Concepts", V.S. Gurfinkel, M.E. Ioffe, J. Massion and J.P. Roll, eds., Plenum Press: New York.

Fulton, J.F., 1949, "Physiology of the Nervous System", 3rd ed., Oxford Univ. Press, New York.

Gahéry, Y., Ioffe, M.E., Massion, J., and Polit, A., 1980, The postural support of movement in cat and dog, Acta Neurobiol. Exp., 40: 741-756.

Gorska, T., 1974, Functional organization of cortical motor areas in adult dogs and puppies. Acta Neurobiol. Exp., 34: 171-203.

Ioffe, M.E., 1973, Pyramidal influences in establishment of new motor coordinations in dogs, Physiol. Behav., 11: 145-153.

Ioffe, M.E., 1975, "Cortico-spinal mechanisms of instrumental motor reactions" (in Russian), Nauka Publishing House, Moscow.

Ioffe, M.E., Frolov, A.A., Gahéry, Y., Frolov, A.G., Coulmance, M., and Davydov, V.I., 1982, Biomechanical study of the mechanisms of postural adjustment accompanying learned and induced limb movements in cats and dogs, Acta Neurobiol. Exp., 42: 469-482.

Ioffe, M.E., Frolov, A.A., Konnov, M.I., Kiseljova, N.V., and Birjukova, E.V., 1985, The learned rearrangement of an inborn postural pattern, Physiol. Bohemoslov., 34 (suppl.): 63-66.
Korjakin, M.F., 1958, Contribution of postural excitation in patterning the conditioned defensive motor reflex in the dog (in Russian), Fiziol. Zh. SSR, 44: 393-403.
Lawrence, D.G., and Kuypers, H.G.J.M., 1968, The functional organization of the motor system in the monkey. I. The effects of bilateral pyramidal lesions, Brain, 91: 1-14.
Martin, J.P., 1967, "The basal ganglia and posture", Pitman Med. Publ.: London.
Massion, J., 1979, Role of motor cortex in postural adjustments associated with movement, in: "Integration in the Nervous System" (H. Asanuma and V.J. Wilson, eds), pp. 239-260, Igaku-Shoin: Tokyo-New York.
Massion, J., 1984, Postural changes accompanying voluntary movement. Normal and pathological aspects. Human Neurobiol., 2: 261-267.
Phillips, C.G., and Porter, R., 1977, "Corticospinal Neurons. Their Role in Movement", Academic Press: New York.
Shumilina, N.I., 1949, On the participation of pyramidal and extrapyramidal system in motor activity of deafferented limb (in Russian), in: "Problems of Higher Nervous Activity" (P.K. Anochin, ed.), pp. 196-207, Publishing House of USSR Academy of Medical Sciences, Moscow.
Tsukahara, N., 1986, Cellular basis of classical conditioning mediated by the red nucleus in the cat, in: "Neural Mechanisms of Conditioning" (D.L. Alkon and C.D. Woody, eds), pp. 127-139, Plenum Press: New York.
Wiesendanger, M., 1981, The pyramidal tract. Its structure and function, in: "Handbook of Behavioral Neurobiology", v. 5, (A.L. Towe and E.S. Luschei, eds), pp. 401-491, Plenum Publishing Corporation: New York.
Zhuravin, J.A., and Bures, J., 1986, Operant slowing of the extension phase of the reaching movement in rats, Physiol. Behav., 36: 611-617.

ON THE INFLUENCE OF MOVEMENT KINEMATICS ON THE SUPPORT PRESSURE PATTERN DURING POSTURAL ADJUSTMENT OF QUADRUPEDS

A.A. Frolov, E.V. Birjukova and M.E. Ioffe

Institute of Higher Nervous Activity and Neurophysiology
USSR Academy of Sciences, Moscow, USSR

Summary

The biomechanical aspects of the postural adjustments accompanying limb lifting were studied. A mechanical model of a quadruped animal is proposed. It consists of 10 links for dogs and 12 links for cats. Good agreement between experimental and model data is shown to exist so that the rigid body mechanical model is suitable for use in the analysis of postural adjustment.

The main result of this study is the finding that the biomechanical model can be used to determine correlations between all kinds of body movements involving a shift in the center of gravity, and the support pressure pattern. It can be used to derive all possible kinds of information about the movement kinematics from the vertical pressure force components usually recorded in experiments. Two different cases are investigated: the one corresponding to static conditions and the other to dynamic ones. For the static case, a clear geometrical interpretation of the connection between pressure force patterns and the center of gravity (CG) shift is presented. It is shown that the kinematics in themselves do not set any limits on the type of support pressure pattern.

Introduction

Postural adjustments are necessary components of many kinds of movement. The main behavioral function of these adjustments seems to be to ensure the body stabilization of a standing animal during the movement of one of the limbs. Usually, when an animal is standing, its CG is shifted slightly forwards. If the forelimb is lifted without any postural adjustment, the CG projection onto the support plane will leave the support triangle and the animal will fall. The main purpose of postural adjustment is to replace the animal's CG within the area defined by the three supporting limbs. In order to make this shift in the CG, some body movement occurs. Most investigations on postural adjustments have not analysed the body movement but only the vertical components of support pressure forces on the four limbs. This may be due to the difficulties encountered in directly recording body movement and to the existence of simple methods for obtaining indirect information about these movements by recording only the vertical components of support pressure forces. Some details on the calculation of vertical and horizontal CG shifts by vertical pressure force measurements have already been described (Gahéry et al., 1985; Ioffé et al., 1982).

The aim of this study was to analyse the correlations between all kinds of body movements and the support pressure patterns using our biomechanical model of the animal body. It is possible in this way to derive all possible kinds of information about the movement kinematics from the vertical pressure recordings.

Model description and verification

Biomechanical models of the dog and cat are shown in Fig. 1. They consist of 10 links for the dog and 12 links for the cat. Each link is a rigid component, connected to the

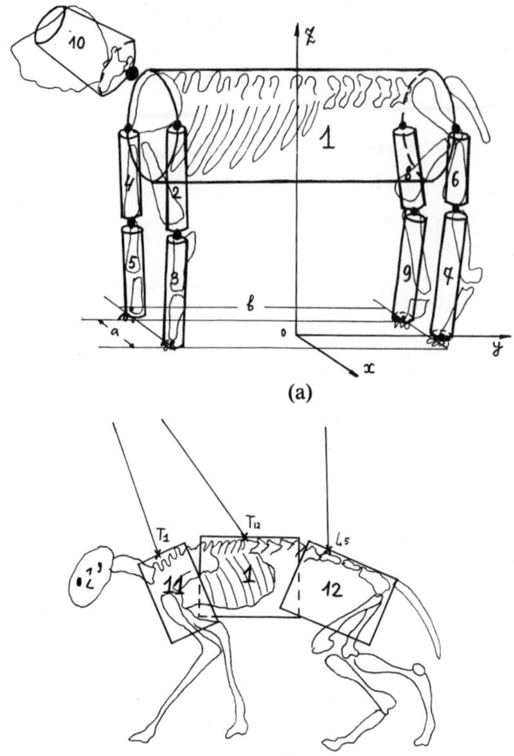

Fig. 1. Mechanical model, coordinate system and link numbers for a dog (top) and a cat (bottom). The numbers of the cat model links are the same as for the dog except for the two additional links. 3 needles, mounted on markers in vertebrae Tl, Tl2 and L5, are shown.

next by joints which are assumed to be pin joints with frictionless coupling. Muscles are simulated in the form of weightless threads with variable strength. In the case of a standing dog, provision was made for 10 kinematical degrees of freedom: 3 rotational and 3 translational body movements, up-down and left-right head movements and movements at humeral and elbow joints in the forelimb being lifted. In the case of a standing cat, we added 6 degrees of freedom to account for the cat's greater body flexibility. The relative masses of all parts are given in Table l. To calculate their moments of inertia we assumed them to be uniform cylinders except for the head, which was assumed to be a truncated cone. The positions of the muscle attachment points were based upon anatomical data (Miller, 1964; Nozdrachev, 1973). The number of muscles was 23 for the dog model and 29 for the cat model. Muscle forces were calculated using second order Lagrange equations in the tensor convolution form (Korenev, 1979). After the muscle force, the vertical pressure components were calculated.

It is essential to note that the number of Lagrange equations was equal to the number of degrees of freedom, i.e. they were much fewer than the number of muscles. Each movement can therefore be performed by several muscle activity patterns. The difference between number of muscles and number of degrees of freedom can be used as a mathematical measure of the muscular apparatus redundancy. The choice of one among the several possible muscle activity patterns for performing a given movement is determined by neurophysiological principles which have not yet been elucidated. However, as shown previously (Frolov and Birjukova, 1986), the vertical components of support pressure forces do not depend on the muscle activity pattern, but are completely determined by the movement kinematics. So for our calculations we required only the set of muscles necessary for solving the Lagrange equations for the given movement. Some details of the calculations have been described elsewhere (Frolov and Birjukova, 1986).

Table 1: Relative masses of model links. i is the link number (Fig. 1), m_i is its mass, M is the total mass. Index "d" for a dog (Frolov and Birjukova, 1986), "c" for a cat (Langer and Hemami, 1985).

i	1	2	3	4	5	6
$(m_i/M)_d$	0.60	0.02	0.03	0.02	0.03	0.07
$(m_i/M)_c$	0.19	0.08	0.05	0.08	0.05	0.04

i	7	8	9	10	11	12
$(m_i/M)_d$	0.03	0.07	0.03	0.10		
$(m_i/M)_c$	0.03	0.04	0.03	0.12	0.05	0.24

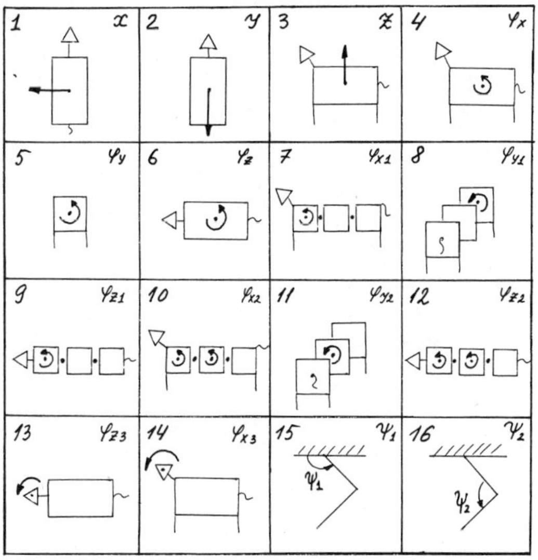

Fig. 2. Generalized coordinates, i.e. the values which determine all model degrees of freedom: X, Y, Z - cartesian coordinates of CG position of the first link; φ_x, φ_y, φ_z - its Eulerian angles, which correspond to the rotations about axes X, Y, Z; φ_{x1}, φ_{y1}, φ_{z1}; φ_{x2}, φ_{y2}, φ_{z2} - Eulerian angles of link 11 and 12 relative to the link 1; φ_{x3} φ_{z3} - Eulerian angles of head relative to the link 1; Ψ_1, Ψ_2 - angles in the lifting limb joints. Coordinate numbers are indicated.

We had a good opportunity of testing the validity of the model, thanks to the kindness of J. Massion and M. Dufossé who provided us with movies showing cats performing postural adjustments while lifting the forelimb and corresponding recordings of vertical components of the support pressure forces. Their experiments were performed on cats which were trained to lift the right forelimb above a certain level (2 cm) after an auditory CS.

To measure movements of the cat's body, markers were implanted in selected vertebrae (Tl, T12 and L5) (Fig. lb). Just before the experiment, 3 needles each about 20 cm long were mounted on these markers thus producing a rigid extention of the vertebrae. Film was used to record the cat's movement from lateral and posterior viewpoints. The method of these experiments has been described in detail by Dufossé et al. (1982). An analysis of these data is part of our ongoing co-operative research project.

3 coordinates of each needle base and tip as well as both ear tips were obtained from the movies using frame by frame analysis. With these coordinates, it is possible to calculate all the generalized coordinates that correspond to all degrees of freedom. The numbers of generalized coordinates used are presented in Fig. 2. Calculated positions of cat's body links

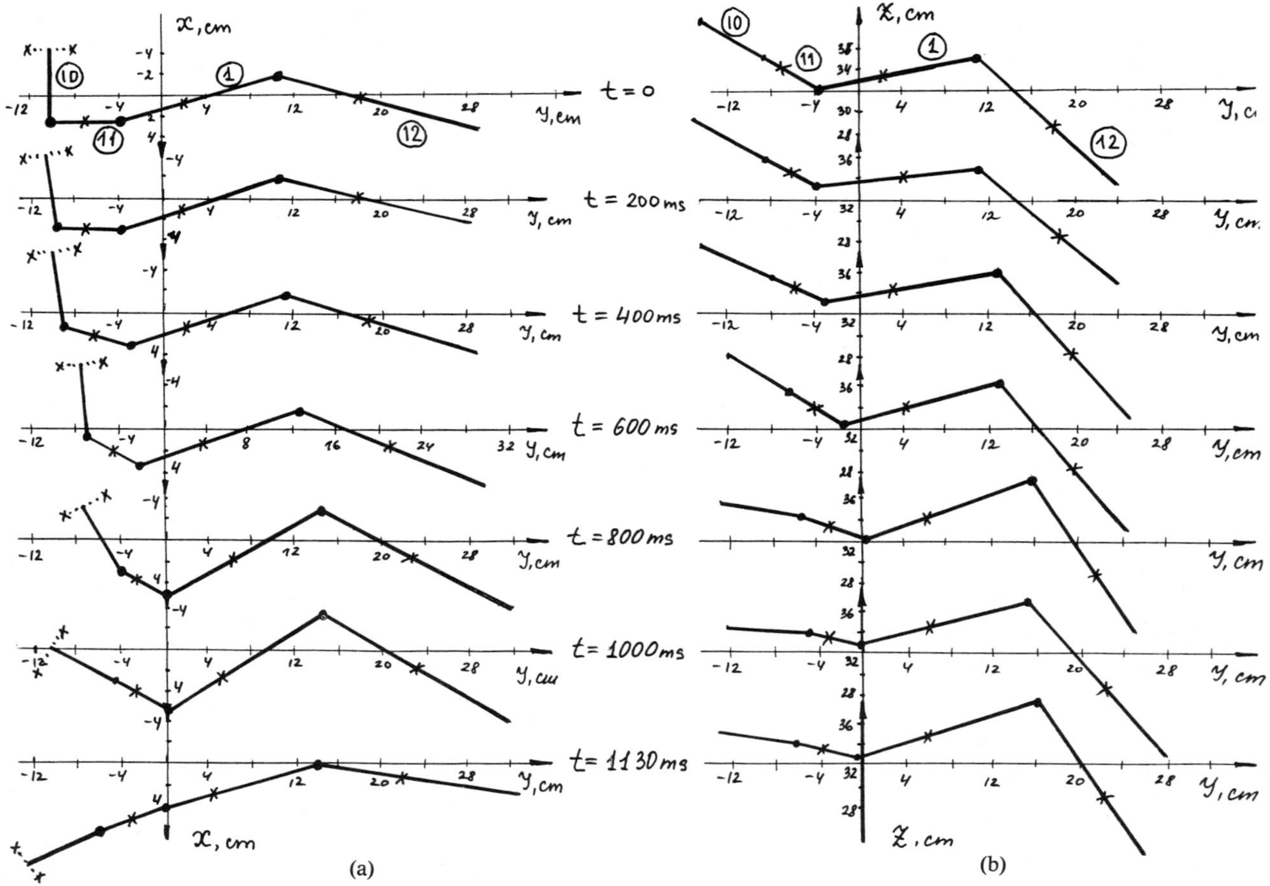

Fig. 3. Cat's body movement during right limb lifting following an auditory CS. Reconstruction by movies (every 200 ms). Link numbers (in circles) are the same as in Fig. 1. Solid line - mechanical model of the cat spine; circles are the joints between links; crosses are the points of marker implantation and ear tips, coordinates of these points are taken from the movie. a, top viewpoint; b, lateral viewpoint.

in one of the trials and the corresponding generalized coordinates are shown in Fig. 3 and Fig. 4a. It can be noted that the method of needle coordinate determination from the frames is not precise enough: it was therefore necessary to smooth the date prior to the calculation of pressure forces using the biomechanical model. They were smoothed by minimizing the square root deflections, assuming that their first derivatives were equal to zero at the initial and final moments of the movement. This means that a movement was assumed to start and end in a static state. The results of such smoothing are shown in Fig.4a for certain coordinates. Calculated and experimental values of the vertical components of the pressure forces are given in Fig.4b. The good agreement between these two sets of values indicates that the accuracy of the model was satisfactory.

Static solution

As has been shown previously (Frolov and Birjukova, 1986), we can use the following general representation of the vertical components of the pressure forces :

$$P_i = P_i^S + P_i^D \tag{1}$$

where P_i^S are static and P_i^D are dynamic components of the pressure forces, i - the number of animal's limbs (i = 1, 2, 3, 4) (Fig. 1 and 5). We assume further that limb 1 is lifted. The values of P_i^S are completely determined by the current CG position :

$$\begin{aligned} P_2^S &= - P_1 + P\,(1/2 - y_c / b) \\ P_3^S &= - P_1 + P\,(1/2 + x_c / a) \\ P_4^S &= P_1 + P\,(yc / b - x_c / a) \end{aligned} \tag{2}$$

where x_c, y_c are the CG coordinates, P is the animal's weight, and a, b distances between the limbs (Fig. 1). The value of P_i^D depends on the velocities and accelerations of moving links of an animal. According to equation (1) for very slow movement components, P_i^D are negligible in comparison with P_i^S.In this case it is possible to give a simple geometrical description of the support pressure patterns as depending on the CG displacement (Fig. 5a). The following quantitative parameters are commonly used for pattern description (Gahéry et al, 1980) :

(Index of diagonality) $$D = 1 - [(| \Delta P_1 - \Delta P_4 | + | \Delta P_2 - \Delta P_3 |) / (\sum_{i=1}^{4} | \Delta P_i |)]$$

(Index of laterality) $$L = (| \Delta P_2 + \Delta P_4 | - | \Delta P_1 + \Delta P_3 |) / (\sum_{i=1}^{4} | \Delta P_i |)$$

Table 2: Coefficients α_{ij} for calculating the dynamic components of pressure forces.j - generalized coordinate number, for 1,2,3 dimensionless, for all the others in cm. i - limb number.

i\j	1	2	3	4	5	6	7
2	0.0	4.7	1.6	-50.0	0.0	0.0	-29.0
3	9.3	0.0	1.8	0.0	33.0	-1.0	-6.0
4	-9.3	-4.7	0.2	50.0	-33.0	1.0	-23.0

i\j	8	9	10	11	12	13	14
2	0.0	0.0	-57.0	0.0	0.0	0.0	-15.0
3	1.2	-25.0	-21.0	2.0	12.0	15.0	-2.0
4	-1.2	-25.0	37.0	-2.0	-12.0	-15.0	13.0

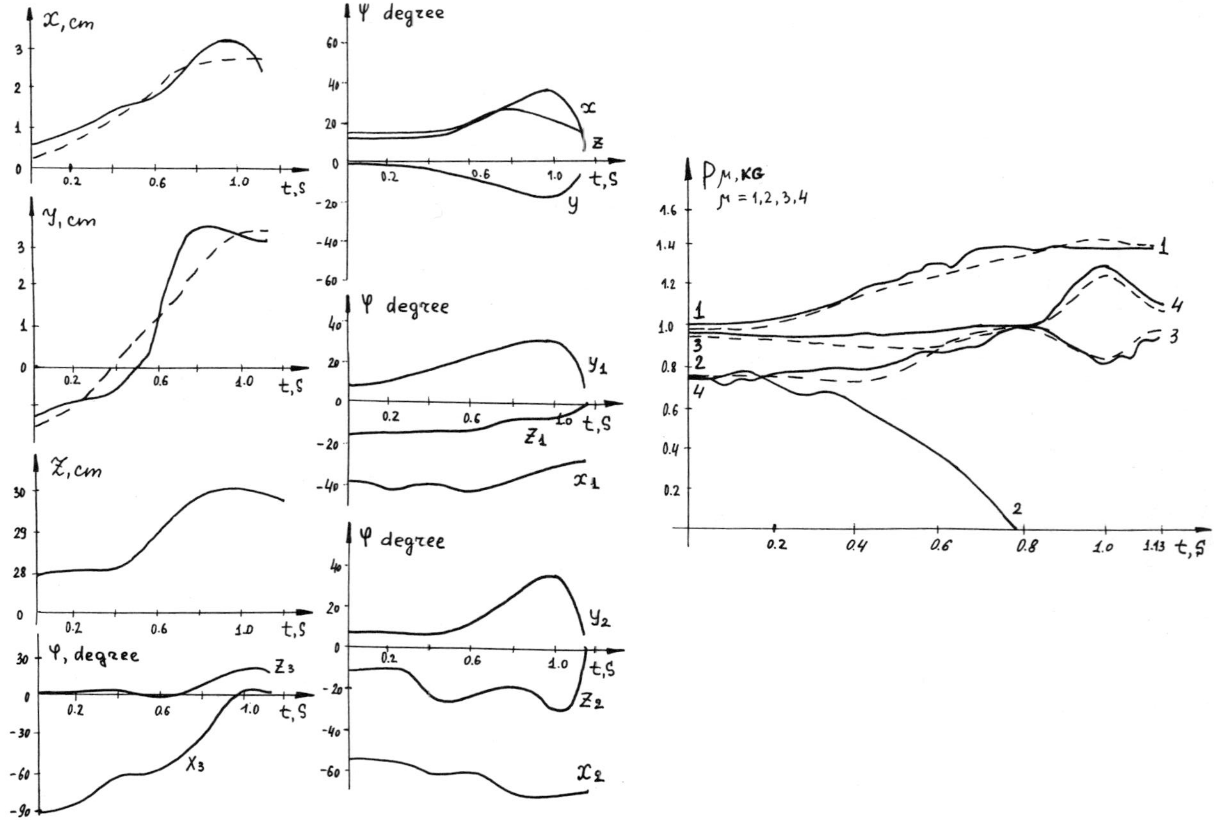

Fig. 4. Generalized coordinates (a) and vertical support pressure forces (b) for the movement shown in Fig. 3. Generalized coordinate numbers are the same as in Fig. 2. In a: solid lines are coordinates before, and dashed lines after smoothing. In b: solid lines are experimental and dashed lines calculated values.

Table 3: Limits on the amplitudes of movement owing to the dog's construction. i - generalized coordinate numbers according to Fig. 2. A^+_i, A^-_i - amplitude limits: for 1,2,3 in cm, for 4,5,6 in degrees; a = 21 cm, b = 62 cm, Z_c = 40 cm.

i	1	2	3	4	5	6
A^+_i	10.5	21.0	45.0	23.0	32.0	40.0
A^-_i	-10.5	-21.0	22.0	-7.0	-32.0	-40.0

From equations (2) it follows that:

$$\Delta P_2 / P = -\Delta P_1 / P - \Delta y_c / b$$

$$\Delta P_3 / P = -\Delta P_1 / P + \Delta x_c / a$$

$$\Delta P_4 / P = \Delta P_1 / P + \Delta y_c / b - \Delta x_c / a$$

It is therefore easy to find the areas on the support plane where CG must be shifted in order to obtain various patterns of changing P_i. These areas are presented in Fig. 5, a in coordinates $(\Delta x / a)$ and $(\Delta y / b)$. They are separated by the lines where $\Delta P_2 = 0$ (line l_2: $\Delta y_c / b = -\Delta P_1 / P$), $\Delta P_3 = 0$ (line l_3: $\Delta x_c / a = \Delta y_c / b$) and $P_4 = 0$ (line l_4: $\Delta x_c / a - \Delta y_c / b = -\Delta P_1 / P$). When the CG displacement is within the area R_D then $\Delta P_2 > 0$, $\Delta P_3 > 0$, $\Delta P_4 < 0$, D = 1. When it is within the area R_L then $\Delta P_2 > 0$, $\Delta P_3 < 0$, $\Delta P_4 > 0$, L = 1.

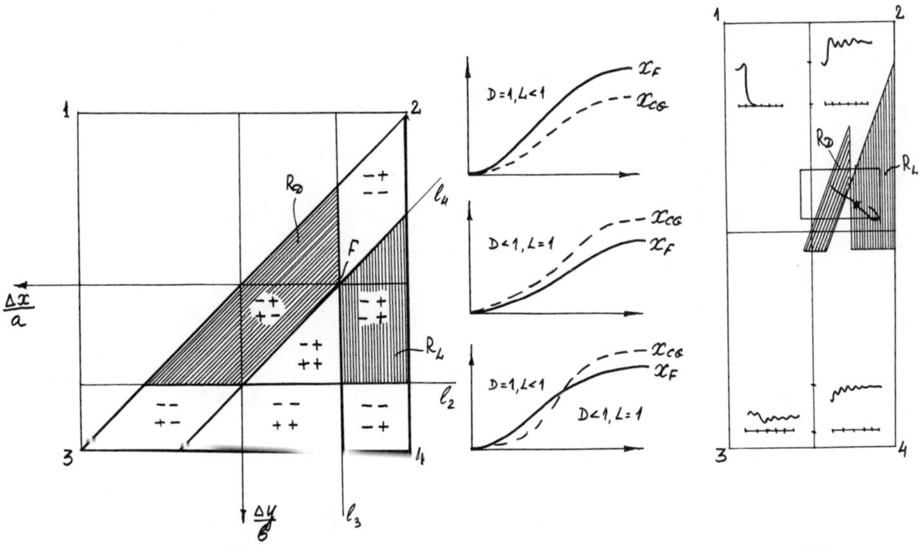

(a) (b) (c)

Fig. 5. Support pressure patterns depending on the CG position in a static case. a, support rectangle in normalized coordinates $\Delta x / a$, $\Delta y / b$, where Δx, Δy are the changes in CG coordinates; 1 and 2 points of forelimb support, 3 and 4, the same for hindlimbs, 1 - lifting limb; the areas with signs + and - inside the rectangle are the regions in which the CG must be shifted if the alterations of support pressure forces have a given sign ("+" an increase and "-" a decrease in a pressure force); the areas are formed by the coordinate axes and lines l_2, l_3, l_4; these lines correspond to the constant values of some support pressure force: l_2: $\Delta P_2 = 0$, l_3: $\Delta P_3 = 0$, l_4: $\Delta P_4 = 0$; R_D, R_L - areas of diagonal and unilateral postural pattern (see signs of pressure force alterations); b, different cases of CG (solid line) and the point F (dash line) shifting resulted in different pressure patterns; abscissa: time; ordinate: normalized change of CG and point F coordinate; D, L: indexes of diagonality and laterality. c, example of pressure pattern depending on the CG shift during postural adjustment of a dog (see Fig. 3, Ioffe et al., this issue); the rectangle corresponds to the area of support; force traces for the corresponding limbs are at the corners of the rectangle; diagonal 2-3 separates the support triangle, occurring after the left forelimb has been lifted; areas R_D and R_L correspond to the moment when P_1 drops to zero. This movement is marked by the asterisk in the CG trajectory.

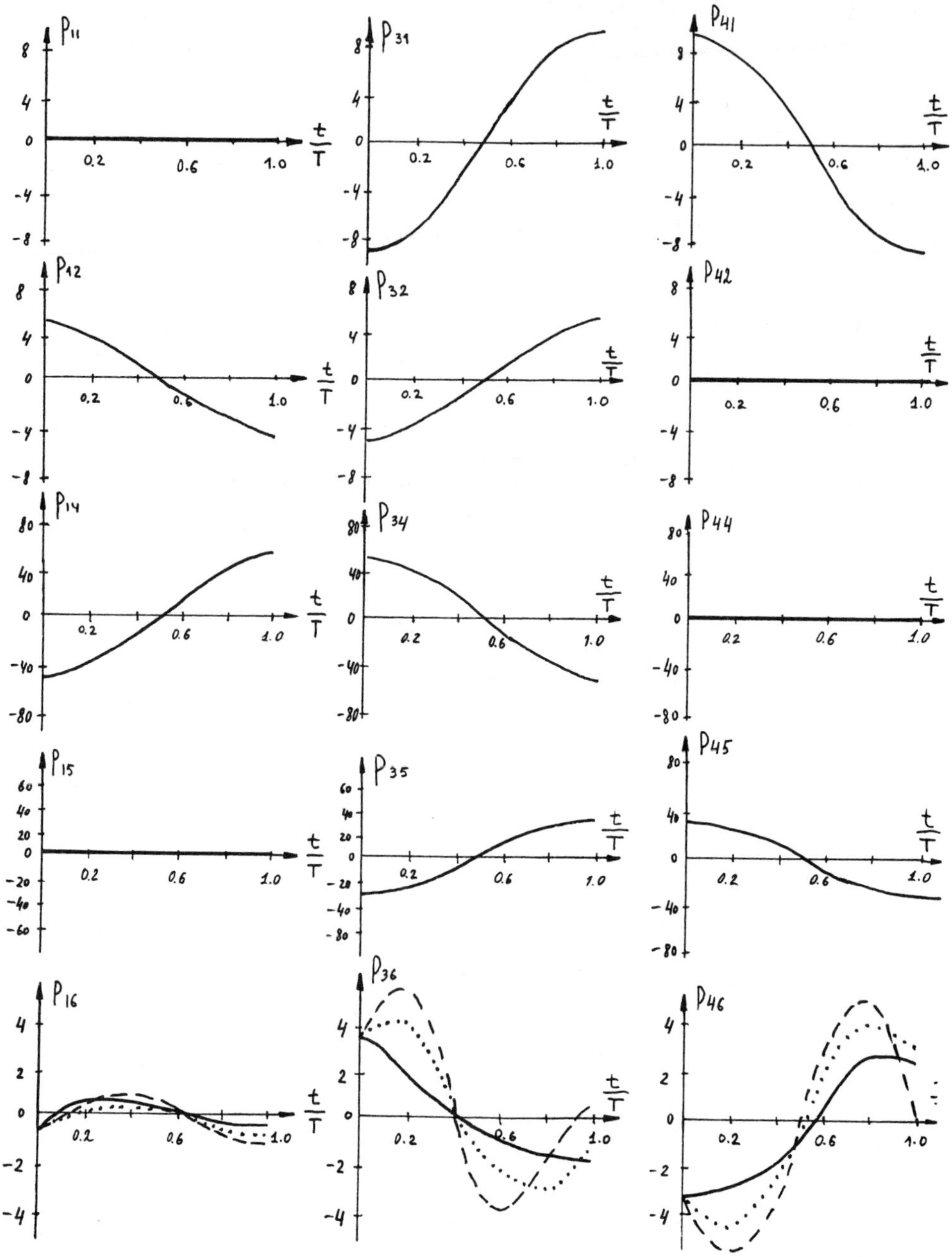

Fig. 6. Dynamic components of support pressure forces; abscissa, time, divided by period of movement; ordinate, $k_{ij} = P_{ij}^D T_j^2 (A_j M)$: dimensionless values of dynamic components for all possible amplitudes of movements (see Table 3); P_{ij}^D: dynamic components of pressure forces for i-th limb due to the change in the j-th generalized coordinate; A_j: amplitude and T_j: period of this change. For example, if the animal's body moves back along axis Y (coordinate number 2), then the dynamic changes in pressure forces will be: for lst limb: ΔP_{12}^D, for 2nd limb: ΔP_{22}^D for 3th limb: ΔP_{32}^D, and for 4th limb: ΔP_{42}^D. Values of k_{ij} do not depend on A_j and T_j with any generalized coordinates except for the rotation around the axis Z; with this coordinate the value of k_{ij} is shown for $A_j = 2°$ (solid line), $A_j = 10°$ (dotted line), $A_j = 20°$ (dashed line).

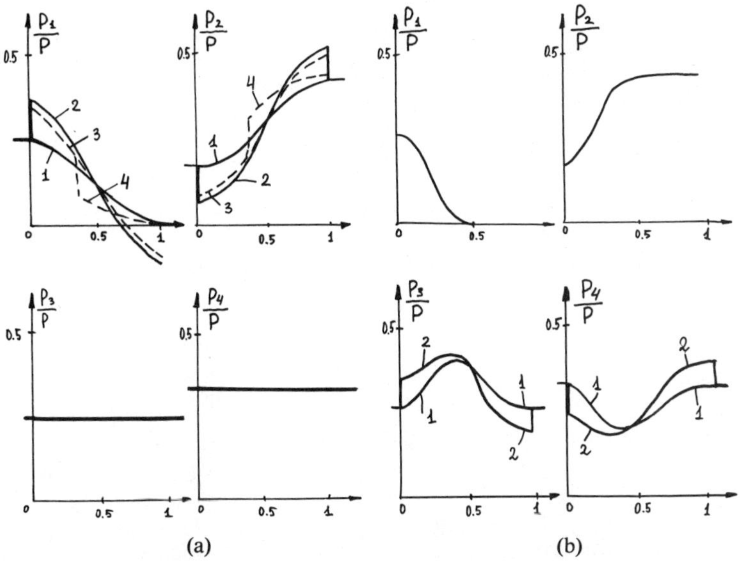

Fig. 7. Influence of the movement dynamics on pressure patterns derived from the model; abscissa, time divided by period of movement; ordinate, Pi/P, where i is the limb number; 1 is the static case; 2 is the movement along the axis X without rotation around the axis Y, T = 0.5 s; 3 is the additional rotation around the axis Y with amplitude 20 and period T = 0.5 s; 4 is the same movement with rotation, but T = 0.3 s. a, translational and flexional movements with the same CG shift as with forelimb pressure pattern, results coincide ; b, translational movement with diagonal pattern.

For all cases it is assumed that $\Delta P_1 < 0$. Hence R_D is the area of diagonality and R_L is the area of unilaterality. Diagonal pressure patterns are typical of inborn postural adjustments in dogs (Gahéry et al., 1980). After special training, dogs can learn to perform movements with the unilateral pressure pattern (Ioffé et al., this issue). When the CG position coincides with point F (the crossing point of the lines l_3 and l_4) then $\Delta P_3 = \Delta P_4 = 0$. In this case only the forelimb pressure forces change during the movement performance, so we can call it the "forelimb" pressure pattern. This pattern can be obtained after conditioning in cats (Dufossé et al., 1982). It can be seen from Fig. 5a that this pattern is intermediate between the diagonal and unilateral patterns. With a diagonal pattern the CG shift is minimal. This may be the reason for the inborn preference for this pattern and for the difficulty encountered in changing it during conditioning (Ioffé et al., this issue). The same type of pattern must be obtained during the fast limb lifting caused by any abrupt perturbation (Dufossé et al., 1982); here again, the CG shift is obviously small as compared to the shift in point F. It should be noted that lines l_2, l_3, l_4 (Fig. 5a) shift with the decrease in P_1. Hence to keep a given type of pressure pattern, the CG shift must follow the displacement of these lines. For example, let us consider a CG shifting only in the lateral direction (along the axis x). Various cases of such a movement in relation to the movement of point F are schematized in Fig. 5b. If CG is behind point F (the x-coordinate only is important), the pressure pattern will be diagonal. If it is in front of this point, the pattern will be unilateral. If it is initially behind and then in front of point F, the pattern will change from a diagonal to a unilateral one. The latter case is typical of dogs after conditioning (Fig. 5c or Fig. 3d in Ioffé et al., this issue). In this case, fast lifting of the forelimb produced a fast shift of point F during the initial phase of the movement and at that moment D = 1 (Fig. 5b). With the forelimb pattern, shifts of point F and CG must coincide. As the CG movement is rather inertial and consequently slow, with the forelimb pattern the P_1 decrease must also be slow. This can be observed in cats after conditioning (Dufossé et al., 1982).

Dynamic solution

For the general case we can use the following representation for dynamic components of the vertical support pressure forces :

$$P_i^D = M \sum_{j=1}^{L} \alpha_{ij} \ddot{\eta}_i + (M/b) \sum_{j=1}^{L} \sum_{k=1}^{L} \beta_{ijk} \dot{\eta}_j \dot{\eta}_k \tag{3}$$

where η_j is the value of the j-th generalized coordinate, L is their number, coefficients α_{ij} and β_{ijk} depend only on the current values of generalized coordinates. To calculate these coefficients we undertook computer simulation based on the biomechanical models. The change in each coordinate is given by the formula :

$$\eta_j(t) = A_j/2 \begin{cases} 0 & , t < 0 \\ 1 - \cos \pi t / T_j & , 0 \le t \le T_j \\ 2 & , t > T_j \end{cases}$$

where A_j is the amplitude and T_j is the period of the movement in question. For example, this characteristic is typical of arm (Long et al., 1961) and eye (Jarbus, 1965) movements.

Some results derived from computer simulation are shown in Fig. 6. It can be seen that: 1) the greatest influence on the values of vertical components of the support pressure forces is exerted by translational movements of the body and its rotational movements around the axes X and Y; 2) for most of the generalized coordinates the second term in equation (3) is negligible as compared to the first one; 3) for these components coefficients α_{ij} can be approximated by constants. The values of these coefficients are given in Table 2.

Two examples of the influence of the dynamic components on the pressure pattern are shown in Fig. 7. It can be seen that forelimb pattern may occur in the case not only of slow but also to fast movements and that the diagonality increases with the decrease in the movement time. The influence of the translational movement may be compensated for by the rotational movement. So the forelimb pressure pattern may occur without any lateral flexion of the vertebral column, and the occurrence of this flexion during the movement in cats (Dufossé et al., 1982) is not the reason for this pressure pattern although the flexion correlates with it.

The same pressure pattern may result from a different movement. An example is shown in Fig. 7a where the translational movement and the body flexion can be seen to have produced the same pressure pattern.

Bearing in mind that translational movements of an animal's body and its rotations around axes X and Y exert the greatest influence on pressure pattern we can write the following system :

$$\begin{aligned} z_c M \ddot{x}_c &= x_c \sum_{i=1}^{4} P_i - a(-P_1 + P_2 - P_3 + P_4) + M k_1 \ddot{\varphi}_y a \\ z_c M \ddot{y}_c &= y_c \sum_{i=1}^{4} P_i - b(P_1 + P_2 - P_3 - P_4) + M k_2 \ddot{\varphi}_x b \\ M \ddot{z}_c &= \sum_{i=1}^{4} P_i - M g \end{aligned} \tag{4}$$

where coefficients k_1 and k_2 can be taken from Table 2:

$$(k_1 = \alpha_{35} - \alpha_{45} = 66, k_2 = \alpha_{24} - \alpha_{44} = -100).$$

When rotations are absent or negligible ($\ddot{\varphi}_x = \ddot{\varphi}_y = 0$) this system coincides with that given in the previous study (Ioffé et al., 1982) which makes it possible to describe the CG displacement in a closed form. This system must be solved for cases where the initial and final positions of the animal approximate the static ones. This corresponds to the following boundary conditions (for example for x_c (t)):

$$x_c(0) = x_p(0), \ \dot{x}_c(0) = 0, \ x_c(T) = x_p(T), \ \dot{x}_c(T) = 0$$

where x_p is a coordinate of the center of pressure.

Our experience in solving this system with experimental records of P_i has shown that in many cases all these conditions are not satisfied simultaneously. This shows that rotations were not negligible (Frolov and Birjukova, 1986). It is evident that we can choose such a rotation of the body (φ_x and φ_y in equation (4)) as to correct the solution of the system (4) and make the solution satisfactory for all boundary conditions.

So biomechanics themselves do not set any limits on the type of support pressure pattern and for any pattern it is possible, having solved the system (4), to find movements (and not just one) that produce it. This is true for animals whether they have more flexible bodies (such as cats) or more rigid bodies (such as dogs). And the difference between the experimental patterns obtained with trained dogs and cats cannot be explained by the differences in the biomechanical properties of their bodies.

Discussion

In dealing with the problem of motor control, it is very important to take into account the biomechanical aspects of movements. It is evident that many features of central and peripheral movement adjustments must depend on the biomechanical properties of an animal's body such as its construction, the inertia of different links, etc. The most complete description of these properties can be obtained by means of the biomechanical model of an animal's body. Such models for quadrupeds have been developed for analyses of gating (Alexander, 1980) and jumping (Alexander, 1974; Langer and Hemami, 1985). Here we considered a similar model for analysing postural adjustment. All these models are based on the same assumptions, namely that it is possible to separate the animal's body into rigid links, to describe their coupling by frictionless joints and to simulate the muscles as weightless threads. The good agreement obtained between experimental and calculated values of the vertical components of support pressure forces with some examples of body movement (Fig. 4b) shows the validity of these assumptions. This provides a good foundation for all models of this type.

The mathematical formalism used in this study, which was that developed by Prof. Korenev (1979), is the most convenient for computer programming. It makes it possible to analyse arbitrary construction by the same universal set of programmes. It has already been used to analyse arm (Birjukova, 1983) and eye (Alexandrovich, 1980) movements.

Generally movement kinematics are primarily determined by the nervous system. But in some cases biomechanical properties may play an important role. This is true for example of abrupt perturbations of posture when the nervous system does not have time intervene and the animal's body movement resembles the movement of a passive mechanical construction (Dufossé et al., 1982). The role of the central nervous system is obviously predominant in the case of learned and voluntary movements. In the latter case two types of redundancy are to be found. The first is kinematical redundancy, i.e. the possibility of reaching the final body position by several movement trajectories. The second is muscular redundancy, i.e. the possibility of producing the same movement by several patterns of muscle activity. Mathematically, this means that the number of muscles is more than the number of degrees of freedom. Muscle forces thus determine the movement kinematics completely but the converse is not true. Here we demonstrate a similar situation with support pressure patterns and movement kinematics. Equations (1) and (2) completely determine the vertical components of support pressure forces for a given movement kinematics, but the same pressure pattern may be produced by different movements (equation (4)).

Although these equations do not yield a single solution for movement kinematics, they can be very usefully in combined with any method of qualitative movement analysis (for example filming). Furthermore the biomechanical model can be used to find some limits on the type and amplitude of movements. The simplest ones are the kinematic limits due to body construction. The boundaries of a possible region for several generalized coordinates are presented in Table 3. The second type of limits relates to the the sign of vertical components of support pressure forces. If the animal cannot catch at the support platform, their sign must be positive. So it can be seen from Fig. 7 that in this case the forelimb pressure pattern is impossible without rotational movement around axis Y. Likewise it is easy to find from equations (4) the upper limits of the translational movement accelerations along the axes X and Y. Here we did not consider the third type of kinematic limits which are those due to the muscle force limits.

An understanding of all the consequences from the biomechanical point of view is the necessary prerequisite for successful investigation of the motor control effected by the central nervous system

References

Alexander, R. McN., 1974, Mechanics of jumping by a dog, J. Zool. (London), 173: 549-573.

Alexander, R. McN., 1980, Optimum walking techniques for quadrupeds and bipeds, J. Zool. (London), 192: 97-111.

Alexandrovich, E.V., 1980, Some results of the mathematical modelling of the human eye muscular apparatus (in Russian), Biophysics, 25: 902-906.

Birjukova, E.V., 1983, Modelling of human hand movements (in Russian), Biophysics, 28: 715.

Dufossé, M., Macpherson, J., and Massion, J., 1982, Biomechanical and electromyographical comparison of two postural supporting mechanisms in the cat, Exp. Brain Res., 45: 38-44.

Frolov, A.A., and Birjukova, E.V., 1986, Biomechanical model of a quadruped animal body in postural adjustment process (in Russian), Biophysics, 31: 908-912.

Gahéry, Y., Ioffé, M., Massion, J., and Polit, A., 1980, the postural support of movement in cat and dog, Acta Neurobiol. Exp., 40: 741-756.

Gahéry, Y., Ioffé, M.E., Frolov, A.A., and Coulmance, M., 1985, Analysis of vertical displacement of the center of gravity in cat during limb flexion induced by cortical stimulation, J. Motor Behavior, 17: 463-479.

Ioffé, M.E., Frolov, A.A., Gahéry, Y., Frolov, A.G., Coulmance, M., and Davydov, V.I., 1982, Biomechanical study of the mechanisms of postural adjustment accompaning learned and induced limb movements in cats and dogs, Acta Neurobiol. Exp., 12: 469-482.

Ioffé, M.E., Ivanova, N.G., Frolov, A.A., Birjukova, E.V., Kiseljova, N.V., 1988, On the role of motor cortex in the learned rearrangement of postural coordination, in: "Stance and Motion: Facts and Concepts", (J. Massion, V.S. Gurfinkel, M.E. Ioffé,and J.P. Roll, eds), Plenum Press, New York.

Jarbus, A.L., 1965, "Role of eyes movement in vision" (in Russian), Nauka Publishing House, Moscow.

Korenev, G.V., 1979, "Goal-oriented mechanics of guided manipulators" (in Russian), Nauka Publishing House, Moscow.

Langer, D.F., and Hemami, H., 1985, Dynamic control and simulation of the jump of a quadruped, Mathemat. Biosci., 75: 257-277.

Long, C., Brown, M.E., and Weiss, G., 1961, Electromyographic kinesiology of the hand, Arch. Phys. Med. and Rehab., 42: 559-565.

Miller, M.E., 1964, "Anatomy of the dog", Sounders, Philadelphia, London.

Nozdrachev, A.D., 1973, "Anatomy of the cat" (in Russian), Nauka Publishing House, Leningrad.

THE EFFECT OF REPETITIVE STIMULATION ON FIRING MOTONEURONS. COMPUTER SIMULATION BASED ON A MOTOR UNIT STUDY

M. Piotrkiewicz *, R. Person ** and L. Kudina **

* Institute of Biocybernetics and Biomedical Engineering
Polish Academy of Sciences, Warsaw

** Institute for Problems of Information Transmission
USSR Academy of Sciences
Moscow

Summary

A computer model for the repetitive stimulation of firing motoneurons (background activity) was based on experimental data derived from the response characteristics of human triceps surae motoneurons to single stimuli. Extrapolated data of response probability in the interspike interval were used in the model. The parameters of motoneuronal firing (firing rate and standard deviation of interspike interval duration) were based on experimental data.

The results of this simulation make it possible to extend our knowledge about the control of motoneuronal firing rate in the low firing-rate range observed during voluntary contraction of human muscles. In the model, stimulation at a mean frequency of 200 st/s shifted the motoneuronal firing rate from 3-12 imp/s to 11-16 imp/s. With decreasing background firing rates below 6-7 imp/s, the excitability of firing motoneurons depending on the mean firing rate increment increased. The mean decrement of interspike intervals decreased with the increase in background firing rate within the whole range of firing rates studied. The data obtained in the experiments and by simulation may help to clarify the controversy about the ramp slopes of membrane potential trajectories of firing motoneurons. Our results suggest that these slopes increase with the firing rate in the range investigated.

Introduction

One of the essential questions in understanding the integrative function of motoneurons is the interaction between different synaptic inputs in the control of motoneuronal firing rate. This problem was investigated by Granit and his coworkers who introduced additional synaptic excitation (muscle stretch or high-frequency nerve stimulation) to cat hindlimb motoneurons activated by intracellularly injected current (Granit et al., 1966a, b). They defined two ranges of firing rates having different slopes of rate-current intensity characteristics: the "primary" and "secondary" ranges corresponding to lower and higher intensities, respectively. In the "primary" range a shift of the rate-intensity characteristics was usually observed without any change in the slope: i.e. an increment in the firing rate which did not depend on background activity. This phenomenon was described by these authors as an "algebraical summation". On the other hand, repetitive stimulation of descending fibres evoking in motoneurons the monosynaptic EPSPs (Shapovalov, 1972) resulted in an increment in firing rate as well as an increase in the slope.

In these and other studies on intracellularly activated motoneurons the lower limit of motoneuronal firing rate was usually 15-20 imp/s and was only occasionally as low as 7-10 imp/s (cf. Kernell, 1965). On the other hand, it is well known that human motor unit firing rates during voluntary muscle contraction may have much lower values. For example, firing rates recorded during weak and moderate contractions of the human muscle triceps surae fell within the range of 4 to 14 imp/s (Khushivadse, 1979). Even with such low rates the distribution of interspike intervals in single spike trains was monomodal, indicating rhythmical firing. In man, low firing rates of some motoneurons have been recorded in most muscles at different levels of contraction (e.g. Freund, 1983; Person, 1985).

The effect of excitatory inputs on discharging motoneurons has been investigated in humans (Kudina, 1976, 1980, 1983, 1987; Ashby and Labelle, 1977; Ashby and Zilm, 1982b; Fournier et al., 1986). In some of these studies monosynaptic excitatory input on motoneurons during voluntary steady state contractions has been induced by stimulation of Ia afferents (evoking an H-reflex) and the responses of single motor units have been evaluated using statistical methods. The single stimuli were presented randomly during the interspike interval of firing motoneurons. Ashby and Zilm (1982b) studying tibialis anterior motoneurons showed that with an increase of background motoneuronal firing rate the probability of response to the stimulus did not change or slightly decreased. Kudina (1983, 1987) reported an inverse relationship between human triceps surae motoneuronal excitability and background firing rate in the range of rates lower than 8 imp/s.

In all of the above human studies single stimuli were used. However, to come nearer to the physiological conditions of motoneuron activation it seemed interesting to study the effects of repetitive stimulation on motoneurons, especially on motoneurons firing within the physiological range of rates. Though the effect of single stimuli is known, the quantitative results of repetitive stimulation are hard to predict: in the former case the timing of an excitatory volley with respect to the beginning of the interspike interval is random, whereas in the latter case the timing of each volley within an interval is influenced by the previous one. Effects of repetitive stimulation cannot be studied with the method described above since the monosynaptic reflex evoked by each stimulus is followed by many undesirable side effects (including inhibitory ones) caused by skin stimulation and muscle contraction. Therefore these effects were investigated by computer simulation. The results of this study provide us with more information about the interaction between two excitatory inputs in motoneurons.

Methodology

Experimental basis of the computer model

The model was based on experiments described in detail elsewhere (Kudina, 1987). Briefly, the test conditions consisted of single stimuli applied randomly to the tibial nerve during slight voluntary activation of triceps surae in human subjects. The motor unit potentials were recorded. The amplitude of the evoked H-reflex (according to surface EMG) was no more than 10% of maximum-H and the interstimulus interval was no less than 5 s. Twenty-five single active motoneurons were investigated, the total number of stimuli being about 4000. Mean motoneuronal background firing rate varied from one test to another between 3.4 and 11.6 imp/s. All tests were subdivided into three groups depending on the mean firing rate and the data obtained in each group were summarized (Table 1.). As a statistical criterion of motoneuronal excitability a summed response index was used (RIs, the number of responses to the stimuli as a percentage of the total number of stimuli applied).

Table 1. Background firing rates of human triceps surae motoneurons classified into three ranges with their corresponding summed response indices (adapted from Kudina, 1987).

Firing rate, imp/s			Summed response index (RI_S), %
Range		Weighted mean	
1.	3.4 - 5.9	4.7	38
2.	6.0 - 7.9	6.7	27
3.	8.0 -11.6	8.4	29

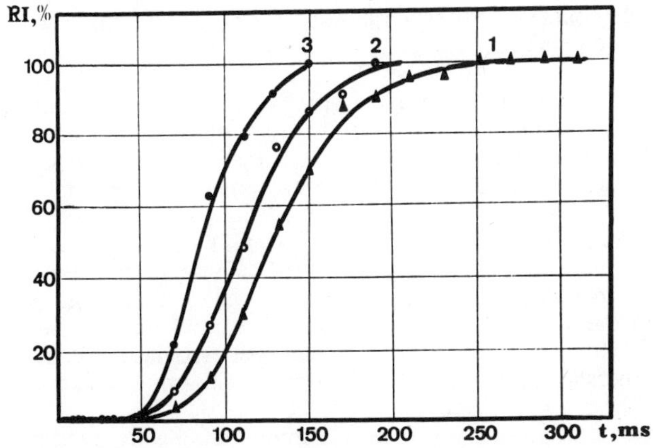

Fig. 1. The change in the response index, RI, during motoneuronal interspike intervals in the 3 groups of tests corresponding to different firing rates of motoneurons (from Table 1): group 1 - 4.7 imp/s, group 2 - 6.7 imp/s, group 3 - 8.4 imp/s; t - time from the last background motoneuronal spike to the arrival of the excitatory volley (adapted from Kudina, 1987).

The change in response probability after spikes of background activity was tested by calculating the response index for different parts of the interspike interval (RI, number of responses as a percentage of the number of stimuli arriving at a given time after the last motoneuronal spike). This change is shown in Fig. 1. Each curve represents one of the above mentioned groups of mean firing rates.

The results of these experiments were used to develop the computer model as follows. Experimental values of RIs were plotted against their corresponding weighted mean interspike intervals, $\overline{X}$ (Fig. 2a, points). The relationship was then approximated by a smooth curve and extrapolated throughout the range of motoneuronal firing rates under investigation. There was no attempt to fit any analytical function to these data. The values of RIs for individual evaluation of $\overline{X}$ were taken from the curve shown in Fig. 2a for further evaluation. The change in RI within the interspike interval was taken to be the change in the response probability (P) and approximated as shown in Fig. 2b. Each impulse was followed by a period of inexcitability (t_1) which was assumed to be constant for all firing rates and equal to 55 ms. The second parameter, t_m, for each mean value of interspike interval, $\overline{X}$, was computed from the equation:

$$RIs = ((\overline{X}-t_m)/\overline{X}).100\%$$

where RIs is the corresponding value of the summed response index, taken from Fig. 2a. Since the relationship P(t) was derived from experimental data, we therefore assumed in the model that the stimulus strength was the same as in the experiments.

Computer Model

In each simulation cycle a time sequence of output spikes was obtained using the following procedure (see Fig. 3):

1. The time of occurrence of consecutive spikes of background activity (trace X) was computed according to the normal distribution of interspike intervals with parameters $\overline{X}$ (mean) and SDx (standard deviation). The range of mean values, $\overline{X}$, was the same as in the above described experiments (87-300 ms). Standard deviations, SDx, were taken from an experimental study on human triceps surae motor units (Khushivadse, 1979) and varied from 7 to 35 ms depending on the mean interspike interval. In each simulation cycle a time sequence of output spikes was obtained using the following procedure (see Fig. 3).

2. The time-course of the response probability (trace P) was determined as described in the previous section (see Fig. 2b).

3. The time sequence of stimuli (trace Y) was generated according to the normal distribution of interstimulus intervals with parameters $\overline{Y}$ (mean) and SDy (standard deviation). The value of $\overline{Y}$ varied from 5 to 20 ms, which corresponded to a range of stimulus rates between 50 and 200 st/s; SDy was assumed to be 5% of $\overline{Y}$.

4. The spike in the output train (trace Z) would be generated at the same time as the stimulus if 1) the latter fell into the excitable part of the interspike interval ($P \neq 0$) and 2) the number (generated in a pseudorandom fashion over a uniform distribution between 0 and 1) was smaller than the corresponding value of the response probability. If there was no response to any stimulus before the time of the next background spike, the subsequent output spike would be generated concurrently with the same background spike.

5. After generation of a response to stimulus the background spike sequence was interrupted and the next spike of the backbround sequence was generated in the appropriate interval with respect to the last output spike.

The sequence 2 to 5 was repeated until the total number of spikes in the background train was greater than 200.

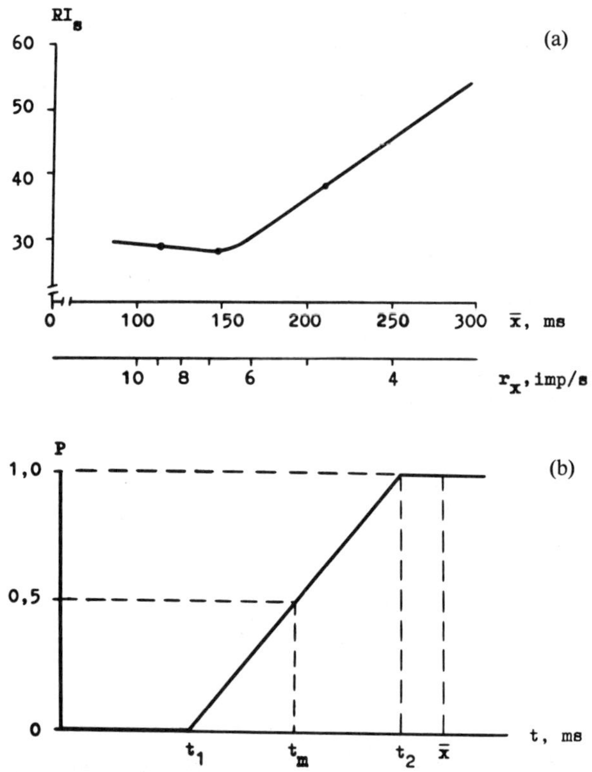

Fig. 2. Basic characteristics of the model: a - the curve fitted to 3 experimental points determined by RIs (see text) as a function of mean interspike interval ($\overline{X}$) and mean firing rate (r_x); b - the model change in motoneuronal response probability (P) during interspike interval: t - time after last impulse, t_1 - inexcitable part of the interval, t_m - time when P = 0.5, $\overline{X}$ - duration of the mean interspike interval.

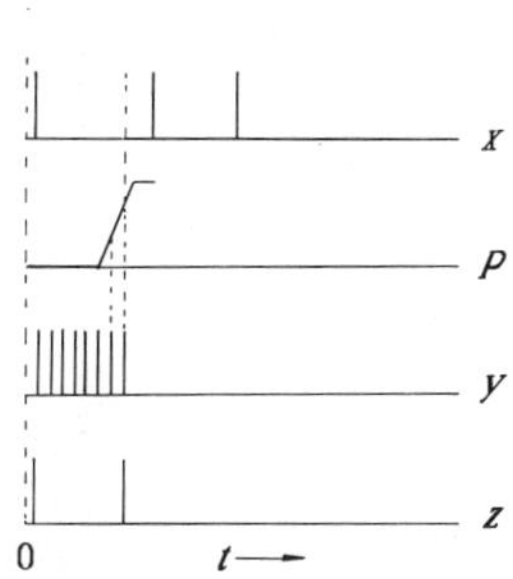

Fig. 3. Principles of simulation: trace X - background train; trace P - response probability; trace Y - stimulus train; trace Z - output train. See text for explanation.

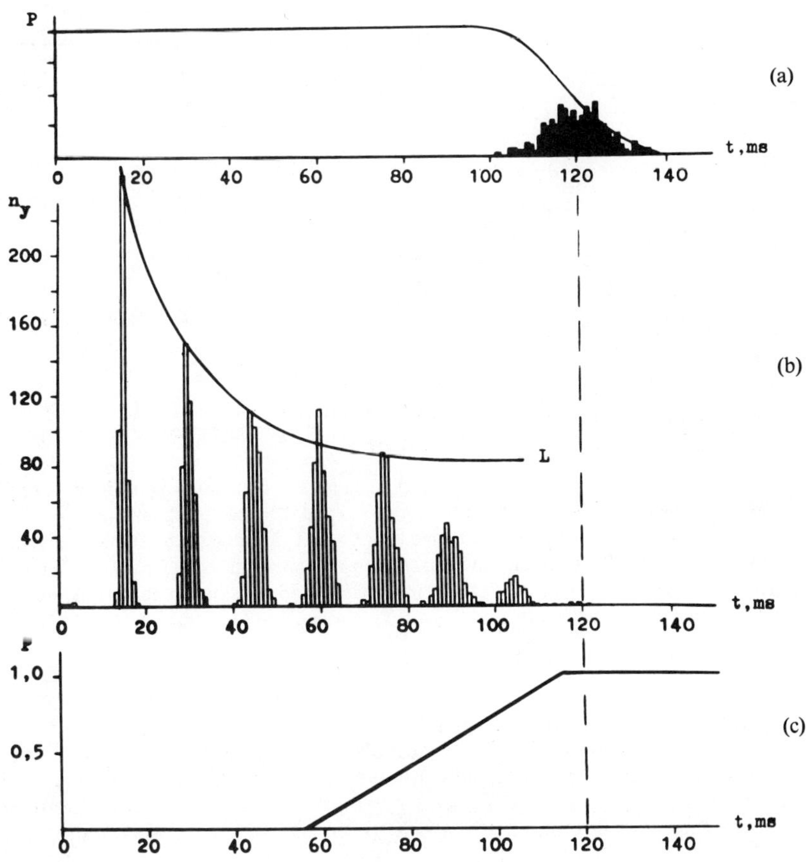

Fig. 4. Comparison between the probability, P, of the arrival of an excitatory volley in the motoneuronal interspike interval during single and repetitive stimulation. In this example simulation parameters were: r_x = 8.3 imp/s ($\overline{X}$ = 120 ms), r_y = 67 st/s: a - theoretical curve of the probability of stimulus arrival during random single stimulation (black columns - interval duration distribution of the background activity); b - stimulus arrival during repetitive stimulation (nY - number of stimuli, L - predicted curve of the decrease in peak number of stimuli arriving during the interspike intervals, based on variability of interstimulus interval alone); c - response probability during the interspike interval in the model: t - time from the last background spike.

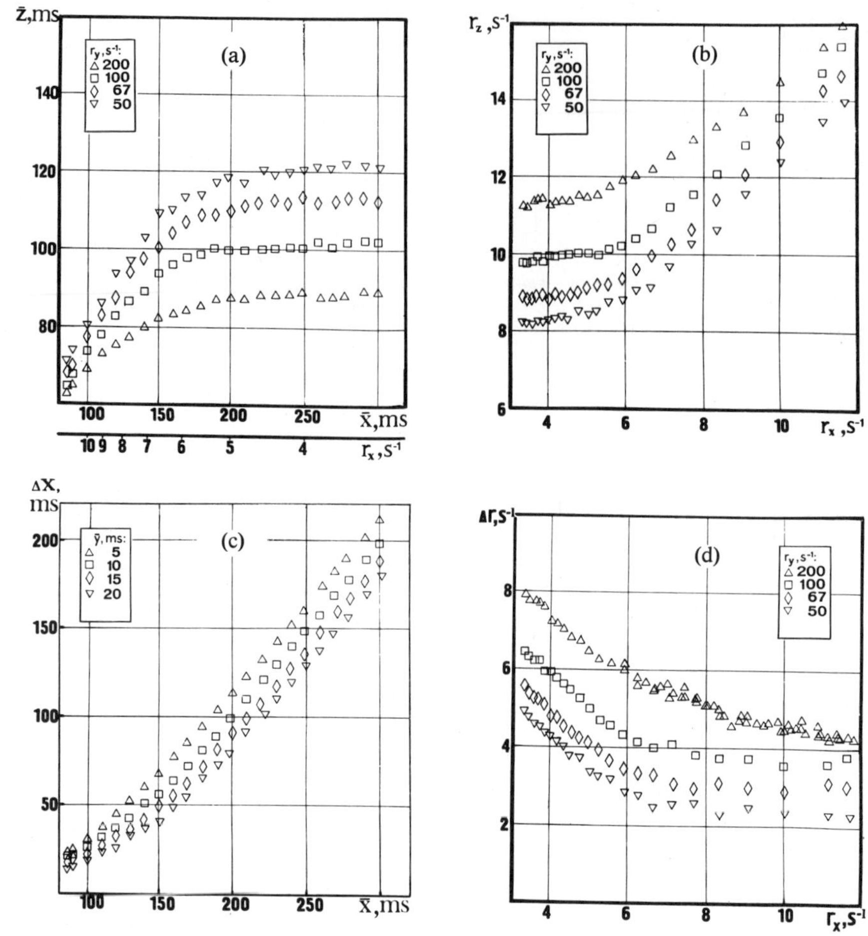

Fig. 5. The effect of repetitive stimulation on firing motoneurons: a - mean interspike interval of the output spike train ($\overline{Z}$) plotted against the mean interspike interval ($\overline{X}$) and the mean firing rate (r_X) of the background activity; b - mean firing rate of the output spike train (r_Z) plotted against the mean firing rate of background activity (r_X); c - decrement of the mean interspike interval (ΔX) plotted against the mean interspike interval of the background activity ($\overline{X}$): d - increment of the firing rate due to the stimulation (Δr) plotted against the mean firing rate of background activity (r_X).

After each simulation cycle the parameters of the resulting output train were computed ($\overline{Z}$ - mean value of interspike intervals, $r_Z = 1/\overline{Z}$ - firing rage) and related to the parameters of the background and stimulus trains (respectively: $\overline{X}$, $r_X = 1/\overline{X}$; $\overline{Y}$, $r_y = 1/\overline{Y}$). In addition, the resulting decrement, ΔX, of the mean interspike interval and increment, Δr, of the mean firing rate were computed as $\Delta X = \overline{X} - \overline{Z}$ and $\Delta r = \Delta r - r_y$, respectively.

Results of simulation

The probability of the arrival of an excitatory volley within the motoneuronal interspike interval was compared with single versus repetitive stimulation. With a single random stimulation the probability of arrival of the volley was uniform throughout the greater part of the mean interspike interval (Fig. 4a). It began to decrease only at the end of the shortest interval and this decline continued throughout the range of these intervals. With repetitive stimulation (Fig. 4b) the probability pattern was entirely different. Due to the short interstimulus interval the arrival of the next stimulus with respect to the last spike was influenced by the previous stimulus. This pattern became more obvious as the ratio of stimulation rate to background firing rate increased. As can be seen in Fig. 4b, c, the decrease in the peak number of stimuli arriving during the inexcitable part of the interval (as a result of the variability in interstimulus intervals) considerably accelerated in the excitable part of the interval, and due to the high stimulation rate the probability of a stimulus reaching the end of the interval was very low.

The results of the computer simulation are summarized in Fig. 5 a-d. The parameters of the output spike train are plotted against the mean rates and mean intervals of the background motoneuronal activity. In each figure these relationships are demonstrated by families of curves for the four different stimulation rates.

The stimulation caused a decrease in the interspike intervals (Fig. 5a), i.e. an increase in firing rates (Fig. 5b) throughout the whole range of background firing rates. Stimulation with the highest frequency (200 st/s) shifted the motoneuronal firing rate from 3-12 imp/s to 11-16 imp/s. The whole range of background firing rates was rather narrow but the slope of curves was different with frequencies above and below 6-7 imp/s (Fig. 5a and b). From Fig. 5d, it can be seen that the firing rate increment due to stimulation (Δr) was inversely related to background firing rates over the range of low rates. At rates above 6-7 imp/s Δr changed only slightly or remained constant. The parameter ΔX, i.e. interval decrement, decreased with the decreasing duration of background interspike intervals throughout the whole range of firing rates studied (Fig. 5c).

The summed response index, RIs, was very low throughout the whole range of r_x and decreased with increasing stimulation rate (with r_y = 200 st/s this index was less than 10%). This result reflected the fact that under simulation conditions a significant part of the total number of stimuli were redundant since they arrived at the motoneuron during the early, inexcitable part of the interspike interval (see Fig. 4b). Thus, this index was regarded as an inadequate measure of stimulation efficiency. Instead, the number of responses to the stimuli was related to the total number of impulses in the output train. This index reached very high values and with stimulation rates of 100 st/s, it was equal to 100%, i.e. all the impulses in the output spike train were responses to stimuli. Furthermore, RIs could not be used as a measure of motoneuronal excitability in our model of repetitive stimulation as it was with single stimuli. The firing rate increment, Δr, can be regarded as a more appropriate measure of this excitability.

Discussion

Certain limitations in the model should be noted. First, the model was based on experimental data relating only to the triceps surae muscle, i.e. to a muscle with slow motor units and a long duration of motoneuronal after-hyperpolarization (Eccles et al., 1958; Buchthal and Schmalbruch, 1970). To overcome this limitation our model is at present being tested on data obtained from flexor carpi ulnaris which contains faster units than those recorded in triceps surae.

Secondly, as noted above, the stimulus strength used in the model was as low as that used in the experiments. In these experiments, a low stimulus strength was chosen since only under these conditions was the identification of individual motor unit potentials possible. We believe the stimulus strengths used in the simulation to be adequate from the physiological point of view, however, since the range of firing rates obtained in the model (up to 16 imp/s) was approximately the same as that of human triceps surae motor units recorded during weak and moderate voluntary contractions (Khushivadse, 1979).

A third limitation was that temporal summation of excitatory volleys was neglected in the model. However, in animal experiments temporal summation of EPSPs during repetitive stimulation of Ia afferents at a frequency of up to 250 st/s only slightly affected EPSP amplitude (Honig et al., 1983). Therefore in our study stimulus strength was not changed even at the highest stimulation frequencies (up to 200 st/s).

The parameter Δr can be regarded as an adequate index to the firing motoneuronal excitability during repetitive stimulation. This parameter can be used as a comparitive measure between the results of animal experiments and the results of simulation. This

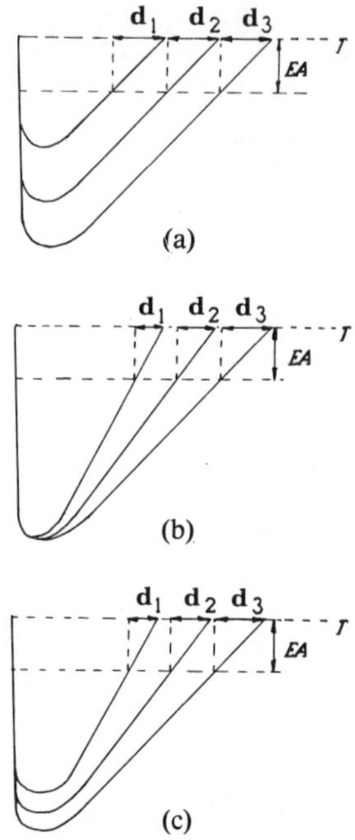

Fig. 6. Schematic representation of motoneuronal membrane potential trajectories when firing with different rates: a - trajectories of "primary" and b - "secondary" range of firing rates according to Schwindt and Calvin (1972); c - trajectories according to Baldissera and Gustafsson (1973) and Kostyukov et al. (1981): T - firing threshold; EA - assumed EPSP amplitude; d_1, d_2, d_3 - parts of interspike intervals in which the EPSP is within reach of threshold.

comparison is reasonable since the activation of motoneurones by synaptic input (in our case by voluntary muscle contraction) can be taken to be equivalent to that by injected current (Kernell, 1969; Schwindt and Calvin, 1973). In animal experiments, as mentioned above, only the range of background firing rates higher than 7-8 imp/s was studied. Within this range, the results of the simulation did not contradict the results of animal experiments by Granit et al. (1966, a, b) as Δr in the model was almost constant.

In the lower range of motoneuronal firing rates the Δr obtained by simulation increased with a decrease in background firing rate. Thus, these results (using Δr as a criterion of excitability) were consistent with those of single stimuli in human experiments (using probability of responses as a measure of excitability; Kudina, 1983, 1987). It has been shown in many studies (for a review see Person, 1985) that the variability of motoneuronal interspike intervals decreases with increasing firing rate. Therefore, the finding that motoneuronal excitability increases at the lower firing rates may reflect a property of a motoneuron which facilitates the transfer into the range of higher, thus more stable rates.

At stimulation rates of 100 and 200 st/s the increment of motoneuronal firing rate, Δr, was only 4-8 imp/s (see Fig. 5d). This increment, though low as compared to the stimulation rate, is very valuable in terms of physiological efficiency. The range of firing rates obtained from our model was the same as that corresponding to the steepest slope of the force-rate relationship with slow motor units in the cat gastrocnemius muscle (Kernell, 1984). Thus, even a small increase in firing rate in this range may cause a substantial increase in motor unit force. The upper level of motoneuronal firing rate is suppressed by an after-spike decrease in excitability (after-hyperpolarization) which determines the functionally preferable range of motoneuronal firing rates.

The effects of additional excitatory inputs to active motoneurons depend on the trajectory of the membrane potential during the interspike interval (Schwindt, 1973; Baldissera and Gustafsson, 1974; Ashby and Zilm, 1982a, b). However, the existing reports on the shapes of these trajectories are controversial. Schematic representations of possible trajectories are shown in Fig. 6.

Schwindt and Calvin have performed several studies on membrane potential trajectory shapes in firing cat motoneurons activated by injected current (Schwindt and Calvin, 1972; Schwindt, 1973; Calvin, 1975). They concluded that the trajectories at different motoneuronal firing rates within the "primary" range (with the exception of the lowest initial firing rates) were parallel: i.e. had a "stereotyped ramp slope" (Fig. 6a) whereas in the "secondary" range, the ramp slopes increased with the increase in firing rate (Fig. 6b).

On the other hand, Baldissera and Gustafsson (1974) analyzed this problem by means of a model of a motoneuron based on the after-hyperpolarization conductance time course. The conclusion from their analysis was that within the whole range of steady firing the ramp slope continuously increased with increasing firing rate (Fig. 6c). A similar result was obtained by Kostyukov et al. (1981) on cat motoneurons firing within the range of 9-22 imp/s.

Ashby and Zilm (1982a, b) investigated this problem by computer simulation and in an experimental study on firing motoneurons responding to single stimuli. Stereotyped ramp slopes and linear summation of EPSP and membrane potential similar to Fig. 6a were assumed in the model. In their simulations the probability of responses depending on the value of the cross-correlogram peak was significantly larger when the mean interspike interval was shorter. This result can be easily explained by the behaviour of the parameters d and X (interspike interval) when the trajectories are parallel (Fig. 6a).

As can be clearly seen, parameter d is constant at a given EPSP amplitude. Therefore, with shorter interspike intervals the ratio d/X is larger and the motoneuron in the model is excitable for a larger part of its interspike interval. However, these authors' experimental study on human tibialis anterior motor units showed that the probability of motoneuronal responses to single stimuli was the same or smaller with higher firing rates. Ashby and Zilm therefore concluded that a trajectory characterized by a variable rate of rise to threshold would be more appropriate for explaining the behaviour of motoneurons in experiments. L. Kudina's (1983, 1987) data agree with this conclusion (cf. Table 1).

The results of our simulations of repetitive stimulations are also consistent with variable ramp slopes rather than with stereotyped ones since ΔX in the model is not constant, but increases with interspike interval (see Fig. 5c). This conclusion fits the experimental data reported by Granit et al. (1966b) and Shapovalov (1972) which have been recalculated and expressed in terms of the dependence of ΔX on X in Fig. 7.

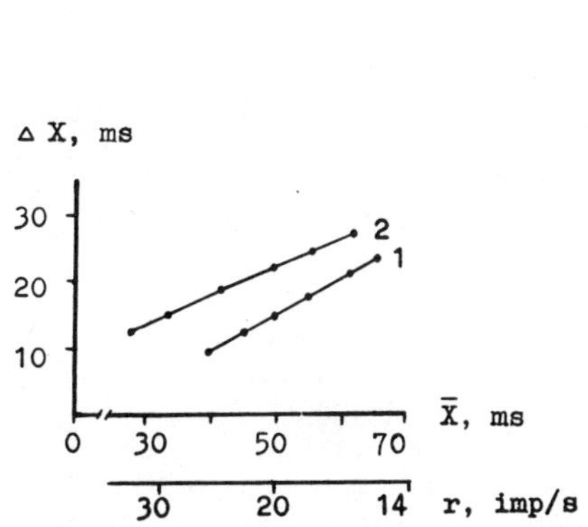

Fig. 7. Dependence of the decrement in interspike interval (ΔX) during additional excitation on the interspike interval of background motoneuronal activity ($\overline{X}$), calculated from: 1 - Granit et al. (1966b, Fig. 3B); 2 - Shapovalov (1972, Fig. 7).

Acknowledgements

The authors wish to thank Dr. V. S. Gurfinkel and Dr. W. Zmyslowski for their valuable discussions. Thanks are also due to Ms. Mindy Levin for her extensive help in revision of the manuscript.

References

Ashby, P. and Labelle K., 1977, Effects of extensor and flexor group I afferent volleys on the excitability of individual soleus motoneurones in man. J. Neurol. Neurosurg. Psychiatry, 40: 910-919.

Ashby, P. and Zilm, D., 1982a, Relationship between EPSP shape and cross-correlation profile explored by computer simulation for studies on human motoneurons. Exp. Brain Res., 47: 33-40.

Ashby, P. and Zilm, D., 1982b, Characteristics of postsynaptic potentials produced in single human motoneurons by homonymous group I volleys. Exp. Brain Res., 47: 41-48.

Baldissera, F. and Gustafsson, B., 1974, Firing behaviour of a neurone model based on the afterhyperpolarization conductance time course and algebric summation. Adaptation and steady state firing. Acta Physiol. Scand., 92: 27-47.

Buchthal, F. and Schmalbruch, H., 1970, Contraction times and fibre types in intact human muscle. Acta Physiol. Scand., 79: 435-452.

Calvin, W., 1975, Generation of spike trains in CNS neurons. Brain Res., 84: 1-22.

Eccles, Y.C., Eccles, R.M. and Lundberg, A., 1958, The action potentials of the alpha motoneurones supplying fast and slow muscles. J. Physiol. (London), 142: 275-291.

Fournier, E., Meunier, S., Pierrot-Deseilligny, E. and Shindo, M., 1986, Evidence for interneuronally mediated Ia excitatory effects to human quadriceps motoneurones. J. Physiol. (London), 377: 143-169.

Freund, H.J., 1983, Motor unit and muscle activity in voluntary motor control. Physiol. Rev., 63: 387-436.

Granit, R., Kernell, D. and Lamarre, Y., 1966a, Algebraical summation in synaptic activation of motoneurones firing within the "primary range" to injected currents. J. Physiol. (London), 187: 379-399.

Granit, R., Kernell, D. and Lamarre,Y, 1966b, Synaptic stimulation superimposed on motoneurones firing in the "secondary range" to injected currents. J. Physiol. (London), 187: 401-415.

Honig, M.G., Collins, W.T. and Mendell, L.M., 1983, -Motoneuron EPSPs exhibit different frequency sensitivities to single Ia-afferent fiber stimulation. J. Neurophysiol., 49: 886-892.

Kernell, D., 1965, The limits of firing frequency in cat lumbo-sacral motoneurones possessing different time course of after-hyperpolarization. Acta Physiol. Scand., 65: 87-100.

Kernell, D., 1969, Synaptic conductance changes and the repetitive impulse discharge of spinal motoneurones. Brain Res., 15: 291-294.

Kernell, D., 1984, The meaning of discharge rate: excitation-to-frequency transduction as studied in spinal motoneurones. Arch. Ital. Biol., 122: 5-15.

Khushivadse, T., 1979, Comparison of motor unit activity of gastrocnemius and soleus human muscles. Fiziol. Cheloveka (Moscow), 5: 102-109 (in Russian).

Kostyukov, A.J., Ivanov Yu.N. and Kryzhanovsky, M.V., 1981, Probability of neuronal spike initiation as a curve-crossing problem for Gaussian stochastic processes. Biol. Cybern., 39: 157-163.

Kudina, L.P., 1976, Effect of tendon reflex on motoneurons of human antagonistic muscle. Neurofiziologiia (Kiev), 8: 624-631 (in Russian).

Kudina, L.P., 1980, Reflex effects of muscle afferents on antagonist studied on single firing motor units in man. Electroencephalogr. Clin. Neurophysiol., 50: 214-221.

Kudina, L.P., 1983, Firing motoneuron: relationship between frequency and excitability. Electroencephalogr. Clin. Neurophysiol., 56: S119.

Kudina, L.P., 1987, Excitability testing of human firing motoneurons during interspike interval. Neurofiziologiia (Kiev), 19: 210-216 (in Russian).

Person, R.S., 1985, Spinal mechanisms of control of muscle contraction, 183 p., Nauka, Moscow (in Russian).

Schwindt, P., 1973, Membrane-potential trajectories underlying motoneuron rhythmic firing at high rates. J. Neurophysiol., 36: 434-449.

Schwindt, P. and Calvin, W., 1972, Membrane-potential trajectories between spikes underlying motoneuron firing rates. J. Neurophysiol., 35: 311-325.

Schwindt, P. and Calvin, W., 1973, Equivalence of synaptic and injected current in determining the membrane potential trajectory during motoneuron rhythmic firing. Brain Res., 59: 389-394.

Shapovalov, A.I., 1972, Extrapyramidal monosynaptic and disynaptic control of mammalian alpha-motoneurons. Brain Res., 40: 105-115.

TOPOLOGICAL ASPECTS OF SENSORI-MOTOR CONTROL

J. Droulez

Laboratoire de Physiologie Neurosensorielle
CNRS, 15, rue de l'Ecole de Médecine
75006 Paris, France

Introduction

Recent investigations in the field of sensori-motor interaction suggest that the main concepts and models which were operational in the past when restricted experimental conditions were used, have to be largely reevaluated when the geometrical constraints and the multidimensional aspects of sensori-motor control are taken into consideration (Gurfinkel and Shik, 1973; Pellionisz and Llinas, 1979; Droulez et al., 1985).

The goal of this paper is first to review some of the experimental data that we obtained at our laboratory which show that movement control cannot be completely understood using models based on servo-control mechanisms. The second part will deal with to the "conceptual story" which has led us to develop new ideas in the field of motor control.

System analysis has been extensively used to explain postural balance and movement responses to various stimuli and is still an essential tool (see a review in Agarwal and Gottlieb, 1982, 1984, 1985). It is also obvious that the underlying concepts of system analysis can be and have been extended to the multidimensional aspects of sensori-motor control (Ormsby, 1974; Robinson, 1982; Hain, 1986). However, the "dimensionality" of the problem itself has not been taken into consideration: schematically, the constraints, the number and the status of the various signals are not explicitly controlled in such models.

One of our favorite games consists of looking for invariant laws which govern the process analyzed. When studying movement control, we are often faced with the problems associated with observing considerable changes: immediate response to a sudden change in stimulation, fast modification of the observed behavior following repetitive experiments, slow and progressive adaptation to a modified or unusual environment. From these observations some quantitative invariants have been deduced, such as, for instance, the dynamic properties of the vestibulo ocular reflex. However, many authors have shown that these quantitative "invariants" are relative: they are goal-dependent, they could be completely modified by purely mental effort (Melvill-Jones et al., 1984), they are subject to considerable variations depending on the environmental conditions under which the experiment takes place. Then, these "quantitative invariants" appear to be themselves subjected to a higher level control in which the laws are expressed in a more "qualitative" language.

A well-known example of these qualitative invariants is provided by the observation of a very precise topographical arrangement of the connections between the various parts of the central nervous system involved in sensori-motor control. One of the central questions that could be asked to-day is: to what extent can this topographical invariant observed at the anatomical level be expressed in terms of functional topological invariants?

I - Experimental results

Sensori-motor adaptations have been shown to occur when people or animals are continuously exposed to a modified environment. Since the original experiment by Helmholtz (1866), who used laterally displacing prisms, and Stratton (1897), who wore

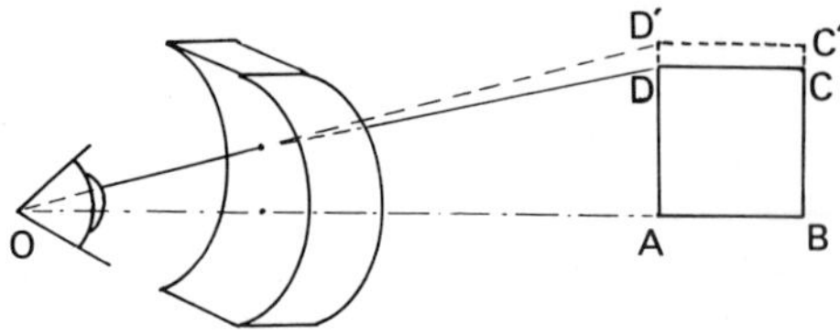

Fig. 1. Optical effect of the lenses. Both faces of each lense were horizontal cylinders with the concave face towards the eye. The square ABCD viewed through these lenses by the observer O appeared as a vertical rectangle ABC'D'.

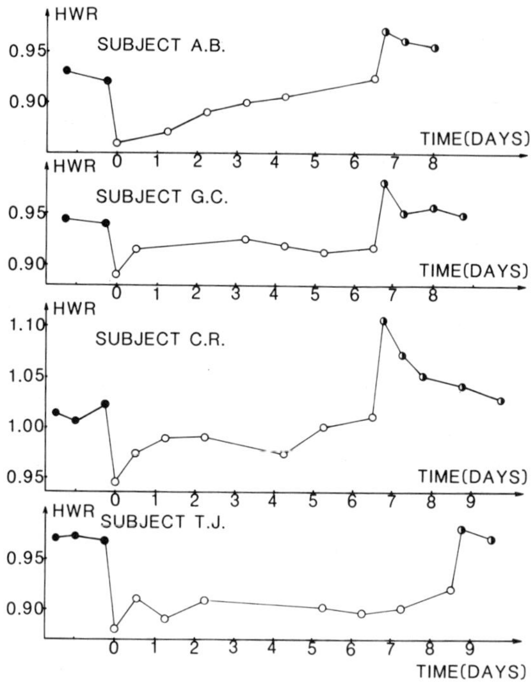

Fig. 2. Perceived-square test. Combined results on the subjects wearing the 8% lenses (A.B., C.R. and T.J.) and the 5% lenses (G.C.). Abscissa: time of exposure (days), starting at the beginning of lens wearing. Ordinate: the mean height/width ratio (HWR) of a rectangle perceived as a square by the subjects, before (solid circles), during (open circles) and after (half-solid circles) lens wearing.

up-down inverting spectacles for several days, many authors have observed significant visuomotor adaptation (Gonshor and Melvill-Jones, 1973; Gauthier and Robinson, 1975; Miles and Fuller, 1974). For instance, vestibulo-ocular reflex (VOR) adaptation was initially interpreted in terms of gain and phase modifications, due to retinal error signals; but since such changes have been proved to be plane specific (Berthoz et al., 1981) and can be elicited by purely mental effort in darkness (Melvill-Jones et al., 1984), more complex mechanisms of adaptation have been suggested (Robinson, 1982; Pellionisz, 1985; Droulez et al., 1985).

The first experiment was aimed at finding evidence of adaptive mechanisms acting on the metrical operators involved in perception as well as in sensori-motor coordination. In this study, we used distorting lenses, which induce anisotropic optical deformation of the subject's visual field. Instead of overall modification such as inversion or magnification, these lenses produce selective vertical lengthening, thus leaving the horizontal direction as a reference for the measurement of vertical changes. As the plane-specificity of VOR adaptation has been demonstrated in previous studies, the modifications induced by distorting lenses are expected to be direction-specific and not reducible to a simple parametric control.

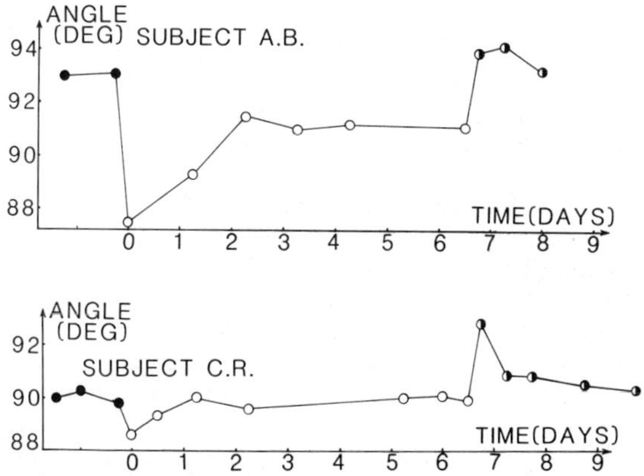

Fig. 3. Perceived orthogonality test. The two subjects (A.B. and C.R.) wore the 8% lenses for 7 days. Abscissa: time of exposure (days), starting at the beginning of lens wearing. Ordinate: mean angle (in degrees) between two oblique lines which were perceived as orthogonal by the subjects, before (solid circles), during (open circles) and after (half-solid circles) lens wearing.

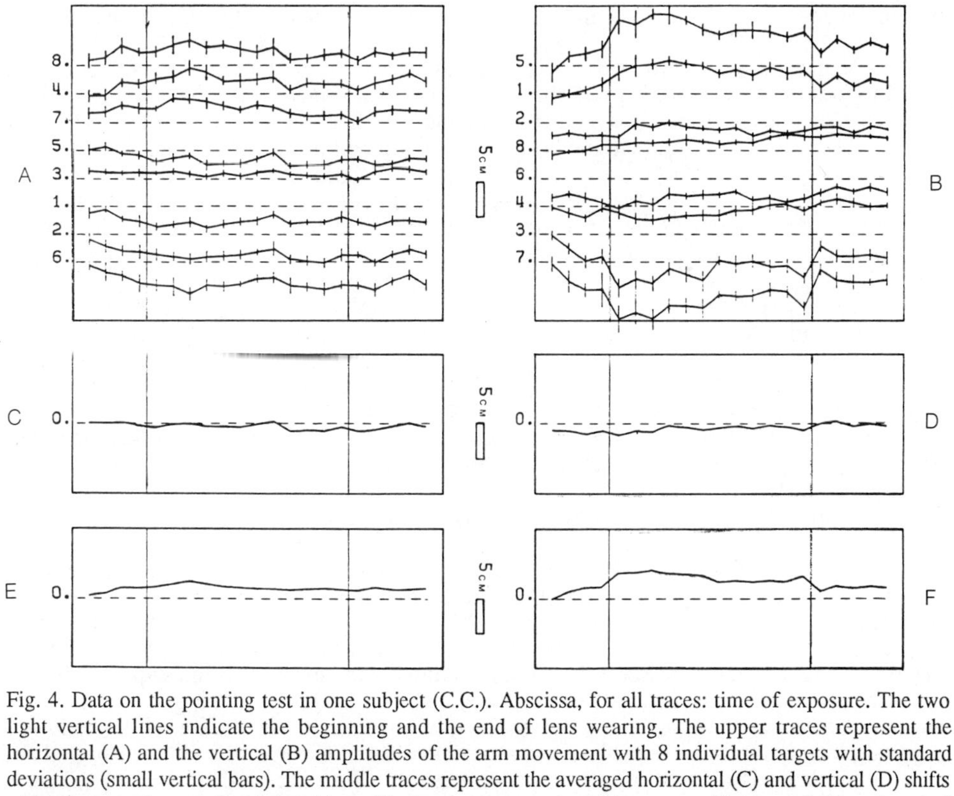

Fig. 4. Data on the pointing test in one subject (C.C.). Abscissa, for all traces: time of exposure. The two light vertical lines indicate the beginning and the end of lens wearing. The upper traces represent the horizontal (A) and the vertical (B) amplitudes of the arm movement with 8 individual targets with standard deviations (small vertical bars). The middle traces represent the averaged horizontal (C) and vertical (D) shifts with respect to the actual target positions. The lower traces represent the averaged horizontal (E) and vertical (F) spreads with respect to the actual target positions.

We used three separate pairs of cylindrical lenses, with the concave side facing the eyes. The total optical power of each lense was zero and the alteration of the retinal image was due only to the refraction phenomenon (Figure 1). The axes of the cylindrical faces of the lenses were in the horizontal plane of the eyes, so that there was little or no disturbance of the visual input in the horizontal direction; in particular, changes in binocular convergence and disparity were very slight. Because of the curvature of the lenses, the vertical disturbance was a magnification of 5%, 8%, or 26%, and varied weakly depending on the distance of the object seen, while the ratios of vertical to horizontal lengthening were 5, 8 and 25%, respectively. Because of the selective lengthening in the vertical direction, oblique lines appeared to be slightly rotated towards the vertical, and any rotation of an object around the visual axis induced a perceived deformation of the object. Moreover, the vertical VOR had to adapt to maintain the retinal stabilization of images during pitch head movement.

We tested 12 subjects for metrical judgements and visuo-manual coordination before, during and after continuous exposure to these distorting lenses (Droulez and Cornilleau, 1986). The metrical judgements were quantified in the "perceived-square" and "perceived-orthogonality" tests. In the perceived-square test, the subjects were presented with a series of rectangles with various height-width ratios (the area of the rectangles remained constant), and they were asked to answer, by force-choice, whether they perceived the rectangle as horizontally or vertically oriented. The procedure was the same in the perceived-orthogonality test, except that two oblique lines with various angles were displayed and subjects were asked to adjust them up to the perfect ratio (HWR) between rectangles perceived as perfect squares by the subjects (Figure 2), and the mean angle between the two lines perceived as orthogonal (Figure 3).

As shown in figure 2, all subjects exhibited a progressive adaptation in their metrical judgements, and the lense removal was followed by a marked after-effect. These results can be interpreted as the consequences of slow adaptive changes in metrical operators such as metrical tensor (Pellionisz and Llinas, 1979) or Riemann connection (Droulez, 1986). The perceptual changes tend to restore the metrical invariance of the object in the physical world.

Visuo-manual coordination was tested in an open-loop pointing test; subjects were asked to point with a pencil on a digitizing tablet to various visible targets with no visual feedback. The lenses worn introduced considerable modifications in the arm movements: the ratio of the vertical amplitude to the horizontal component of the arm movement is plotted against time of exposure in figure 4. Individual traces of horizontal and vertical components of each target show that this change in visuo-manual coordination was only observable on the vertical component of the arm movement, which is in agreement with the optical distortion induced by the lenses.

Both perceptual responses and visuo-manual coordination exhibited similar adaptation and after-effect time courses. This adaptation process selectively affects one direction of space. The overall motor command was changed accordingly. Therefore, these effects cannot be explained by a simple parametric adjustment acting on a particular visuo-motor pathway. On the contrary, it suggests the existence of an internal representation of space, common to perceptual processes and motor planning, in which metrical operators are subjected to slow adaptive mechanisms.

II - Mental set and goal-dependent control mechanisms

Contrary to the slow behavioral changes described above, recent observations show that motor patterns could be reversed or drastically modified depending on the goal or set of postures or available points of "anchoring". These observations are consistent with the idea that determination of the "programs" underlying goal-oriented behavior are centrally determined as suggested by Bernstein (1966). An interesting example of these processes was given by a recent study of acrobatic movements. During jumps made on a trampoline, athletes performed somersaults which consisted of either a simple rotation of the body in a saggital plane (backward somersault), or a combination of two rotations in the saggital and horizontal planes (full twisting backward somersault). We have studied the athlete's head trajectory during various visual situations, in collaboration with D. Rezette (INSEP: Institut National Supérieur de l'Education Physique).

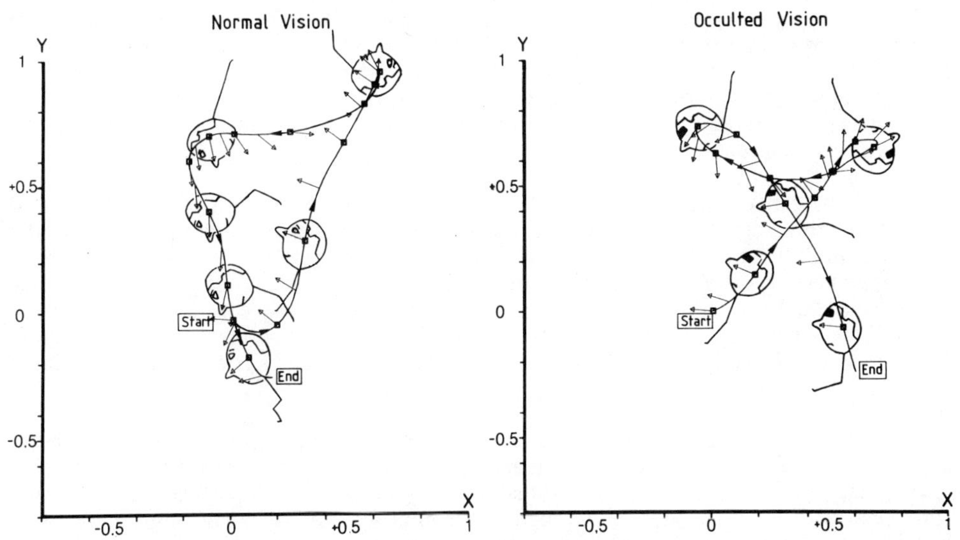

Fig. 5. Head trajectory and orientation during a simple backward somersault. Left part: normal vision. Right part: occulted vision. Horizontal (X) and vertical (Y) head coordinates are expressed in meters. Arrows indicate the subject's frontal head direction every 50 milliseconds.

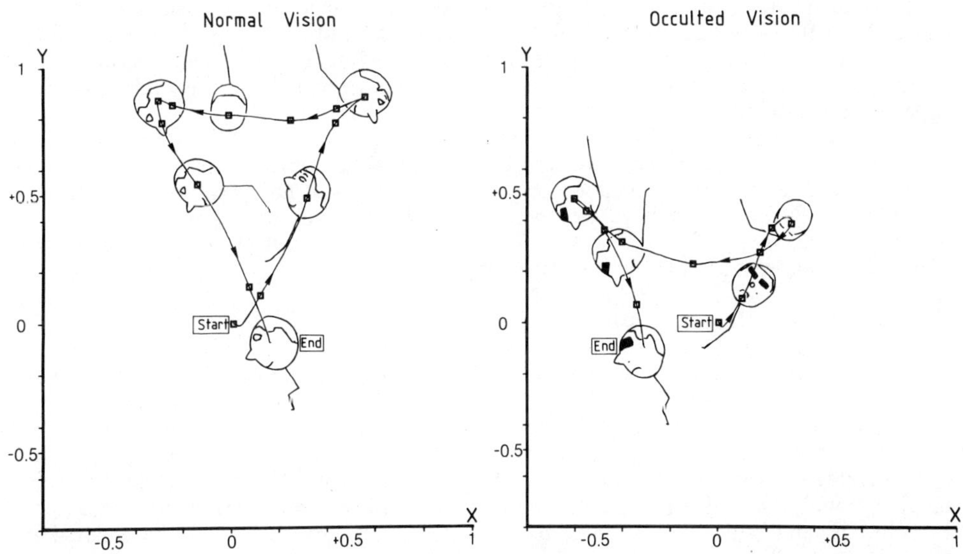

Fig. 6. Head trajectory and orientation during the full twisting backward somersault. Left part: normal vision. Right part: occulted vision. The schematic drawings of the head indicate its successive orientations in space. Horizontal (X) and vertical (Y) head coordinates are expressed in meters.

During the somersault (simplest one plane movement) performed in the light, we observed two periods of head angular stabilization (see figure 5): at the beginning of the jump with the head upright, and at the end of the jump, during the preparation to land phase, with the head pitched forward (nose-down). This particular orientation and stabilization of the head is clearly required by the visual system to gain precise information on landing point distance. When the same jump is performed in darkness, these head angular stabilization phases disappear, indicating a strong visual dependency. It should be stressed that the simple somersault, while easily performed and early learned in the course of an athlete's training, is badly executed in darkness or perturbed vision, particularly during the reception phase. Moreover, the subjects frequently reported a sensation of spatial disorientation after the experiments in darkness.

However, during a more complex movement such as the full twisting backward somersault which combines a vertical 360 degree rotation and a 360 degree rotation around the longitudinal axis of the subject, darkness does not prevent the prelanding head stabilization (see figure 6). We have hypothesized that during this complex movement, vision cannot be used, due to high rotational velocities and poor performance of the vestibulo-ocular reflex in the torsional plane, and that therefore the subject has learned the movement differently: in this sault, the head angular stabilization was mainly controlled by vestibular signals.

During complex motor activities, some specific inputs were selected and used to control specific phases of the movement, and the selection of one sensori-motor program or another depends upon the overall task that the subject is asked to perform. In particular, some phases of the movement are required by the multisensory data acquisition, and have to be integrated at some precise time in the motor program.

There also exist other experiments which show that motor activity cannot be correlated uniquely with immediate, contemporary sensory inputs, but are also influenced by the overall context of the experiment and even by some kind of projection into the future. Lacour et al. (1978) have described very nicely during a succession of at least two activities in the extensors of limb and neck free fall in the monkey: an initial reactive burst of EMG with a short latency (50-100 ms) which is dependent upon the presence of the vestibular receptors and is probably induced by the otolithic stimulation, and a second more labile EMG component which disappears when there is no actual landing, and which is therefore related to a postural preparation for landing. The latency of this second burst is dependent upon the expected time of landing.

Along the same lines of investigation, P.P. Vidal (unpublished results) has observed in man a considerable reorganization of postural reactions to unpredictable perturbations depending on the experimental environment. The motor activities following the perturbation, which were normally recorded in the inferior limbs, were observed in the superior limbs when a fixed bar was placed in front of the subject.

III - Theoretical interpretation

How are we to interpret these results? First, it is our opinion that in posture and movement control, the variables which are under control cannot be local parameters such as single joint angles, but rather global parameters such as the position of center of gravity or the distance between hand or head to a given target point. As these global parameters depend upon several sensory signals, we must assume that the control of multiarticular systems involves a multidimensional space of representation. This has been particularly emphasized by Pellionisz and Llinas (1979) who introduced the tensorial formalism. We have called this mode of motor control: the conservative mode, in the sense that whatever the initial state of the system, the conservative mode tends to bring it back to its final equilibrium state. Clearly, the equilibrium point or final state and the actual state of the system belongs to a multidimensional space. A picture of this conservative mechanism could be given by a well-shaped potential function in this multidimensional space. The bottom of the well defines the equilibrium state. Whatever the actual state of the system, the potential function brings the system back to the vicinity of the bottom of the well. A generalization of such well-shaped functions is the notion of the attractor (Thom, 1972). It should be mentioned also that the word "equilibrium" should be taken in the general sense: in particular, it is not strictly limited to static conditions.

The potential function may exhibit several "wells" or attractors. Moreover, the shape of the wells and their number are dependent upon the space in which the motor activity is

Fig. 7. Reading and walking task. Note the position of the arms and the elbows near the trunk.

described. For instance, in the case of a single joint movement, the potential function can be defined either in the 1D space of the joint angle, or in the 2D space of the couple of antagonist muscle activities. In the case of a complex multi-joint movement controlled by several sensory inputs, there are clearly many spaces in which the potential function could be defined. Therefore, the shape of the potential function may be rather complex and may change depending on the task and the constraints, as soon as the movement is assumed to be controlled at a high level of integration.

One important notion is that in this mode of control the relevant information about the environment and the system itself is simply expressed with a small number of localization variables. Why is this conservative mode not sufficient to explain posture and movement control? First, it cannot account for changes in strategies, as shown by Nashner (1985) or different utilization of synergies as defined by Bernstein (1966), depending on the goal or the context of the action. Secondly, because the mode of representation, and the space in which the variables and the relevant signals are defined does not allow the coding of the contextual data, nor even the simultaneous representation of several states.

These considerations led us to propose another mode of control in which what is controlled is not the state of the system, but the space in which the conservative mode operates. We have called it the projective or topological mode of control. The term "projective" was used because we assumed that it is at this level of control that the movement is "projected", that is programmed and internally simulated. The term "topological" was used because the operators acting on the spaces of representation may behave like some topological operators.

This point can be illustrated by a simple postural example. Two apparently similar postural tasks were given to naive subjects. Both tasks consisted of carrying an object with both hands while walking. The first task was to read a book, as shown in figure 7, while the second task consisted of carrying a glass full of water. Due to the well-known limitations of ocular smooth pursuit performances in velocity and frequency, the reading task required a considerable reduction in the relative movement between head and hands, particularly in the high temporal frequency domain. The attitude of the superior limbs was clearly well adapted to the constraints of the task: the arms were held close to the trunk, and the overall posture tended to reduce the number of degrees of freedom.

On the contrary, when the subject was required to carry a full glass, the opposite tendency was observed (see figure 8): the arms and the elbows were detached from the trunk and were partly raised forward. This posture increases the number of degrees of freedom. As a matter of fact, the second task did not require subjects to reduce the relative intersegmental

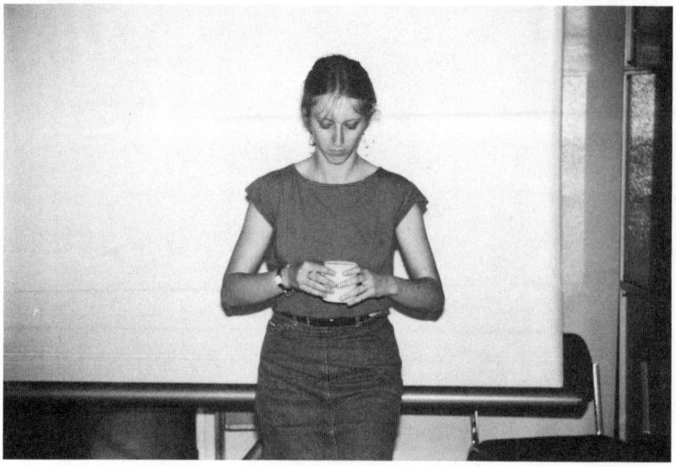

Fig. 8. Carrying a full glass while walking. See text for discussion.

movement, but rather to reduce the final segment displacements with respect to the gravito-inertial frame of reference, particularly in the low frequency domain.

Our interpretation of these postural changes is as follows: let us consider that the whole posture can be completely described with n variables. The conservative mode of control has then to operate in an n-dimensional space. In the reading task, the subjects adopted a "rigid" attitude, which corresponds to the so-called trivial topology of the n-dimensional space: all the joint angles involved in the postural control had to be independently fixed and stabilized. On the contrary, in the glass carrying task, the subject adopted a different attitude which freed the hands from the perturbations. This posture corresponds to the so-called quotient topology: here only the final segment orientation and position are controlled, independently of the various joint angles which are no more relevant to this task. To accomplish this, the role of synergies could be essential: the joint angles are not individually controlled, but are set by global signals resulting from the combination of several adjacent joint angles and motor commands. These combinations allow the simultaneous co-contractions of flexor and extensor muscles of different groups which are required to stabilize the hands independently of the trunk movements.

Analyzing each task separately, it is possible to interpret the postural reactions within the conservative mode, using a particular multidimensional model, and a particular potential function. As discussed above, the change of task or constraints imposed on the system may imply a drastic change in the shape of the potential function. We used the term "topological space transformation" instead of "potential function transformation" because for the sake of simplicity, assuming that the shape of the potential function remains globally simple and invariant, but is applied to different spaces of representation. This is well known in physics that a given problem can be highly simplified by an appropriate choice of reference system. Actually, the idea that some geometrical operators in the central nervous system can change the space of representation is not new: for instance, Robinson (1982) has postulated that a visual target position is successively coded in a retinotopic and a craniotopic frame of reference. More generally, the integration of several kinds of sensory information originating from different segments (eye, head, joints) certainly requires the use of geometrical operators to perform space transformations. However, these operations were generally assumed to be purely sensory (acting on purely sensory signals) and limited to coordinate system transformations. We generalized this idea to sensori-motor loops (not only to purely motor information), and to more drastic space transformations, including the change of dimensionality - the projection of 2D retinal images in the 3D perceptual space is an example of a complex transformation - or the introduction of discontinuities.

Moreover, the goal, the global parameters, the range of validity, and the role of sensory inputs and the interaction between sensory modalities appear to be very different in the two tasks. Therefore, a higher level of motor control, which assigns the goal, defines the parameters and the limits of the "conservative" mode, has to be postulated. Taking into account the importance of the contextual information (task, available sensory channels, global activity), we propose that the higher "projective" mode is based on a different internal representation of data (topological maps instead of vectors) and uses different operators (topological operators instead of tensors). As these maps convey a large amount of data on the system itself and the environment in which the action takes place, the motor activities can be planned and executed in these maps without any sensory feedback or with only the intermittent and goal-oriented sensory inputs described above.

References

Agarwal, G.C. and Gottlieb, G.L., 1982, Mathematical modelling and simulation of the postural control loop. CRC Critical Reviews in Biomedical Engineering. Part I (1982), Vol. 8-2, p.93-134; Part II (1984), vol. 11-2, p. 113-154; Part III (1985), vol. 12-1, p. 49-93.

Bernstein, N., 1966, The coordination and regulation of movement. Pergamon Press. Oxford.

Berthoz, A., Melvill-Jones, G. and Begue, A.E., 1981, Differential visual adaptation of vertical canal dependent vestibulo-ocular reflexes. Exp. Brain Res., 44: 19-26.

Droulez, J., 1986, Neurosensory processing of moving images. Biol. Cybern. (subm. to pub.).

Droulez, J., Berthoz, A. and Vidal, P.P., 1985, Use and limits of visual vestibular interaction in the control of posture. In: Vestibular and Visual Control on Posture and Locomotor Equilibrium, Igarashi, M. and Black, F.O., eds, Karger, Basel, pp. 14-21.

Droulez, J. and Cornilleau, V., 1986, Adaptive changes in perceptual responses and visuomanual coordination during exposure to visual metric distortion. Vision Res., 26-11: 1783-1792.

Gauthier, G.M. and Robinson, D.A., 1975, Adaptation of the human vestibulo-ocular reflex to magnifying lenses. Brain Res., 92: 331-335.

Gonshor, A., and Melvill-Jones, G., 1973, Changes of human vestibulo-ocular response by vision reversal during head rotation. J. Physiol. Lond., 234: 102-103.

Gurfinkel, V.S. and Shik, M.L., 1973, The control of posture and locomotion. In: Gydikov et al.: Motor Control. Plenum Press New York: 217-236.

Hain, T.C., 1986, A model of the nystagmus induced by off-vertical axis rotation. Biol. Cybern., 54: 337-350.

Helmholtz, H. von, 1866, Handbuch der physiologischen Optik. Vos: Leipzig.

Lacour, M., Xerri, C. and Hugon, M., 1978, Muscle responses and monosynaptic reflexes in falling monkeys. Role of the vestibular system. J. Physiol., Paris, 74: 427-438.

Melvill-Jones, G., Berthoz, A. and Segal, B., 1984, Adaptive modification of the vestibulo-ocular reflex by mental effort in darkness. Exp. Brain Res., 56: 149-153.

Miles, F.A., and Fuller, J.H., 1974, Adaptive plasticity in the vestibulo-ocular responses of the Rhesus monkey. Brain Res., 80: 512-516.

Nashner, L.M., 1985, Strategies for organization of the human posture. In: Vestibular and Visual Control on Posture and Locomotor Equilibrium, Igarashi, M. and Black, F.O., eds, Karger, Basel: pp.1-8.

Ormsby, C., 1974, Model of human dynamic orientation. Ph.D. Thesis, MIT, Cambridge.

Pellionisz, A., 1985, Tensorial aspects of the multidimensional approach to the vestibulo-ocular reflex and gaze. Reviews of oculomotor research. In: Adaptive Mechanisms in Gaze Control, Berthoz, A., and Melvill-Jones, G., eds, Elsevier, Amsterdam.

Pellionisz, A., and Llinas, R., 1979, Brain modelling by tensor network theory and computer simulation. The cerebellum: distributed processor for predictive coordination. Neuroscience, 4: 323-348.

Robinson, D.A., 1982, The use of matrices in analyzing the three dimensional behavior of the vestibulo-ocular reflex. Biol. Cybern., 46: 53-66.

Stratton, G., 1897, Upright vision and the retinal image. Psychol. Rev., 4: 182-187.

Thom, R., 1972, Stabilité structurelle et morphogénèse, Ediscience, Paris.

RAPID ONE-JOINT MOVEMENTS: A QUALITATIVE MODEL AND ITS EXPERIMENTAL VERIFICATION

R.M. Abdusamatov, S.V. Adamovich, M.B. Berkinblit,
A.V. Chernavsky, and A.G. Feldman

Institute of Problems of Information Transmission
Academy of Sciences, Moscow 101447, USSR

Summary

One of two versions of the equilibrium-point hypothesis is used to provide a qualitative explanation for the control of fast one-joint movements. It is suggested that discrete unidirectional movements are produced by a rapid gradual shift in an equilibrium point to a final position, in combination with another gradual command which produces agonist-antagonist coactivation and ensures the stiffness of the joint. The monotonic central commands and the reflex effects occurring in the course of movement give rise to the characteristic three-burst pattern, as is qualitatively illustrated using the concept of muscle activation area. This form of central command also makes it possible to explain the observable correlation between the duration of the first agonist burst and movement amplitude. Two predictions of the model are experimentally corroborated in the present study: (1) movement trajectories with terminal oscillations can be transformed into aperiodic trajectories by means of transitory perturbations; (2) the onset time, magnitude and duration of the first agonist burst are reflex-dependent. The results are consistent with the assumption that neither the form of movement trajectory nor the form of EMG bursts (time periods and amplitudes) are centrally programmed.

Introduction

The equilibrium-point (EP) hypothesis concerning motor control (Feldman, 1974, 1986) has been used to formulate a minimal model for calculating the kinematic and electromyographic (EMG) patterns underlying simple one-joint movements (Abdusamatov and Feldman, 1986; Abdusamatov, Adamovich and Feldman, in press). In particular, the model includes a triphasic EMG pattern similar to that observable during fast arm movements to a final position: an initial burst of activity in the agonist muscle followed by an antagonist burst and then a second small agonist burst (e.g., Hallet et al., 1975, Lestienne, 1979). On the other hand, the origin of the EMG pattern can be directly explained in qualitative terms on the basis of the EP hypothesis, without complex calculations. This type of approach, which was used in the present study, can promote an understanding of the system's dynamic behavior. For this purpose the concept of muscle activation area (AA) will be used. AA is defined as the domain of kinematic and control variables in which the muscle undergoes activation (Feldman, 1986). Next we will experimentally test predictions made with the qualitative model concerning the form of movement trajectories and the dependence of EMG parameters on kinematic variables and external perturbations.

Triphasic EMG pattern: a qualitative model

Here we describe how, using a representation of AAs and movement trajectories on the phase plane, a triphasic EMG pattern emerges. In the phase plane, the axes are the angular displacement (φ) and its velocity (ω). The displacement and velocity values act as coordinates of a point in two-dimensional space. As time varies, the point (φ, ω) describing the motion of the joint moves along a certain path on the phase plane. Fig. 1 shows flexor and extensor AAs on the phase plane in the case of fixed central commands descending to alpha and gamma motoneurons. The commands are assumed to specify the positions (β1 and β2) and the slopes on the border lines of the AAs (for detail see Feldman, 1986). Physiologically, the position of AA border line is associated with the threshold of the tonic stretch reflex whereas the slope determines the sensitivity of the threshold to velocity of muscle stretching or shortening. For the sake of simplicity, the slopes are assumed to be fixed at certain values. The flexor AA is situated to the right and the extensor AA to the left of appropriate border line. It is reasonable to suggest the following rule for EMG bursts and silent periods. If the phase movement trajectory enters a given AA, activity of corresponding muscle occurs and increases the more the trajectory penetrates inside the AA. When the trajectory leaves the AA, the EMG activity disappears. The concept of AA takes into account both central and reflex modes of muscle activation. Central commands manifest themselves in the motion of the border lines due to changes in β1 and β2. Consequently, the distances from the current point of the movement trajectory to the border lines alters and, in accordance with the above rule, the EMG varies correspondingly.

The next point touches upon the question of the coordination of central commands on flexor-extensor coupled muscles. It has been suggested that the brain controls the muscles cooperatively, as a coherent unit (cf. Kugler et al., 1980), but that there exist a variety of cooperative commands. Each command is expressed as a coupled modification of reflex thresholds or, geometrically, as co-ordinated shifts of AA borders. Fig. 2 A shows in these terms central commands on a rapid flexor movement from an initial to a final position against a zero load. Dashed lines in Fig. 2 A show an initial position of AA borders. There is a small overlap between the two AAs, and a slight coactivation of the muscles provides an initial equilibrium position ($\varphi = R$). To perform a flexor movement, the nervous system shifts the AA border lines to the left (solid lines), and the equilibrium position biases to R'. As a result, reciprocal changes in the activity of flexor and extensor muscles occur (see below). This is why this cooperative central command (R → R') is called "reciprocal". This command is combined with another, coactivation command (C) that gives rise to an increase in the overlapping between the two AAs. The final equilibrium position R' is situated inside the new coactivation zone.

The reciprocal and coactivation commands are functionally different: the first specifies a final equilibrium position whereas the second ensures the system's stiffness to provide the appropriate muscle torques for the execution of rapid movement. In addition, muscle damping is known to increase with muscle activity (Gasser and Hill, 1924) and thus the coactivation command promotes the system's stability. Obviously, the slopes of AAs as

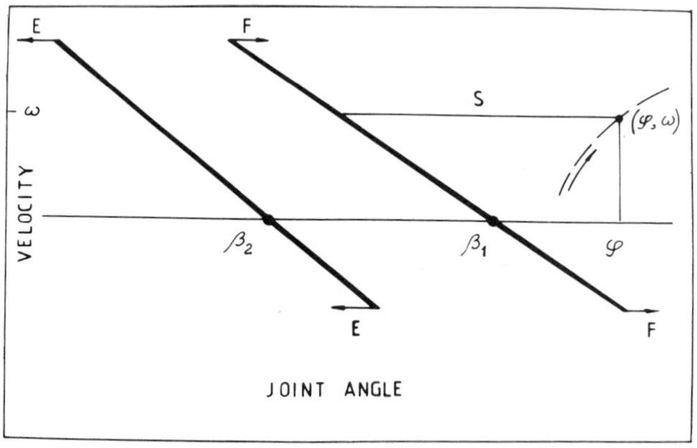

Fig. 1. Activation areas (AAs) for the flexor (to the right of the borderline FF) and extensor muscles (to the left of the borderline EE) on the phase plane (joint angle φ versus its velocity ω). Recruitment of motor units of a given muscle occurs when the current point (φ,ω) of movement trajectory (dashed) enters the appropriate AA and increases as a function of horizontal distance S from that point to the borderline. Central commands specify the position (β_1,β_2) of the borderlines and thus participate in the control of muscle activity. When $\beta_2<\beta_1$, as here, the muscles act apart, otherwise there is a coactivation area for the muscles (Fig. 2A).

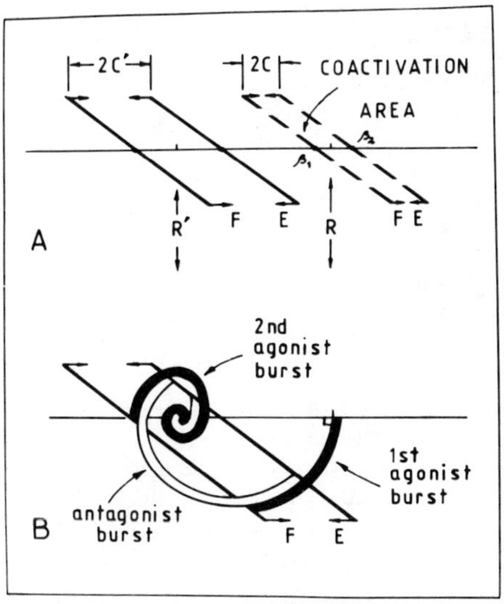

Fig. 2. The generation of EMG burst during rapid flexor movements against a zero load: schematic representation using the concept of muscle AAs. A: cooperative central commands for the movement, represented in terms of shifts of AA borders. Coactivation command $C = (\beta_1 - \beta_2) / 2$ specifies the magnitude of the overlapping between the AAs (coactivation area) and gradually increases ($C \rightarrow C'$) during the performance of the movement. Reciprocal command $R = (\beta_2 + \beta_1) / 2$ specifies the appropriate equilibrium position of the joint, $\varphi = R$. Geometrically, modifications of the command ($R \rightarrow R'$) are associated with unidirectional shifts of the AA borders to the left. B: phase movement trajectory (oval curve) and periods of muscle activation resulted from the central commands presented in A. Axes as in Fig. 1.

speed factors also determine the system's stability. On the basis of previous experimental data, we have suggested that the shifts of AA borders and equilibrium position during fast unidirectional movements are executed rapidly but gradually (Adamovich et al., 1984). However, to simplify the qualitative explanation for the triphasic EMG pattern, we assume in the meanwhile that the shifts are executed in a step-like manner. As has been shown by numerical solution of the above-mentioned minimal model, a three-burst EMG pattern remains if the central commands sweep gradually (Abdusamatov and Feldman, 1986).

Fig. 2 B shows the kinematics and "EMGs" resulting from the commands presented in Fig. 2 A. When the AA border lines are shifted to the left, the phase point (R, 0) characterizing the initial system state occurs deeper in the flexor AA and outside the extensor AA. As a result, additional flexor activity and a "silent period" in extensor activity emerge, which provide an initial torque for the movement. As the movement velocity is negative ($\omega < 0$), the phase point moves downwards and to the left along an oval curve (Fig. 2B) and soon approaches the new border lines, leaves the flexor AA and enters the extensor AA. Thus, the first agonist burst ends giving place to a burst of activity in the antagonist muscle which begins to decelerate the movement. The phase movement trajectory begins to go upward, to the abscissa axis. In this phase, the trajectory may enter the flexor AA again and then agonist muscle activity resumes (a second agonist burst). The movement ends in equilibrium position (R') with a coactivation of agonist and antagonist muscles, generating considerable stiffness. The movement may end with terminal fading oscillations or without them (aperiodically), depending on the overall system damping, which is specified in particular by the coactivation command and the slopes of AA border lines. No support is necessary for the considerable stiffness after the end of the movement. Correspondingly, it has been suggested that the brain in this post-movement phase decreases the distance between the two border lines (this post-movement phase is not shown in Fig. 2 B). As a result, the amount of muscle coactivation gradually diminishes while the final position remains, as has been observed experimentally (Feldman, 1980; Vincken, 1983).

Experimental testing of the model

The model suggests that the terminal oscillations usually observed during fast one-joint movements to a final position are not present in the central control signals which specify shifts of a system. Actually, the model hypothesizes that central control signals have a monotonic form (Fig. 3). They elicit an intensive inflow of energy for the motion due to a rapid shift in the equilibrium state. The system is possibly unable, despite its strong damping, to dissipate the energy entirely while coming to the final position and does so only through a terminal overshoot or fading oscillations. This means that the oscillatory form of movement trajectory can be transformed into an aperiodic form if part of the kinetic energy is absorbed in the course of the movement by an external system via its interaction with the limb. This suggestion is tested in Experiment 1.

The model predicts some correlations between EMG parameters and kinematic variables. For example, the form of central control signals (Fig. 3) indicates that the movement distance is directly proportional to the duration of central command signals. Insofar as the commands elicit the first agonist burst, there must be a positive correlation between its duration and the movement distance as has been shown (Berardelli et al., 1984; Brown and Cooke, 1984).

On the other hand, the dependence of the first agonist burst on the central control signals is not absolute as they elicit the EMG burst indirectly by modulating the stretch reflex threshold. Consequently, while the control signals have a predominant role in the generation of the first agonist burst, there remains a possibility that reflex modification of the burst amplitude, duration and onset time may occur if movement perturbations give rise to threshold changes in kinematic variables. This is especially so if they precede the movement onset, a suggestion that is tested in Experiment 2.

As far as the antagonist burst is concerned, it is possibly generated when the changes in the central control signals cease, and hence the burst will be more sensitive to movement perturbations in the course of movement than the first agonist burst. This suggestion is consistent with well-documented observations (e.g., Ghez and Martin, 1982; Soechting et al., 1976; Terzuolo et al., 1973; Lestienne, 1979) and is not tested here.

Methods

4 normal right-handed subjects participated in Experiment 1 and 3 in Experiment 2. They were seated in a dental chair, grasping a vertical rod attached to a manipulandum handle which was free to move horizontally. The subject's right arm was positioned with the elbow beneath the point of rotation of the handle. The arm was supported just distally to the elbow. A vertical beam was projected from a light source fastened to the distal end of the handle onto a horizontal screen just under the manipulandum to indicate its position to subjects. In Experiment 1 subjects were required to perform flexor movements from an initial (about 140°) to a final position (60-70°). The movements had to be performed as fast as possible in response to a sound signal switched on after an interval of 4-6 s. The amplitude was not specified precisely - the subjects were asked to reproduce more or less the same amplitude

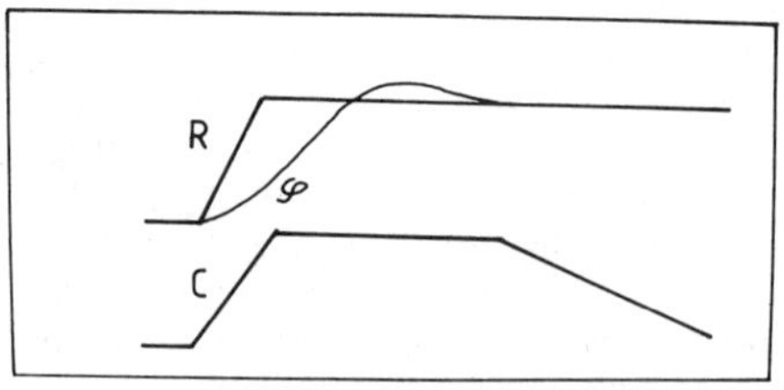

Fig. 3. The timing of central commands R and C (specified in Fig. 2A) for a rapid one-joint movement (φ).

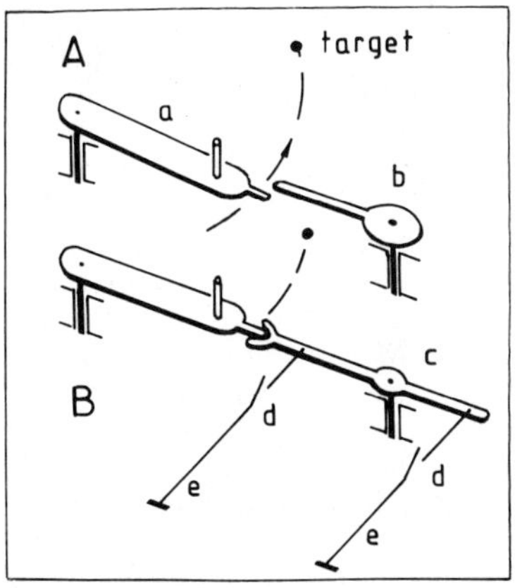

Fig. 4. Arrangements for perturbation of one-joint movements by a sudden inertial load (A) and impulsive torques (B). a: manipulandum handle, b: inertial disk, c: lever for applying perturbations, d: electromagnetic locks, e: rubbers.

with their eyes closed (they could open their eyes in inter-trial intervals to check on their performances). In some trials extensor movements were performed.

In Experiment 1, subjects were asked to make a series of 30 movements. In 10 trials randomly selected out of the 30 trials the movements were perturbed soon after their onset (after about one third of their full run); the manipulandum knocked against a lever bonded with an 0.01 kg.m2 inertial disk and caused it to rotate (Fig. 4 A). The interaction lasted over a distance of 8-10° and then the movement was free again. The subjects were asked to ignore the perturbation and move the arm to the intended final position. The primary data recorded were the angular position and acceleration of the manipulandum handle and the surface EMG activities from the biceps caput longum and the lateral head of the triceps muscles. Handle position and acceleration were measured by commercially available electroinductive transducers. The EMG signals were detected, rectified and slightly smoothed (time constant 10 ms). All data were digitized with an effective sampling interval of 2 ms, displayed on a four-beam oscilloscope with digital memory and filmed for later analysis.

In Experiment 2 subjects were asked to make a series of 50 flexor movements with the same instructions but the movement amplitude was reduced to about 45-50°. Torque perturbations were applied to the handle in randomly selected movements. Perturbations began simultaneously with the sound signal. Their initial magnitude ranged from 1.0 to 5.0 Nm and either opposed or assisted the forthcoming movement. Subjects were asked to disregard whatever limb displacement might occur because of the perturbation and to concentrate on moving to the intended final position. Perturbations were delivered by a light lever which engaged the distal end of the handle (Fig. 4 B). Two rubbers were attached with the help of two electromagnetic locks to the lever in points symmetric about its rotational axis. In the initial position, rubber torques were balanced and the load on the handle was zero. In a given trial, either one or both the rubbers could be switched off simultaneously with the sound signal, and thus the forthcoming movement was either perturbed or non-perturbed. Engagement of the lever ceased when the handle had been rotated 8-10° in any direction from its initial position. In Experiment 2, the amplitude, duration and onset time of the initial agonist burst were examined.

Results

Experiment 1. In Fig. 5 each set of traces shows arm position, acceleration and EMG activity associated with individual flexion movements. Control, non-perturbed movement trajectories (A) had terminal oscillations. A transitory interaction with an inertial load (B) which absorbed part of the kinetic energy of the arm resulted in near-complete suppression of the terminal oscillations. This effect was significant (p < .01, t-test) in all the trials and in all the subjects that participated in the experiments, although the suppression of terminal oscillations was not always complete.

The effect of the inertial load on the final position was insignificant (p >.O5), according to the property of equifinality of motor systems predicted by the EP hypothesis (e.g., Feldman, 1966; Kelso and Holt, 1980). EMG activity was not analyzed in detail in Experiment 1. However, the load perturbation was often observed to give rise to reciprocal EMG burst activity in the flexor and extensor muscles just prior to and after the end of the movement (Fig.6). These rapid EMG oscillations (about 12 Hz) had a non-significant influence on the form of movement trajectory, possibly due to the low frequency bandwidth of the muscle (Partridge and Benton, 1983). The EMG rhythmicity can be explained by a synchronization of the spikes of active motor units elicited by the sudden load and accentuated by the reciprocal reflex inhibitory interaction between the muscles.

Experiment 2. Typical responses to flexion and extension perturbations applied before the onset of movement are shown in Fig.7. Perturbations opposing the forthcoming movement resulted in an earlier initiation of the first agonist burst, an increase in its magnitude and duration, and consequently, in their product (p<.001, t-test, Table 1). By contrast, perturbations assisting the forthcoming movement caused a delay in the initiation of the first agonist burst (p<.01 in two subjects and p>.05 in one subject). Any increase in its magnitude and decrease in its duration was statistically insignificant (p>.05). The mean difference in the onset times with perturbations of opposite directions was 25-30 ms. The effect of the direction of a perturbation on the onset time was significant (p<.01).

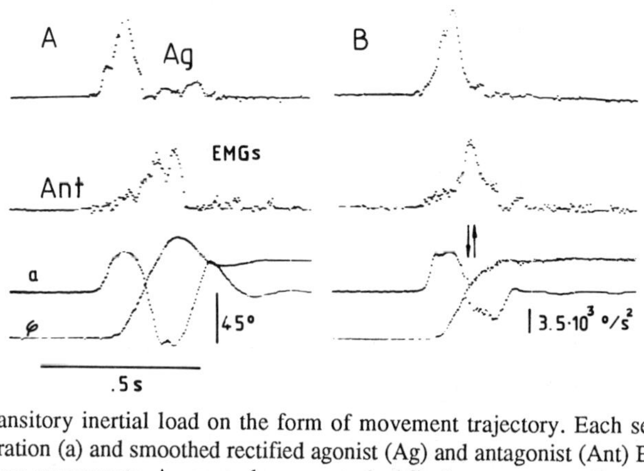

Fig. 5. Effect of transitory inertial load on the form of movement trajectory. Each set of records shows position (φ), acceleration (a) and smoothed rectified agonist (Ag) and antagonist (Ant) EMG activity during individual rapid elbow movements. A - control, non-perturbed flexion movement; B - perturbed movement; perturbation period is indicated by arrows.

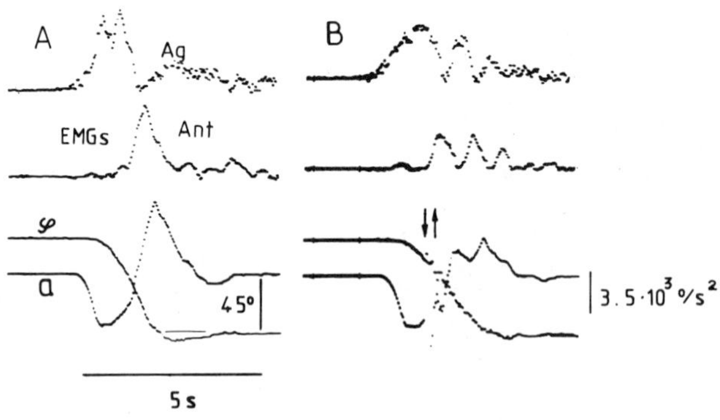

Fig. 6. Another example of effects of inertial load; rapid extension movement. Records as in Fig. 5.

Table 1: EMG parameters in control (zero load) and perturbed (load, unload) movements. Perturbations were applied simultaneously with the signal to move and either opposed (load) or assisted (unload) the forthcoming movement. For measuring the onset time of the first agonist burst (AGl) or antagonist burst (ANT) a reference EMG level was selected (about 1/3 of the mean peak value of the AGl in control trials; see Fig. 7B). Real onset times were about 20 ms earlier than indicated here.

SUBJECT	PERTURBATION	ONSET TIME AG1 (ms)	ONSET TIME ANT (ms)	DURATION AG1 (ms)	MAGNITUDE AG1 (arbitrary units)	DURATION MAGNITUDE AG1 normalized	TRIALS N
A.K.	load	84 ± 13	191 ± 16	95 ± 16	35 ± 6	1.6	16
	zero	98 ± 15	196 ± 16	71 ± 23	28 ± 5	1.0	11
	unloaded	111 ± 16	182 ± 26	66 ± 18	32 ± 8	1.1	18
L.M.	load	86 ± 7	208 ± 16	101 ± 14	33 ± 6	2.3	9
	zero	108 ± 12	206 ± 12	63 ± 19	23 ± 6	1.0	7
	unloaded	111 ± 18	196 ± 20	58 ± 12	25 ± 7	1.0	12
T.M.	load	77 ± 10	-	128 ± 17	65 ± 14	2.5	10
	zero	100 ± 6	-	72 ± 19	44 ± 10	1.0	11
	unloaded	106 ± 10	-	71 ± 26	45 ± 13	0.95	14

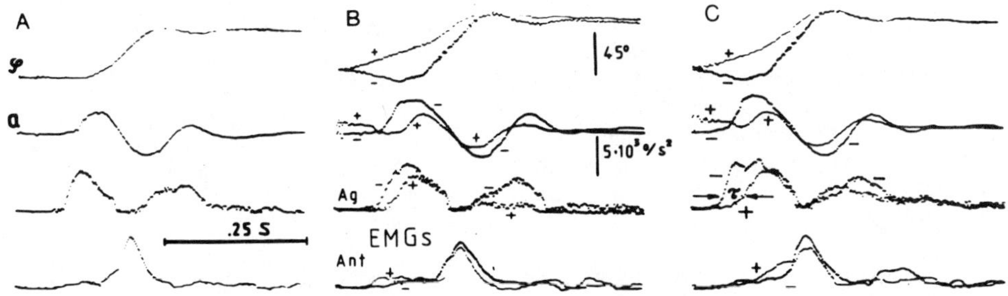

Fig. 7. The influence of perturbations applied prior to movement on the onset time of the first agonist EMG burst. Each set of records shows position (φ), acceleration (a) and smoothed and rectified agonist (Ag) and antagonist (Ant) EMG activity during individual rapid elbow flexion movements. A - non-perturbed movement; B - two sequential movements perturbed in opposite directions (+ and -); τ shows the difference in the onset time of the first agonist burst with the two movements. C - another couple of sequential movements perturbed in opposite directions.

It is interesting to note that in 3 subjects perturbations which assisted the forthcoming movement caused, though irregularly (in about 2-3 trials out of 10), an additional antagonist burst prior to the onset of the first agonist burst. In these cases the overall EMG pattern was similar to that observed during a movement in the opposite direction (Fig.8).

It is known that perturbations applied 50ms after the signal instructing subjects to initiate a movement cause an overall reduction in the reaction time, irrespective of their direction (Wadman et al., 1980a, 1980b; Day et al., 1983; Brown and Cooke, 1986). This effect was not observed in our study. It should be remembered that the perturbations in our study were applied simultaneously with the sound signal instructing subjects to perform the movement, which may have neutralized the effect of the response time reduction.

Discussion

The present discussion will be concerned with the question of the adequacy of various hypotheses or models for rapid one-joint motor control. In this connection, we will

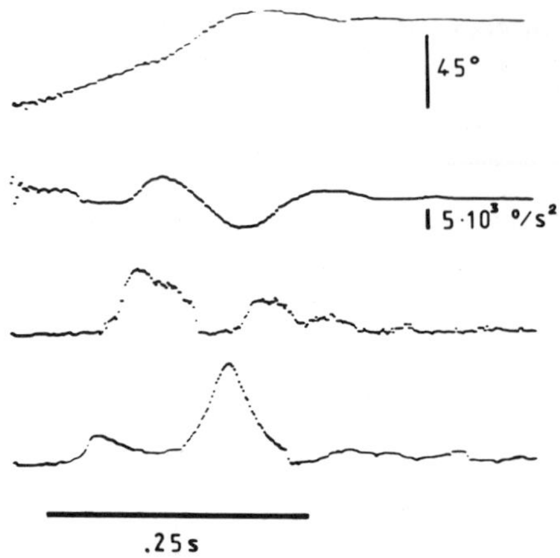

Fig. 8. An example of an additional antagonist burst (bottom trace) preceding the first agonist burst in the case of perturbation applied in the direction of the forthcoming movement. Records as in Fig. 7.

refer, in particular, to the following results of our study. 1) Movement trajectories with terminal oscillations could be transformed into aperiodic trajectories if part of the surplus kinetic energy of the arm was suddenly removed in the course of movement by an external load. 2) The onset time of the first agonist burst was affected by the direction of perturbations: perturbations applied in the direction of the forthcoming movement gave rise to an increase and those in the opposite direction to a decrease in the onset time. 3) The first agonist burst increased in magnitude and duration in the case of perturbations which opposed the forthcoming movement. By contrast, perturbations assisting the forthcoming movement only weakly influenced the first agonist burst. 4) Assisting perturbations sometimes, although irregularly, gave rise to an additional antagonist burst prior to the first one (an inverted EMG pattern).

The first result is consistent with the EP hypothesis which suggests that the form of a movement trajectory, particularly the terminal overshoot, is not present in the central control signals descending to motoneurons (cf. Bizzi, 1980) but rather reflects dynamic transfer processes in an executive part of the motor system which take the load into account (see also Lestienne, 1979).

Observation (2) indicates that not only can central command signals participate in the timing of the first agonist burst but so can afferent signals from the muscle. More specifically, one might suggest that the stretch-reflex mechanism functions efficiently even in an initial phase of rapid movement, shifting the onset time of the first agonist burst as appropriate to the changes in muscle length elicited by perturbations. The same reflex may underlie the modifications in the magnitude and duration of the first agonist burst in the case of perturbations opposing the forthcoming movement. Thus, observations (2)-(3) are also consistent with the EP hypothesis based on the idea that the brain uses the stretch reflex to control movements (Feldman, 1974, 1986). The preceding activation of the antagonist muscle observed in the case of perturbations assisting the forthcoming movement also indicates that the initial phase of rapid movement is reflex-dependent. On the whole, our observations are consistent with the idea that it is not a "central program" which is responsible for EMG bursts or their time periods and amplitudes (Kelso and Saltzman, 1982; Feldman, 1986).

The pattern of the central control signals given in Fig. 3 helps to explain why reflex modifications of the onset time of the first agonist burst were not observed in other studies in which perturbations were applied later, 50 ms after the signal to movement (Wadman et al., 1979, 1980a, 1980b; Brown and Cooke, 1986). The fact is that the control signals underlying rapid movements sweep the EP at a high speed of 500-700°/s (Abdusamatov, Adamovich and Feldman, in press). Perturbations must preset a comparable angular speed of the arm prior to the movement in order to elicit appropriate afferent signals which are capable of modifying the onset time of the burst. Correspondingly, perturbations must begin to speed up the limb long (say, 100 ms or even more) before the first burst in order to counter the limb's inertia and affect the onset time. Otherwise, only the magnitude of the first burst of a later part of the triphasic pattern will be modified, as has been observed in the studies cited above.

Brown and Cooke (1986) have observed that perturbations assisting forthcoming movements which are applied 50 ms after the signal to move increased the magnitude of the first agonist burst and decreased its duration more efficiently than in the present study. A "triggered reaction" (Crago et al., 1976) may underlie this effect and elicit, in addition, an overall reduction in the response time as was observed in the study cited. On the other hand, reflex mechanisms, in particular the Ia reciprocal interaction between flexor-extensor coupled muscles, can also contribute to this effect. Actually, assisting perturbations cause stretching of the antagonist muscle and an activation of its afferents, which can result in hyperpolarization of agonist motoneurons before the movement. In such a case, the central activation of agonist motoneurons will be associated with a postinhibitory rebound (Granit, 1970), that is, the density of spikes of motor units in the first agonist burst will increase whereas its duration will be limited by reinforced recurrent inhibition of the motoneurons. This explanation shows that it is necessary to elaborate the EP hypothesis further to incorporate additional motor mechanisms, particularly, the reflex reciprocal inhibition.

All things considered, we come to the conclusion that the available data provide no substantial evidence supporting the impulse-timing hypotheses on motor control which suggest that "control variables like muscle activation levels and timing of contractions, are pre-set over periods of at least 100 ms (Wadman et al., 1980b). One of two versions of the EP hypothesis, which favours direct central programming of motoneuron activity (Bizzi, 1980; Hogan, 1984), is also in disagreement with the reflex effects of perturbations reported here.

The hypothesis that the triphasic pattern results principally from reflex actions activated in the course of movement and modulated by central control signals (Ghez and Martin, 1982) is, in general, consistent with the EP hypothesis presented here. However, the "pulse-step" form of central command signals postulated in the said hypothesis is not corroborated by the data showing that a gradual form of central control signals (more specifically determined in terms of EP shifts) underlies unidirectional one-joint movements (Bizzi et al., 1982; Adamovich et al., 1984).

References

Abdusamatov, R. M., and Feldman, A. G., 1986, Description of electromyograms by a mathematical model of single joint movements. Biofizika, 31, 503-505 (in Russian).

Abdusamatov, R. M., Adamovich, S. V., and Feldman, A. G. A model for one-joint motor control in man (in press).

Adamovich, S. V., Burlachkova, N. I., and Feldman, A. G. , 1984, On the central wave nature of time-angle trajectory formation in man. Biofizika, 29, 122-125.

Berardelli, A., Rothwell, J. C., Day, B. L., Kachi, T., and Marsden, C. D., 1984, Duration of the first agonist EMG burst in ballistic arm movements. Brain Research, 304, 183-187.

Bizzi, E., 1980, Central and peripheral mechanisms in motor control. In: G. E. Stelmach and J. Requin, eds. Tutorials in Motor Behavior. North-Holland: Amsterdam, pp. 131-144.

Bizzi, E., Accornero, N., Chapple, W., and Hogan, N., 1982, Arm trajectory formation in monkeys. Exp. Brain Research, 46, 139-143.

Brown, S. H., and Cooke, J. D., 1984, Initial agonist burst duration depends on movement amplitude. Exp. Brain Research, 55, 523-527.

Brown, S.H., and Cooke, J. D., 1986, Initial agonist burst is modified by perturbations preceeding movement. Brain Res., 377, 311-322.

Crago, P. E., Houk, J. C., and Hasan, Z., 1976, Regulatory actions of human stretch reflex. J. Neurophysiol., 39, 925-935.

Day, B. L., Rothwell, J. C., and Marsden, C. D., 1983, Interaction between the long-latency stretch reflex and voluntary electromyographic activity prior to a rapid voluntary motor reaction. Brain Res., 270, 55-62.

Feldman, A. G., 1966, Functional tuning of the nervous system with control of movement or maintenance of a steady posture. II. Controllable parameters of the muscle. Biophysics, 11, 565-578.

Feldman, A. G., 1974, Control of the length of a muscle. Biophysics, 19, 771-776.

Feldman, A . G., 1980, Superposition of motor programs. I. Rhythmic forearm movements in man. Neuroscience, 5, 81-90.

Feldman, A. G., 1986, Once more on the equilibrium point hypothesis (δ-model) for motor control. Journal of Motor Behavior, 18, 17-54.

Gasser, H.S., and Hill, A. V., 1924, The dynamics of muscular contraction. Proc. of Roy. Soc. of Lond., B96, 398-437.

Ghez, C., and Martin, J.H., 1982, The control of rapid limb movement in the cat. Exp. Brain Research, 45, 115-125.

Granit, R., 1970, The basis of motor control. London: Academic Press.

Hallett, M., Shahani, B. T., and Young, R. R., 1975, EMG analysis of stereotyped voluntary movements in man. J. Neurology, Neurosurgery and Psychiatry, 38, 1154-1162.

Hogan, N., 1984, An organising principle for a class of voluntary movements. J. Neuroscience, 4, 2745-2754.

Kelso, J.A.S., and Holt, K.G., 1980, Exploring a vibratory system analysis of human movement production. J. Neurophysiol., 43, 1183-1196.

Kelso, J.A.S., and Saltzman, E.L., 1982, Motor control: Which themes do we orchestrate ? Behavioral and Brain Sciences, 5, 554-557.

Kugler, P.N., Kelso, J.A.S., and Turvey, M.T., 1980, On the concept of coordinative structures as dissipative structures: I. Theoretical lines of convergence. In: G.E. Stelmach and J. Requin, eds. Tutorials in Motor Behavior. North-Holland, Amsterdam, pp. 3-45.

Lestienne, F., 1979, Effects of inertial load and velocity on the braking process of voluntary limb movements. Exp. Brain Research, 34, 407-418.

Partridge, L.D., and Benton, L.A., 1983, Muscle, the motor. In: V. Brooks, ed. Motor Control. Handbook of Physiology - The Nervous System. Sec. 1, vol.II, part 1: ` Am. Physiol. Soc., Bethesda, pp. 43-106.

Soetching, J.F., Ranish, N.A., Palminteri, R., and Terzuolo, C.A., 1976, Changes in a motor pattern following cerebellar and olivary lesions in the squirrel monkey. Brain Research, 105, 21-44.

Terzuolo, C. A., Soetching, J.F., and Viviani, P., 1973, Studies on the control of some simple motor tasks. I. Relations between parameters of movements and EMG activities. Brain Research, 58, 212-216.

Vincken, M.H., 1983, Control of limb stiffness. Thesis. Utrecht University.

Wadman, W.J., Denier van der Gon, J.J., Geuze, R.H., and Mol, C.R., 1979, Control of fast goal-directed arm movements. J. Hum. Movement Stud., 5, 3-17.

Wadman, W.J., Denier van der Gon, J.J., and Derksen, R.J.A., 1980a, Muscle activation patterns for fast goal-directed arm movements. J. Hum. Movement Stud., 6, 19-37.

Wadman, W.J., Boerhout, W., and Denier van der Gon, J.J., 1980b, Responses of the arm movement control system to force impulses. J. Hum. Movement Stud., 6, 280-302.

CONTRIBUTORS

Abdusamatov, R.M., Institute of Problems of Information Transmission, Academy of Sciences, Moscow 101447, USSR.

Adamovich, S.V., Institute of Problems of Information Transmission, Academy of Sciences, Moscow 101447, USSR.

Arshavsky, Yu. I., Institute of Problems of Information Transmission, Academy of Sciences, Moscow, Moscow State University, Moscow, USSR.

Batuev, A.S., Department of Higher Nervous Activity, Leningrad State University Leningrad, 199034, USSR.

Beloozerova, I.N., Institute of Biomedical Problems, Ministry of Public Health Moscow, USSR.

Berkinblit, M.B., Institute of Problems of Information Transmission, Academy of Sciences, Moscow 101447, USSR.

Birjukova, E.V., Institute of Higher Nervous Activity and Neurophysiology USSR Academy of Sciences, Moscow, USSR.

Bouisset, S., Laboratoire de Physiologie du Mouvement, U.A. C.N.R.S. 631, Université de Paris-Sud, Orsay, Paris, France.

Bulgokova, N.V., A.A. Bogomoletz Institue of Physiology, Academy of Sciences of the Ukrainian SSR, Kiev, USSR.

Cabelguen, J.-M., Département de Neurophysiologie comparée, Institut des Neurosciences, Université Paris 6, 9 quai Saint-Bernard, 75230 Paris, Cedex 05, France.

Cailler, P., Laboratoire de Physiologie Neurosensorielle du C.N.R.S., 15 rue de l'Ecole de Médecine, 75270 Paris Cedex 06, France.

Chernavsky, A.V., Institute of Problems of Information Transmission, Academy of Sciences, Moscow 101447, USSR.

Chrachri, A., Laboratoire de Neurobiologie et Physiologie comparées, Place Peyneau, 33120 Arcachon, France.

Clarac, F., Laboratoire de Neurobiologie et Physiologie comparées, Place Peyneau, 33120 Arcachon, France.

Crenna, P., Instituto di Fisiologia Umana II, Milano, Itaty.

Deat, A., Unité de Neurosciences Intégratives, LNF 3 CNRS, Marseille, France.

Deliagina, T. G., Institute of Problems of Information Transmission, Academy of Sciences, Moscow, Moscow State University, Moscow, USSR.

Dmitrieva, I., Institute of Biomedical Problems, Moscow, USSR.

Droulez, J., Laboratoire de Physiologie Neurosensorielle, CNRS, 15, rue de l'Ecole de Médecine, 75006 Paris, France.

Fanardjian, V. V., L. A. Orbeli Institute of Physiology, Academy of Sciences of Armenian SSR, 22 Orbeli Bros. Str., 375028, Yerevan, USSR.

Feldman, A.G., Institute of Problems of Information Transmission, Academy of Sciences, Moscow 101447, USSR.

Frigo, C., Centro di Bioingegneria, Milano, Italy.

Frolov, A.A., Institute of Higher Nervous Activity and Neurophysiology USSR Academy of Sciences, Moscow, USSR.

Gelfand, I. M., Institute of Problems of Information Transmission, Academy of Sciences, Moscow, Moscow State University, Moscow, USSR.

Gorodnov, V. L., L. A. Orbeli Institute of Physiology, Academy of Sciences of Armenian SSR, 22 Orbeli Bros. Str., 375028, Yerevan, USSR.

Grigorieva, L., Institute of Biomedical Problems, Moscow, USSR.

Gurfinkel, V.S., Institute for Problems of Information Transmission, USSR Acad. Sci., Moscow, USSR.

Horcholle-Bossavit, G., Laboratoire de Neurophysiologie, Collège de France, 75231 Paris, Cedex 05, France.

Ioffe, M.E., Institute of Higher Nervous Activity and Neurophysiology USSR Academy of Sciences, Moscow, USSR.

Ivanova, N.G., Institute of Higher Nervous activity and Neurophysiology USSR Academy of Sciences, Moscow, USSR.

Jami, L., Laboratoire de Neurophysiologie, Collège de France, 75231 Paris, Cedex 05, France.

Kazennikov, O.V., Institute of Information Transmission Problems, Academy of Sciences of the USSR, Moscow, USSR.

Kirenskaya, A., Institute of Biomedical Problems, Moscow, USSR.

Kiseljova, N.V., Institute of Higher Nervous activity and Neurophysiology USSR Academy of Sciences, Moscow, USSR.

Kozlovskaya, I., Institute of Biomedical Problems, Moscow, USSR.

Kreidich., Yu., Institute of Biomedical Problems, Moscow, USSR.

Kudina, L., Institute for Problems of Information Transmission, USSR Academy of Sciences, Moscow, USSR.

Lestienne, F., Laboratoire de Physiologie Neurosensorielle du C.N.R.S., 15 rue de l'Ecole de Médecine, 75270 Paris Cedex 06, France.

Levik, Yu.S., Institute for Problems of Information Transmission, USSR Acad. Sci., Moscow, USSR.

Liverneaux, P., Laboratoire de Physiologie Neurosensorielle du C.N.R.S., 15 rue de l'Ecole de Médecine, 75270 Paris Cedex 06, France.

Marlinsky, V.V., A.A. Bogomoletz Institute of Physiology, Academy of Sciences of the Ukrainian SSR, Kiev, USSR.

Massion, J., Unité de Neurosciences Intégratives, LNF 3 CNRS, Marseille, France.

Orlov, A.A., Department of Higher Nervous Activity, Leningrad State University Leningrad, 199034, USSR.

Orlovsky, G. N., Institute of Problems of Information Transmission, Academy of Sciences, Moscow, Moscow State University, Moscow, USSR.

Orsal, D., Département de Neurophysiologie comparée, Institut des Neurosciences, Université Paris 6, 9 quai Saint-Bernard, 75230 Paris, Cedex 05, France.

Pachin, Yu. V., Institute of Problems of Information Transmission, Academy of Sciences, Moscow, Moscow State University, Moscow, USSR.

Pavlova, G. A., Institute of Problems of Information Transmission, Academy of Sciences, Moscow, Moscow State University, Moscow, USSR.

Pedotti, A., Centro di Bioingegneria, Milano, Italy.

Perret, C., Département de Neurophysiologie comparée, Institut des Neurosciences, Université Paris 6, 9 quai Saint-Bernard, 75230 Paris, Cedex 05, France.

Person, R., Institute for Problems of Information Transmission, USSR Academy of Sciences, Moscow, USSR.

Petit, J., Laboratoire de Neurophysiologie, Collège de France, 75231 Paris, Cedex 05, France.

Pilyavsky, A.I., A.A. Bogomoletz Institue of Physiology, Academy of Sciences of the Ukrainian SSR, Kiev, USSR.

Piotrkiewicz, M., Institute of Biocybernetics and Biomedical Engineering, Polish Academy of Sciences, Warsaw, USSR.

Popov, K.E., Institute for Problems of Information Transmission, USSR Acad. Sci., Moscow, USSR.

Roll, J. P., Département de Psychophysiologie, Laboratoire de Neurobiologie Humaine, UA CNRS 372, Université d'Aix-Marseille, 13397 Marseille Cedex 13, France.

Roll, R., Laboratoire de Neurobiologie Humaine, U.A. C.N.R.S. 372 - Université de Provence, Avenue Escadrille Normandie Niemen, 13397 Marseille, France.

Selionon, V.A., Institute of Information Transmission Problems, Academy of Sciences of the USSR, Moscow , USSR.

Shaefer, V.I., Department of Higher Nervous Activity, Leningrad State University Leningrad, 199034, USSR.

Shapovalov, A. I., I.M. Sechenov Institute of Evolutionary Physiology and Biochemistry, USSR. Academy of Sciences, Leningrad, USSR.

Shapovalova, K.B., Pavlov Institute of Sciences, Leningrad, USSR.

Shik, M.L., Institute of Information Transmission Problems, Academy of Sciences of the USSR, Moscow, USSR.

Shlikov, V.Yu., Institute for Problems of Information Transmission, USSR. Acad. Sci., Moscow, USSR.

Sirota, M.G., Institute of Biomedical Problems, Ministry of Public Health Moscow, USSR.

Smetanin, B.N., Institute for Problems of Information Transmission, USSR Acad. Sci., Moscow, USSR.

Tsyntsabadze, Th. I., A.A. Bogomoletz Institute of Physiology, Academy of Sciences of the Ukrainian SSR, Kiev, USSR.

Vasilenko, D.A., A.A. Bogomoletz Institute of Physiology, Academy of Sciences of the Ukrainian SSR, Kiev, USSR.

Vedel, J.P., Laboratoire de Neurosciences Fonctionnelles, CNRS, U2, 31, Chemin Joseph Aiguier, 13402 Marseille Cedex 9, France.

Vejsada, R., Institute of Physiology, Czechoslovak Academy of Sciences, Videnska 1083, 142 20 Prague 4, Czechoslovakia.

Vesselkin, N. P., I.M. Sechenov Institute of Evolutionary Physiology and Biochemistry, USSR Academy of Sciences, Leningrad, USSR.

Yakhnitsa, I.A., A.A. Bogomoletz Institue of Physiology, Academy of Sciences of the Ukrainian SSR, Kiev, USSR.

Yakimovsky, A.F., Pavlov Institute of Sciences, Leningrad, USSR.

Zattara, M., Laboratoire de Physiologie du Mouvement, U.A. C.N.R.S. 631, Université de Paris-Sud, Orsay, Paris, France.

Zytnicki, D., Institute of Physiology, Czechoslovak Academy of Sciences, Videnska 1083, 142 20 Prague 4, Czechoslovakia.

INDEX

Accelerometry, 177, 178
Adaptive mechanisms, 252, 253
Adaptive reorganization, 37, 149, 150
Afferen Ia discharge, 11, 12, 24, 127, 154, 186, 268, 240
Afferent Ia fibres, 16, 17, 49, 50, 55, 56
Afferent Ib discharge, 1, 6, 7
Afferent Ib fibres, 2, 3, 8
Afferent inflow, 46, 133, 135, 137, 165
Afferent joint fibres, 99
Alpha-gamma linkage, 11, 17

Balance, 38, 39, 45, 177, 192, 251
Basal ganglia, 71, 79, 80, 83, 85, 172, 222
Behavioral condition, 17, 18, 24
Behaviour, 83, 91, 98
- instrumental, 203
- orienting, 71

Biomechanic, 37, 46, 94,177, 181, 183, 192, 224, 227, 231, 238
Bulbospinal, 123, 125

Carp, 49, 89, 94, 245
Cat, 1, 61, 123, 133, 239
Center of gravity, 180, 195, 227, 256
Central control, 195, 198, 264, 268, 269
Computer simulation, 236, 240, 245, 247
Context, 23, 29, 33, 79, 177, 185, 225, 256, 257, 259
Cortical stimulation, 61, 64
Cortico rubral system, 61, 68
Cortico spinal, 61, 65
Crayfish, 93, 94, 99, 101, 102, 104
Crustacea, 93, 118

Decapod, 93, 94, 99, 104
Diencephalon, 123
Dog, 133, 203, 213, 228
Dopamine 203, 209, 210
Dorsal root potential, 153, 160

Efferent alpha and gamma fibers, 11, 16
Efferent copy, 24
Efferent neurons, 68, 114, 134, 139
Efferent output, 153, 160, 211
Electromyography, 12, 38, 46, 133, 143, 177, 188, 196, 197, 261
Extraocular proprioception, 23, 24, 32, 33
Extrapyramidal, 61
Eye, 23, 71, 79, 236, 258

Force
- contractile, 1, 8
- inertial, 177, 183
- mechanoreceptors, 93, 101
- muscular, 12, 169
- postural, 27, 28, 185, 192, 213
- pressure, 227, 239

Forebrain, 123
Frog, 49, 50, 51, 53, 54, 57
Fusimotor fibre, 11, 13, 15, 17, 18

Gastropod, 107, 113, 117
Goal-directed, 72, 83, 91
Gravitational, 37, 46, 47, 192
Guinea pig, 143

Imbalance, 28
Instantaneous frequency, 1, 4, 8, 12, 101
Intracellular recording, 50, 104, 111, 126

Kinematic, 39, 196, 227, 237, 261, 264

Lamprey, 49, 54, 55, 57, 124
Locomotion, 43, 54, 93, 107, 123, 133, 143, 153, 163, 185, 203
Locomotor cycle, 107, 134, 145, 153, 175
Locomotor rhythm, 107, 137, 143

Man, 11, 24, 33, 37
Mesencephalic, 123, 127, 133, 143, 204
Microgravity, 37, 39, 40, 41, 44, 45, 46
Microinjection, 203, 211
Model, 39, 49, 93, 118, 137, 185, 192, 204, 227, 239, 247, 261, 267
Monkey, 33, 68, 71, 90, 256
Monosynaptic, 17, 49, 61, 123, 129, 239
Motoneuronal firing, 239, 247
Motoneuron, 2, 49, 58, 149, 150, 246
- gamma motoneuron, 11, 14, 16, 18, 262
- interneurons, 107, 112, 138, 149, 153

Movement
- active, 12
- arm, 33, 71, 74, 78, 79, 196, 198, 200, 254, 261
- axial, 195, 196, 198, 200
- body, 33, 94, 169, 177, 178, 185, 191, 200, 227, 237
- eye, 32, 33, 34, 71, 74, 236
- goal directed, 93

head, 26, 33, 71, 78, 79, 228, 254
illusory, 23, 24, 29, 33, 187
joint, 12, 24, 166, 257, 261, 269
knee, 195, 198
learned, 211, 214, 256
leg, 94, 99
limb, 79, 146, 153, 160, 165, 172, 174, 180, 196, 213, 218
locomotor, 133, 139, 144
passive, 12, 155, 156, 160
posture, 37
preparation, 18
respiratory, 196
rhythmic, 107, 108, 113, 119, 143
spatially oriented, 185, 186
stepping, 119, 125, 163, 166, 169, 170, 172, 174, 175
tracking, 43
trunk, 195, 196, 198, 258
voluntary, 164, 177, 193, 203, 237
Muscle spindle, 11, 14, 16, 24, 25, 49, 55, 57, 100
Musculoskeletal, 37

Neostriatum, 83, 87, 89, 90, 91, 203, 211
Neuromuscular, 37, 46, 99, 104
Nigrostriatal, 203, 209, 210

Otolithic, 37, 256

Pallidum, 83, 84, 85, 87, 89, 91
Parietal, 83, 84, 85, 86, 87, 89, 90, 91
Pattern
afferent, 23, 33
discharge, 85, 86
efferent, 16, 139, 254
EMG, 261, 269
generator, 110, 133
locomotion, 93, 98, 102, 133, 135, 137, 175
pressure, 227, 231, 235, 236, 237
regulation, 143
rhythmic, 104, 116
stimulation, 2, 3
synergy, 37, 45, 46, 101
temporal, 79
Performance, 164, 174, 195, 203, 208, 210, 223, 225, 235, 256
Polysynaptic, 18, 123, 125, 126, 127, 129
Post potential
EPSP, 61, 63, 64, 65, 66, 67, 68, 111, 117, 239, 245, 247
IPSP, 109, 110, 111, 117
Postural adjustment, 79, 177, 185, 192, 196, 203, 213, 227, 235
Postural coordination, 213, 214, 222
Postural pattern, 213, 217, 219, 222, 225
Postural reference, 23
Posture, 24, 37, 43, 149, 177, 185, 195, 203, 219, 237, 254
Prefrontal, 80, 83, 85, 86, 87, 89, 90, 91
Presynaptic, 63, 68, 138, 153, 159, 160
Prime mover, 71, 74, 195, 198, 199, 200
Proprioceptive, 24, 29, 32, 34, 39, 46, 74, 185, 191, 195
Pteropod mollusc, 107
Pyramidal, 61, 68, 163, 165, 174, 175

Receptor
DA, 204, 209
Golgi tendon organs, 1, 7, 8, 16
joint, 99
muscle, 2, 6, 18, 25, 37, 46, 55
otolith, 150
stretch, 55, 93
tactile, 33
vestibular, 150, 256
Red nucleus, 61, 62, 63, 64, 68
Reflex
conditioned, 214
fusimotor, 14, 18
H, 17, 18, 38, 40, 240
phasic, 14, 133
stretch, 150, 155, 185, 262, 268
T, 17, 18, 38, 40
vestibulo ocular, 251, 252, 256

Sensory control, 93, 99, 104
Sensory information, 93, 99, 258
Sensory input, 37, 38, 74, 79, 99, 105, 185, 191, 195, 256, 259
Sensory motor cortex, 61, 63, 64, 65, 68
Servo-control mechanism, 251
Space flight, 37, 38, 40, 41
Spinal cord, 49, 58, 62, 123, 129, 174
Stance, 24, 93, 104, 143, 153, 160, 165, 167, 187, 188
Stretch, 11, 18, 32, 33, 55, 94, 104, 150, 156, 185, 239, 262, 269
Substantia nigra, 71, 209, 211
Swing, 93, 104, 143, 153, 160, 165, 174
Synapse, 46, 50, 51, 53, 54, 55, 57, 58, 124, 211
Synaptic transmission, 46, 49, 55, 126
Synergy, 37, 40, 45, 74, 199

Tegmentum, 123
Topographical, 251
Topological, 251, 257, 258, 259

Vibration, 12, 33, 99, 186, 189
VL, 163, 172, 173, 174, 175

Walking, 93, 98, 100, 102, 123, 163, 257
Weightlessness, 37
Whole body, 23